# Mathematical Studies

## for the IB DIPLOMA

**Second Edition**

**Ric Pimentel**
**Terry Wall**

**Acknowledgements**

Examination questions have been reproduced with kind permission from the International Baccalaureate Organization.

This material has been developed independently by the publisher and the content is in no way connected with, nor endorsed by, the International Baccalaureate Organization.

**Photo credits**
**p.1** *t* © Inmagine/Alamy, *bl* © Feng Yu/Fotolia.com, *br* © Ingram Publishing Limited; **p.16** *t* United States Government Work/http://commons.wikimedia.org/wiki/File:Enigma-Machine.jpg/17April2012, *b* © Bettmann/CORBIS; **p.17** NASA Human Spaceflight Collection; **p.18** Walters Art Museum/http://commons.wikimedia.org/wiki/File:Babylonian_-_Economic_Document_-_Walters_482030_-_View_A.jpg/ http://creativecommons.org/licenses/by-sa/3.0/deed.en; **p.27** © The Art Gallery Collection/Alamy; **p.35** Photodisc/ Getty Images; **p.70** *t* www.purestockX.com, *b* Chris brown/http://www.flickr.com/photos/zoonabar/ 5207075123/http://creativecommons.org/licenses/by/2.5/deed.en; **p.71** © INTERFOTO/Alamy; **p.108** *t* © Jonathan Barton/iStockphoto.com, *b* Official White House Photo by Pete Souza/http://www.flickr.com/photos/anhonorablegerman/6824268224/http://creativecommons.org/licenses/by/2.0/deed.en_GB; **p.109** © Danilo Ascione – Fotolia; **p.111** *t* © The Granger Collection, NYC /TopFoto, *b* © Argus – Fotolia; **p.135** © The Granger Collection, NYC/TopFoto; **p.156** © Bettmann/CORBIS; **p.157** *t* © Rita Jayaraman/iStockphoto.com, *c* © Gregory Sams/Science Photo Library, *b* © TOK56 – Fotolia; **p.202** *t* © scooperdigital – Fotolia, *b* © Fotomas/Topham; **p.228** © Harvey Hudson – Fotolia.com; **p.244** © The Art Gallery Collection/Alamy; **p.280** © Luis Carlos Torres/iStockphoto.com; **p.281** *t* © spe/Fotolia.com, *b* © Imagestate Media (John Foxx); **p.283** *tr* © The Granger Collection, NYC/TopFoto, *br* © Science & Society Picture Library/Getty Images, *l* public domain/http://commons.wikimedia.org/wiki/File:ArsMagna.jpg; **p.324** *t* © Imagestate Media (John Foxx), *b* © Department of Energy/Photodisc/Getty Images; **p.325** *t* © Hulton Archive/ Imagno/Getty Images, *c* © Georgios Kollidas – Fotolia, *b* © Photo Researchers/Alamy; **p.327** © Professor Peter Goddard/Science Photo Library; **p.358** © Nick Koudis/Photodisc/Getty Images; **p.359** Photo by Eric Erbe, digital colorization by Christopher Pooley/ARS/USDA

*t* = top, *c* = centre, *b* = bottom, *l* = left, *r* = right


Orders: please contact Bookpoint Ltd, 130 Milton Park, Abingdon, Oxon OX14 4SB. Telephone: (44) 01235 827720. Fax: (44) 01235 400454. Lines are open 9.00–5.00, Monday to Saturday, with a 24–hour message answering service. Visit our website at www.hoddereducation.com


First published in 2010 by
Hodder Education, an Hachette UK Company,
338 Euston Road
London NW1 3BH

This second edition published 2012

Impression number 5 4 3
Year 2015



Cover photo © V. Yakobchuk – Fotolia
Typeset in 10/12pt Goudy by Pantek Media, Maidstone, Kent
Printed in Dubai

A catalogue record for this title is available from the British Library

**ISBN 978 1444 180 176**

# Contents

Answers to the revision exercises and worked solutions to exercises and student assessments can be found on the website: **www.hodderplus.co.uk/IBMSSL2**.

# Introduction

## The IB Mathematical Studies SL student

The underpinning philosophy of the International Baccalaureate® is described in the attributes of the IB learner profile. These attributes encapsulate the IB mission statement and ideally should be embraced and modelled by students, teachers and the entire school community:

| | |
|---|---|
| **Inquirers** | They develop their natural curiosity. They acquire the skills necessary to conduct inquiry and research and show independence in learning. They actively enjoy learning and this love of learning will be sustained throughout their lives. |
| **Knowledgeable** | They explore concepts, ideas and issues that have local and global significance. In so doing, they acquire in depth knowledge and develop understanding across a broad and balanced range of disciplines. |
| **Thinkers** | They exercise initiative in applying thinking skills critically and creatively to recognize and approach complex problems, and make reasoned, ethical decisions. |
| **Communicators** | They understand and express ideas and information confidently and creatively in more than one language and in a variety of modes of communication. They work effectively and willingly in collaboration with others. |
| **Principled** | They act with integrity and honesty, with a strong sense of fairness, justice and respect for the dignity of the individual, groups and communities. They take responsibility for their own actions and the consequences that accompany them. |
| **Open-minded** | They understand and appreciate their own cultures and personal histories, and are open to the perspectives, values and traditions of other individuals and communities. They are accustomed to seeking and evaluating a range of points of view, and are willing to grow from the experience. |
| **Caring** | They show empathy, compassion and respect towards the needs and feelings of others. They have a personal commitment to service, and act to make a positive difference to the lives of others and to the environment. |
| **Risk-takers** | They approach unfamiliar situations and uncertainty with courage and forethought, and have the independence of spirit to explore new roles, ideas and strategies. They are brave and articulate in defending their beliefs. |
| **Balanced** | They understand the importance of intellectual, physical and emotional balance to achieve personal wellbeing for themselves and others. |
| **Reflective** | They give thoughtful consideration to their own learning and experience. They are able to assess and understand their strengths and limitations in order to support their learning and personal development. |

## IB Mathematics grade descriptors

The following amplifies the grade descriptors from the IB. Students should have a clear understanding of what is expected at each grade level.

- **Grade 7 Excellent performance**
  Demonstrates a thorough knowledge and understanding of the syllabus; successfully applies mathematical principles at a sophisticated level in a wide variety of contexts; successfully uses problem-solving techniques in challenging situations; recognizes patterns and structures, makes generalizations and justifies conclusions; understands and explains the significance and reasonableness of results, and draws full and relevant conclusions; communicates mathematics in a clear, effective and concise manner, using correct techniques, notation and terminology; demonstrates the ability to integrate knowledge, understanding and skills from different areas of the course; uses technology proficiently.
- **Grade 6 Very good performance**
  Demonstrates a broad knowledge and understanding of the syllabus; successfully applies mathematical principles in a variety of contexts; uses problem-solving techniques in challenging situations; recognizes patterns and structures, and makes some generalizations; understands and explains the significance and reasonableness of results, and draws relevant conclusions; communicates mathematics in a clear and effective manner, using correct techniques, notation and terminology; demonstrates some ability to integrate knowledge, understanding and skills from different areas of the course; uses technology proficiently.
- **Grade 5 Good performance**
  Demonstrates a good knowledge and understanding of the syllabus; successfully applies mathematical principles in performing routine tasks; successfully carries out mathematical processes in a variety of contexts, and recognizes patterns and structures; understands the significance of results and draws some conclusions; successfully uses problem-solving techniques in routine situations; communicates mathematics effectively, using suitable notation and terminology; demonstrates an awareness of the links between different areas of the course; uses technology appropriately.
- **Grade 4 Satisfactory performance**
  Demonstrates a satisfactory knowledge of the syllabus; applies mathematical principles in performing some routine tasks; successfully carries out mathematical processes in straightforward contexts; shows some ability to recognize patterns and structures; uses problem-solving techniques in routine situations; has limited understanding of the significance of results and attempts to draw some conclusions; communicates mathematics adequately, using some appropriate techniques, notation and terminology; uses technology satisfactorily.
- **Grade 3 Mediocre performance**
  Demonstrates partial knowledge of the syllabus and limited understanding of mathematical principles in performing some routine tasks; attempts to carry out mathematical processes in straightforward contexts; communicates some mathematics, using appropriate techniques, notation or terminology; uses technology to a limited extent.

- **Grade 2 Poor performance**
  Demonstrates limited knowledge of the syllabus; attempts to carry out mathematical processes at a basic level; communicates some mathematics, but often uses inappropriate techniques, notation or terminology; uses technology inadequately.
- **Grade 1 Very poor performance**
  Demonstrates minimal knowledge of the syllabus; demonstrates little or no ability to use mathematical processes, even when attempting routine tasks; is unable to make effective use of technology.



# The syllabus and *Mathematical Studies for the IB Diploma Second Edition*

This textbook fully covers the IB Mathematical Studies SL guide for first examinations in 2014. The assessment breakdown is as follows:

| Assessment | Type | Duration |
|---|---|---|
| External<br>Paper 1<br>40%<br>90 marks | 15 compulsory short-response questions on entire syllabus | 1 hr 30 min |
| External<br>Paper 2<br>40%<br>90 marks | 6 compulsory extended-response questions based on the entire syllabus | 1 hr 30 min |
| Internal project<br>20%<br>20 marks | The project is an individual piece of work involving the collection of information or the generation of measurements, and the analysis and evaluation of the information or measurements. | 25 hours |

For the external assessments a formulae booklet will be provided.

The topics covered in this resource are in the same order as they appear in the syllabus. Syllabus statements appear at the start of each topic. However, it is not intended that teachers and students will necessarily follow this order. Teachers should use this resource in the order which is appropriate for their school context. Some content has been included which will not be examined but is still useful to be taught. These sections have been indicated with a dashed line down the right hand side of the text. Each topic has explanations, examples, exercises and, at the end of the topic, a number of student assessments and past IB examination questions to reinforce learning.

A resource such as this one is used primarily as preparation for an examination. We have written a book which follows the syllabus and will provide excellent preparation. However we have also tried to make the book more interesting than that limited aim. We have written a twenty-first century book for students who will probably be alive in the twenty-second century.

## Presumed knowledge

We are aware of the difficulties presented to teachers by students who are taking this course and come from a variety of backgrounds and with widely differing levels of previous mathematical knowledge. The syllabus refers to presumed knowledge. At the start of the book, in each area of study – number, algebra, geometry, trigonometry and statistics, we have included a section of assessments which can be used to identify areas of weakness and as revision. These areas can then be studied in more detail by reference to our IGCSE® textbook.

## International mindedness

We are aware that students working from this resource will have begun full time education in the twenty-first century. IB students come from many cultures and have many different first languages. Sometimes, cultural and linguistic differences can be an obstacle to understanding. However, mathematics is largely free from cultural bias. Indeed, mathematics is considered by many to be a universal language. Even a Japanese algebra text book with Japanese characters will include recognisable equations using $x$ and $y$. The authors are very aware of the major and fundamental contribution made to mathematics by all cultures. Arabic, Indian, Greek and Chinese scholars have learned from each other and given a basis for the work of more recent mathematicians. We have introduced each topic with references to the history of mathematics to give a context to students' studies. The people who extended the boundaries of mathematical knowledge are many. We have referred to the major contributors by name, often with a photograph or other image.

## The graphic display calculator

The syllabus places great emphasis on the appropriate use of the graphic display calculator (GDC) in interpreting problems, so many teachers will wish to start with the Introduction to the graphic display calculator. This gives a general overview of the use of the GDC, which will assist students who may be unfamiliar with its use. Throughout the book we have then built upon this foundation work and have provided very clear and concise illustrations of exactly how such a calculator is used (not merely showing how a graph might look.) It must be noted that use of a GDC is not a substitution for authentic mathematical understanding. The two models of calculators used are the Casio *fx*-9860G and the Texas TI-84 Plus. Instructions for the Texas TI-Nspire for a selection of exercises are provided on the website.

We also refer to other computer software (Autograph and GeoGebra) where appropriate.

## Command terms and notation

Many examiners reports have highlighted that students are often uncertain what specific command terms mean. Students should become familiar with all these words as they will be faced with them during the external examinations. One such example is the difference between 'draw' and 'sketch'. This issue is dealt with on pages 309–310 of the textbook. Students should also be familiar with the list of notation that the IB uses in the external examinations. Students will not, however, be penalised in examinations for using different appropriate notation but understanding the command terms is essential. The IB has released a new list of command terms and notation with the new guide and the full lists can be found on the hodderplus website.

## The website

The material on the accompanying website is indicated through the book by coloured icons. These are:

powerpoints,

spreadsheets,

GeoGebra files, and

Personal Tutor presentations, step-by-step audio visual explanations of the harder concepts.

We hope that providing free access to this material will enable all students to engage with this invaluable material. A copy of the GeoGebra installer is also available to download.

The website also contains answers to the revision exercises, worked solutions to all the exercises and student assessments throughout the book and instructions for the use of the Texas TI-Nspire.

## Applications and theory of knowledge

Links to other Diploma subjects are made in blue boxes at the side of the text.

These areas should form a vital part of this course. The textbook therefore has a section at the end of each topic for discussion points such as 'Applications' (yellow), 'Project ideas' (**green**) and 'Theory of Knowledge' (**pink**). Students should not underestimate the importance of this facet of the course which is why we have given it a dedicated section at the end of each topic. It is worthwhile to explore these ideas as discussions in lessons to emphasize and explore one of the core elements of the IB diploma programme.

## Revision exercises

At the end of the book is a revision section, with exercises covering the whole course. It is expected that students will also have access to previous examination papers.

# Internal assessment

A project is chosen by the student and is assessed by their teacher and then is externally moderated using assessment criteria that relate to objectives for Group 5 Mathematics. These criteria are as follows:

| | | |
|---|---|---|
| Criterion A | Introduction | 3 marks |
| Criterion B | Information/measurement | 3 marks |
| Criterion C | Mathematical processes | 5 marks |
| Criterion D | Interpretation of results | 3 marks |
| Criterion E | Validity | 1 mark |
| Criterion F | Structure and communication | 3 marks |
| Criterion G | Notation and terminology | 2 marks |
| | | Total 20 marks |

## Project ideas

Project ideas can be found at the end of each topic (on green notes). They are only ideas and are not intended to be project titles. Where it is suggested that students extend their mathematical knowledge as part of the project it is important to discuss this thoroughly with a teacher, both before starting and as the project progresses.

Many teachers feel that too many students choose statistics projects which are too limited in the scope of the mathematics used. Students can look outside the syllabus; it may even be advantageous to look at areas of maths such as symmetry, topology, optimisation, matrices, advanced probability, calculus, linear programming and even mathematics as it applies to art, music and architecture (but be sure that maths is the emphasis). Your teacher is allowed to give help and guidance, so students should discuss their ideas before starting their project to be sure that it is feasible.

## Possible approach for a project

Before starting to plan your project there are a number of things to consider. Your project is assessed according to a number of headings (criteria) as listed on the previous page; these will be referred to below.

1 The project is part of your assessment. You should aim for the highest possible mark.

2 The project is expected to take more time to prepare, plan and complete than even the longest of the seven topics in the Mathematical Studies syllabus. So allow yourself time to get this work done.

3 This is your mathematics project and is meant to reflect your mathematical ability. You would think this does not need to be stated. However it is a fact that many teachers will have received projects which:
   a have a very interesting title (but little to do with maths)
   b are beautifully presented (but have little mathematical content)
   c have a lot of very repetitive, low-level maths (e.g. lots and lots of bar charts)
   d contain mathematics far above the ability of the student submitting it (be prepared to explain anything that you write).

4 You should try to involve mathematics from as many topics in this course as you can, and at the highest level possible. A sensible choice of project will allow you to show the mathematical knowledge you have gained, but in a different way than in a formal exam. You should plan to use a further mathematical process above that in the Mathematical Studies syllabus (your teacher may help with this).

5 The more general the project title, the easier it is to extend it into a variety of areas.

Below is an indication of how a project might be approached. The mathematics involved will be taught as you go through the course but we have shown where it is in the book should you wish to have a look.

7 The ratio of the interior angles of a pentagon is 2 : 3 : 4 : 4 : 5. Calculate the size of the largest angle.

8 A large swimming pool takes 36 hours to fill using three identical pumps.
  a How long would it take to fill using eight identical pumps?
  b If the pool needs to be filled in nine hours, how many pumps will be needed?

9 The first triangle is an enlargement of the second. Calculate the size of the missing sides and angles.

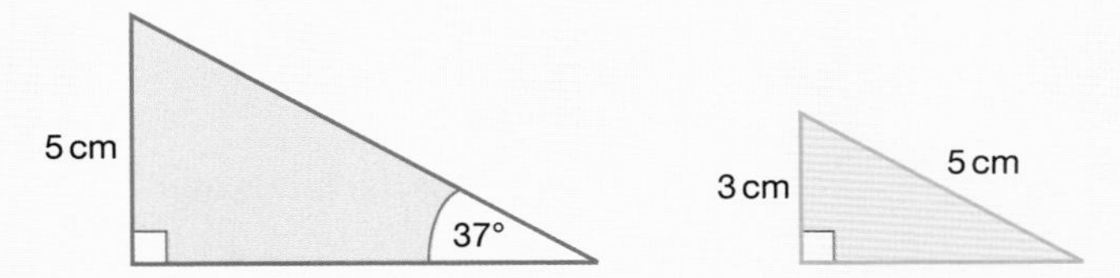


10 A tap issuing water at a rate of 1.2 litres per minute fills a container in 4 minutes.
  a How long would it take to fill the same container if the rate was decreased to 1 litre per minute? Give your answer in minutes and seconds.
  b If the container is to be filled in 3 minutes, calculate the rate at which the water should flow.

## Student assessment 4: Percentages

1 Copy the table below and fill in the missing values.

| Fraction | Decimal | Percentage |
| --- | --- | --- |
| | 0.25 | |
| $\frac{3}{5}$ | | |
| | | $62\frac{1}{2}\%$ |
| $2\frac{1}{4}$ | | |

2 Find 30% of 2500 metres.

3 In a sale a shop reduces its prices by 12.5%. What is the sale price of a desk previously costing €600?

4 In the last six years the value of a house has increased by 35%. If it cost £72 000 six years ago, what is its value now?

5 Express the first quantity as a percentage of the second.
  a 35 minutes, 2 hours
  b 650 g, 3 kg
  c 5 m, 4 m
  d 15 seconds, 3 minutes
  e 600 kg, 3 tonnes
  f 35 cl, 3.5 l

6 Shares in a company are bought for $600. After a year the same shares are sold for $550. Calculate the percentage depreciation.

7 In a sale, the price of a jacket originally costing 17 000 Japanese yen (¥) is reduced by ¥4000. Any item not sold by the last day of the sale is reduced by a further 50%. If the jacket is sold on the last day of the sale, calculate:
  **a** the price it is finally sold for
  **b** the overall percentage reduction in price.

8 Calculate the original price of each of the following.

| Selling price | Profit |
|---|---|
| \$224 | 12% |
| \$62.50 | 150% |
| \$660.24 | 26% |
| \$38.50 | 285% |

9 Calculate the original price of each of the following.

| Selling price | Loss |
|---|---|
| \$392.70 | 15% |
| \$2480 | 38% |
| \$3937.50 | 12.5% |
| \$4675 | 15% |

10 In an examination Sarah obtained 87.5% by gaining 105 marks. How many marks did she lose?

11 At the end of a year, a factory has produced 38 500 television sets. If this represents a 10% increase in productivity on last year, calculate the number of sets that were made last year.

12 A computer manufacturer is expected to have produced 24 000 units by the end of this year. If this represents a 4% decrease on last year's output, calculate the number of units produced last year.

13 A farmer increased his yield by 5% each year over the last five years. If he produced 600 tonnes this year, calculate to the nearest tonne his yield five years ago.

# Student assessment 5: Algebraic manipulation

1 Expand the following and simplify where possible.
  **a** $3(2x - 3y + 5z)$
  **b** $4p(2m - 7)$
  **c** $-4m(2mn - n^2)$
  **d** $4p^2(5pq - 2q^2 - 2p)$
  **e** $4x - 2(3x + 1)$
  **f** $4x(3x - 2) + 2(5x^2 - 3x)$
  **g** $\frac{1}{5}(15x - 10) - \frac{1}{3}(9x - 12)$
  **h** $\frac{x}{2}(4x - 6) + \frac{x}{4}(2x + 8)$

2 Factorize each of the following.
  **a** $16p - 8q$
  **b** $p^2 - 6pq$
  **c** $5p^2q - 10pq^2$
  **d** $9pq - 6p^2q + 12q^2p$

3 If $a = 4$, $b = 3$ and $c = -2$, evaluate the following.
  **a** $3a - 2b + 3c$
  **b** $5a - 3b^2$
  **c** $a^2 + b^2 + c^2$
  **d** $(a + b)(a - b)$
  **e** $a^2 - b^2$
  **f** $b^3 - c^3$

4 Rearrange the following formulae to make the **bold** letter the subject.

a $p = 4m + \mathbf{n}$

b $4x - 3\mathbf{y} = 5z$

c $2x = \dfrac{3\mathbf{y}}{5p}$

d $m(x + y) = 3w$

e $\dfrac{pq}{4\mathbf{r}} = \dfrac{mn}{t}$

f $\dfrac{p + \mathbf{q}}{r} = m - n$

5 Factorise each of the following fully.

a $pq - 3rq + pr - 3r^2$

b $1 - t^4$

c $875^2 - 125^2$

d $7.5^2 - 2.5^2$

6 Expand the following and simplify where possible.

a $(x - 4)(x + 2)$

b $(x - 8)^2$

c $(x + y)^2$

d $(x - 11)(x + 11)$

e $(3x - 2)(2x - 3)$

f $(5 - 3x)^2$

7 Factorize each of the following.

a $x^2 - 4x - 77$

b $x^2 - 6x + 9$

c $x^2 - 144$

d $3x^2 + 3x - 18$

e $2x^2 + 5x - 12$

f $4x^2 - 20x + 25$

8 Make the letter in **bold** the subject of the formula.

a $m\mathbf{f}^2 = p$

b $m = 5\mathbf{t}^2$

c $A = \pi r\sqrt{\mathbf{p} + q}$

d $\dfrac{1}{\mathbf{x}} + \dfrac{1}{y} = \dfrac{1}{t}$

9 Simplify the following algebraic fractions.

a $\dfrac{x^7}{x^3}$

b $\dfrac{mn}{p} \times \dfrac{pq}{m}$

c $\dfrac{(y^3)^3}{(y^2)^3}$

d $\dfrac{28pq^2}{7pq^3}$

# Student assessment 6: Equations and inequalities

**For questions 1–4, solve the equations.**

1 a $x + 7 = 16$

b $2x - 9 = 13$

c $8 - 4x = 24$

d $5 - 3x = -13$

2 a $7 - m = 4 + m$

b $5m - 3 = 3m + 11$

c $6m - 1 = 9m - 13$

d $18 - 3p = 6 + p$

3 a $\dfrac{x}{-5} = 2$

b $4 = \frac{1}{3}x$

c $\dfrac{x + 2}{3} = 4$

d $\dfrac{2x - 5}{7} = \dfrac{5}{2}$

4 a $\frac{2}{3}(x - 4) = 8$

b $4(x - 3) = 7(x + 2)$

c $4 = \frac{2}{7}(3x + 8)$

d $\frac{3}{4}(x - 1) = \frac{5}{8}(2x - 4)$

5 Solve the following simultaneous equations.

a $2x + 3y = 16$
$2x - 3y = 4$

b $4x + 2y = 22$
$-2x + 2y = 2$

c $x + y = 9$
$2x + 4y = 26$

d $2x - 3y = 7$
$-3x + 4y = -11$

6 The angles of a triangle are $x$, $2x$ and $(x + 40)$ degrees.

a Construct an equation in terms of $x$.
b Solve the equation.
c Calculate the size of each of the three angles.

7 Seven is added to three times a number. The result is then doubled. If the answer is 68, calculate the number.

8 A decagon has six equal exterior angles. Each of the remaining four is fifteen degrees larger than these six angles. Construct an equation and then solve it to find the sizes of the angles.

9 A rectangle is $x$ cm long. The length is 3 cm more than the width. The perimeter of the rectangle is 54 cm.

a Draw a diagram to illustrate the above information.
b Construct an equation in terms of $x$.
c Solve the equation and hence calculate the length and width of the rectangle.

10 At the end of a football season, the leading goal scorer in a league has scored eight more goals than the second leading goal scorer. The second has scored fifteen more than the third. The total number of goals scored by all three players is 134.

a Write an expression for each of the three scores.
b Form an equation and then solve it to find the number of goals scored by each player.

## Student assessment 7: Indices

1 Simplify the following by using indices.

a $2 \times 2 \times 2 \times 5 \times 5$
b $2 \times 2 \times 3 \times 3 \times 3 \times 3 \times 3$

2 Write the following out in full.

a $4^3$
b $6^4$

3 Work out the value of the following without using a calculator.

a $2^3 \times 10^2$
b $1^4 \times 3^3$

4 Find the value of $x$ in each of the following.

a $2^{(x-2)} = 32$
b $\frac{1}{4^x} = 16$
c $5^{(-x+2)} = 125$
d $8^{-x} = \frac{1}{2}$

5 Using indices, simplify the following.

a $3 \times 2 \times 2 \times 3 \times 27$
b $2 \times 2 \times 4 \times 4 \times 4 \times 2 \times 32$

6 Write the following out in full.

a $6^5$
b $2^{-5}$

# Student assessment 8: Geometry of plane shapes

1 Calculate the circumference and area of each of the following circles. Give your answers to one decimal place.

a

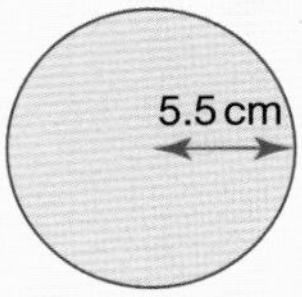

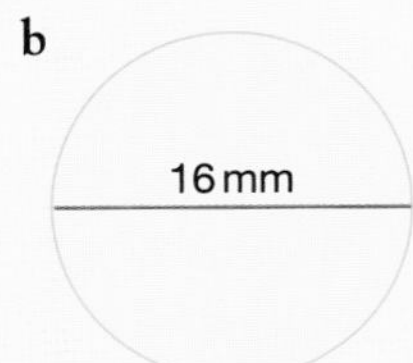


b

2 A semi-circular shape is cut out of the side of a rectangle as shown. Calculate the shaded area to one decimal place.

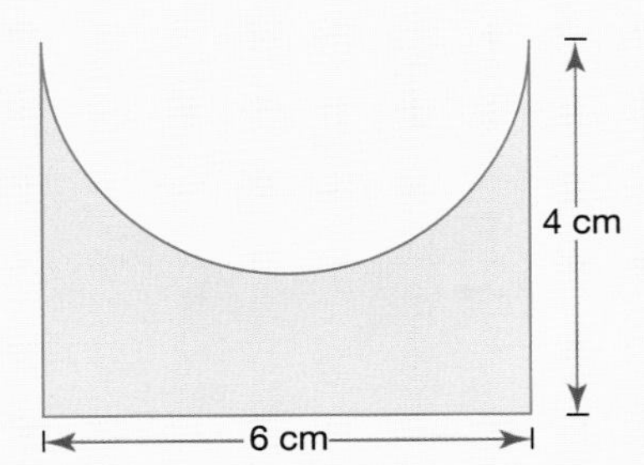


3 For the diagram below, calculate the area of:

a the semi-circle

b the trapezium

c the whole shape.

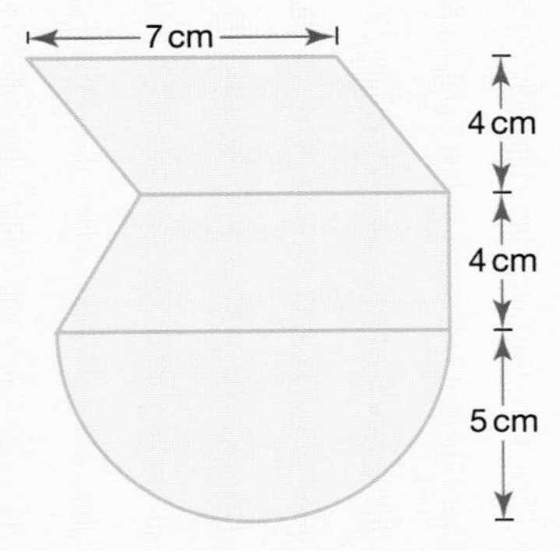


4 Calculate the circumference and area of each of these circles. Give your answers to one decimal place.

a

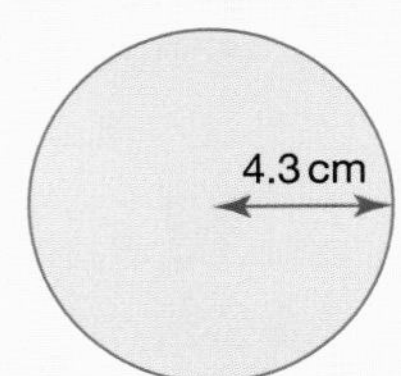


b

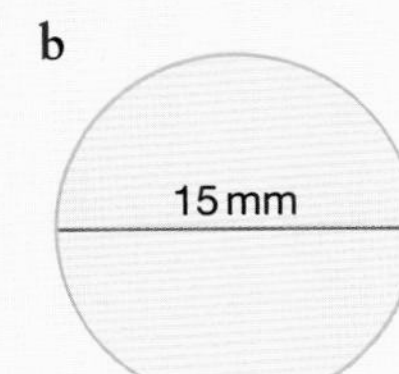

**5** A rectangle of length 32 cm and width 20 cm has a semi-circle cut out of two of its sides as shown. Calculate the shaded area to one decimal place.

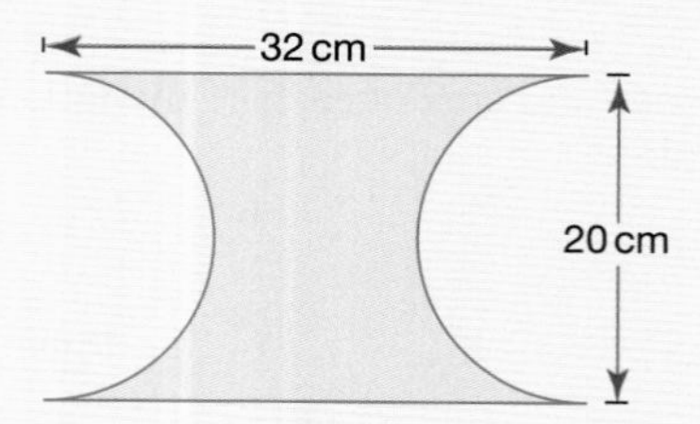


**6** Calculate the area of:

**a** the semi-circle

**b** the parallelogram

**c** the whole shape.

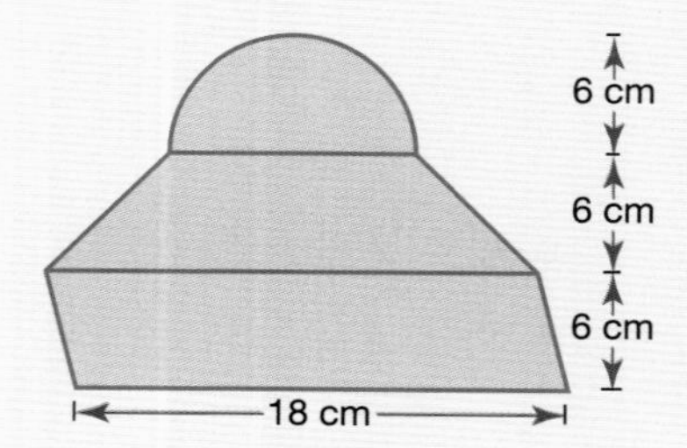

# Introduction to the graphic display calculator

## Introduction

People have always used devices to help them calculate. Today, basic calculators, scientific calculators and graphic display calculators are all available; they have a long history.

an early abacus

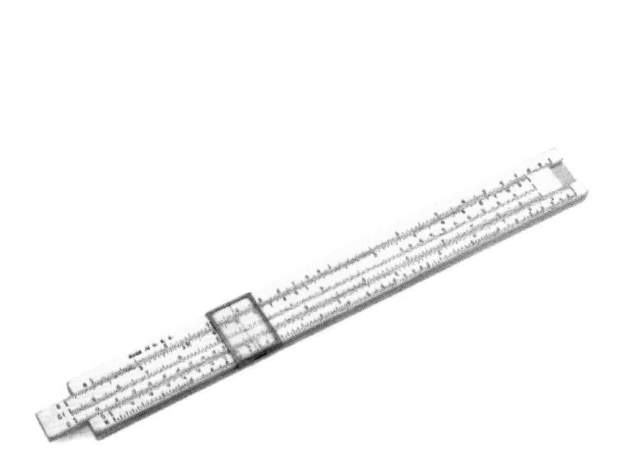

an early slide rule

an early calculator

## Using a graphic display calculator

**Graphic display calculators** (GDCs) are a powerful tool used for the study of mathematics in the modern classroom. However, as with all tools, their effectiveness is only apparent when used properly. This section will look at some of the key features of the GDC, so that you start to understand some of its potential. The two models used are the Casio *fx*-9860G and the Texas TI-84 Plus. Many GDCs have similar capabilities to the ones shown. However, if your calculator is different, it is important that you take the time to familiarize yourself with it.

Here is the home screen (menu/applications) for each calculator.

| Casio | Texas |
| --- | --- |
| MAIN MENU: RUN·MAT 1, STAT 2, e·ACT 3, S·SHT 4, GRAPH 5, DYNA 6, TABLE 7, RECUR 8, CONICS 9, EQUA A, PRGM B, TVM C | ■ |
| The modes are selected by using the arrows key and then pressing EXE, or by typing the number/letter that appears in the bottom right-hand corner of each square representing a mode. Brief descriptions of the seven most relevant modes are given below. | The main features are accessed by pressing the appropriate key. Some are explained below. |
| **1** RUN.MAT is used for arithmetic calculations. | [LIST / STAT] is used for statistical calculations and for drawing graphs of the data entered. |
| **2** STAT is used for statistical calculations and for drawing graphs. | [TEST A / MATH] is used to access numerical operations. |
| **4** S.SHT is a spreadsheet and can be used for calculations and graphs. | [STAT PLOT F1 / Y=] is used for entering the equations of graphs. |
| **5** GRAPH is used for entering the equations of graphs and plotting them. | [TABLE F5 / GRAPH] is used for graphing functions. |
| **6** DYNA is a dynamic graph mode that enables a family of curves to be graphed simultaneously. | |
| **7** TABLE is used to generate a table of results from an equation. | |
| **A** EQUA is used to solve different types of equation. | |

Casio

Texas

## Basic calculations

The aim of the following exercise is to familiarize you with some of the buttons on your calculator dealing with basic mathematical operations. It is assumed that you will already be familiar with the mathematical content.

## Exercise 1

Using your GDC, solve the following.

| Casio | Texas |
|---|---|
| $\sqrt{\ }$ $x^2$ | $\sqrt{\ }$ $x^2$ |

**1** a $\sqrt{625}$

b $\sqrt{324}$

c $2\sqrt{8} \times 5\sqrt{2}$

| Casio | Texas | |
|---|---|---|
| $\sqrt[x]{\ }$ ^ | TEST MATH | 5 |

**2** a $\sqrt[3]{1728}$

b $\sqrt[4]{1296}$

c $\sqrt[5]{3125}$

| Casio | Texas |
|---|---|
| $\sqrt{\ }$ $x^2$ | $\sqrt{\ }$ $x^2$ |

**3** a $13^2$

b $8^2 \div 4^2$

c $\sqrt{5^2 + 12^2}$

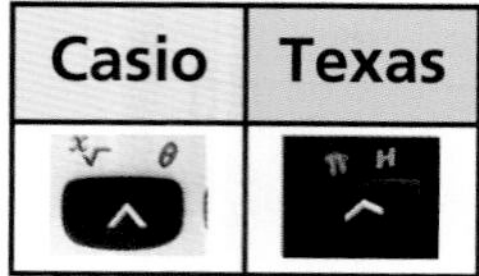

| Casio | Texas |
| --- | --- |
| ^ | ^ |

4 a $6^3$

b $9^4 \div 27^2$

c $\sqrt[4]{\dfrac{4^3 \times 2^8}{8^2}}$

| Casio | Texas | |
| --- | --- | --- |
| EXP | 2ND | EE , |

5 a $(2.3 \times 10^3) + (12.1 \times 10^2)$

b $(4.03 \times 10^3) + (15.6 \times 10^4) - (1.05 \times 10^4)$

c $\dfrac{13.95 \times 10^6}{15.5 \times 10^3} - (9 \times 10^2)$

GDCs also have a large number of memory channels. Use these to store answers which are needed for subsequent calculations. This will minimise rounding errors.

| Casio | Texas |
| --- | --- |
| → ALPHA followed by a letter of the alphabet | STO▸ ALPHA followed by a letter of the alphabet |

## Exercise 2

1 In the following expressions, $a = 5$, $b = 4$ and $c = 2$.
Enter each of these values in memory channels A, B and C of your GDC respectively, and determine the following.

a $a + b + c$

b $a - (b + c)$

c $(a + b)^2 - c$

d $\dfrac{2(b + c)^3}{(a - c)}$

e $\dfrac{4\sqrt{a^2 - b^2}}{c}$

f $\dfrac{(ac)^2 + ba^2}{a + b + c}$

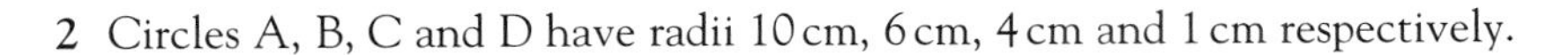

2 Circles A, B, C and D have radii 10 cm, 6 cm, 4 cm and 1 cm respectively.

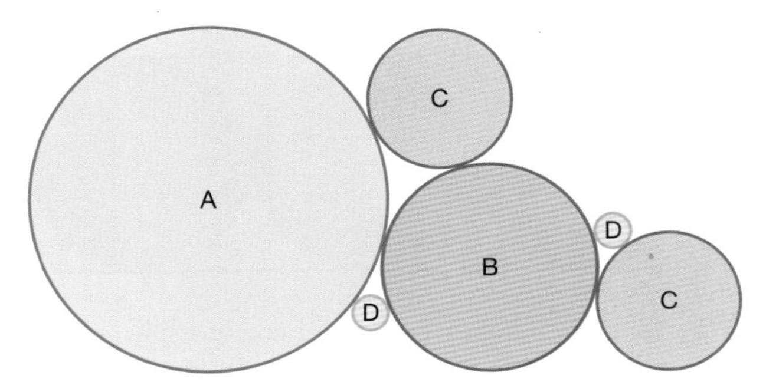


a Calculate the area of circle A and store your answer in memory channel A.
b Calculate the area of circle B and store your answer in memory channel B.
c Calculate the area of each of the circles C and D, storing the answers in memory channels C and D respectively.
d Using your calculator evaluate A + B + 2C + 2D.
e What does the answer to part **d** represent?

3 A child's shape-sorting toy is shown in the diagram. The top consists of a rectangular piece of wood of dimension 30 cm × 12 cm. Four shapes W, X, Y and Z are cut out of it.

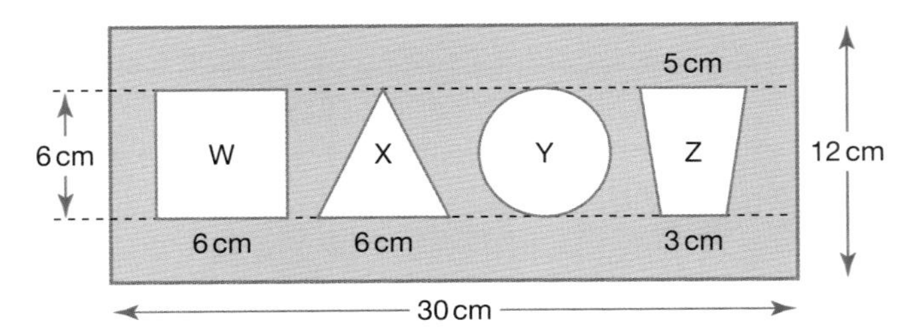


a Calculate the area of the triangle X. Store the answer in your calculator's memory.
b Calculate the area of the trapezium Z. Store the answer in your calculator's memory.
c Calculate the total area of the shapes W, X, Y and Z.
d Calculate the area of the rectangular piece of wood left once the shapes have been cut out.

4 Three balls just fit inside a cylindrical tube as shown. The radius ($r$) of each ball is 5 cm.

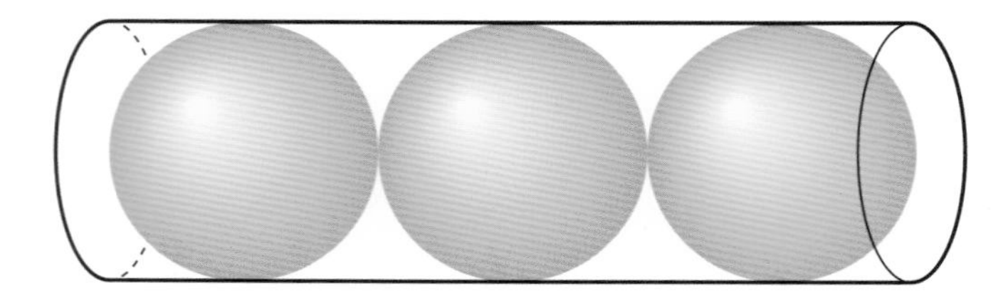

a Using the formula for the volume of a sphere, $V = \frac{4}{3}\pi r^3$, calculate the volume of one of the balls. Store the answer in the memory of your calculator.
b Calculate the volume of the cylinder.
c Calculate the volume of the cylinder *not* occupied by the three balls.

## Plotting graphs

One of a GDC's principal features is to plot graphs of functions. This helps to visualize what the function looks like and, later on, it will help solve a number of different types of problem. This section aims to show how to graph a variety of different functions. For example, to plot a graph of the function $y = 2x + 3$, use the following functions on your calculator.

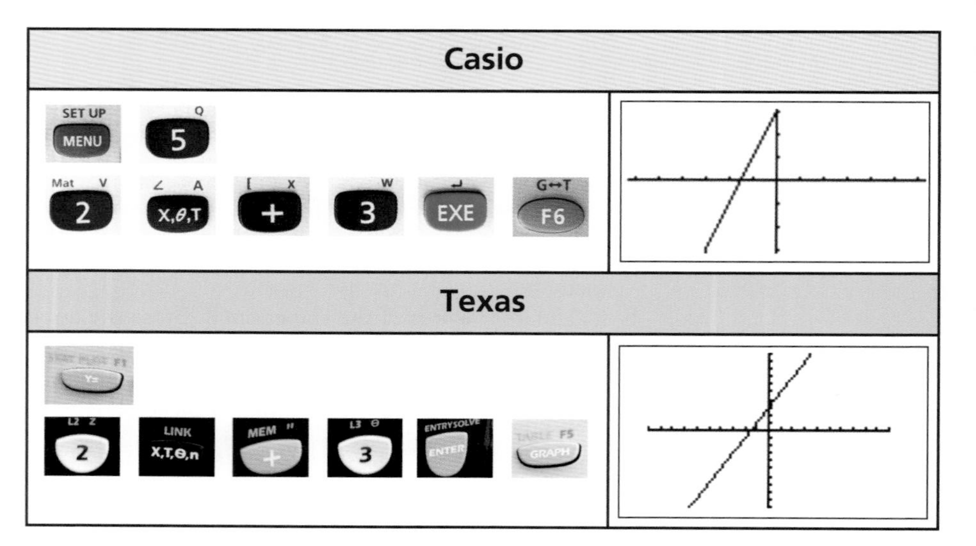


Occasionally, it may be necessary to change the scale on the axes to change how much of the graph, or what part of the graph, can be seen. This can be done in several ways, two of which are described here.

- By using the zoom facility

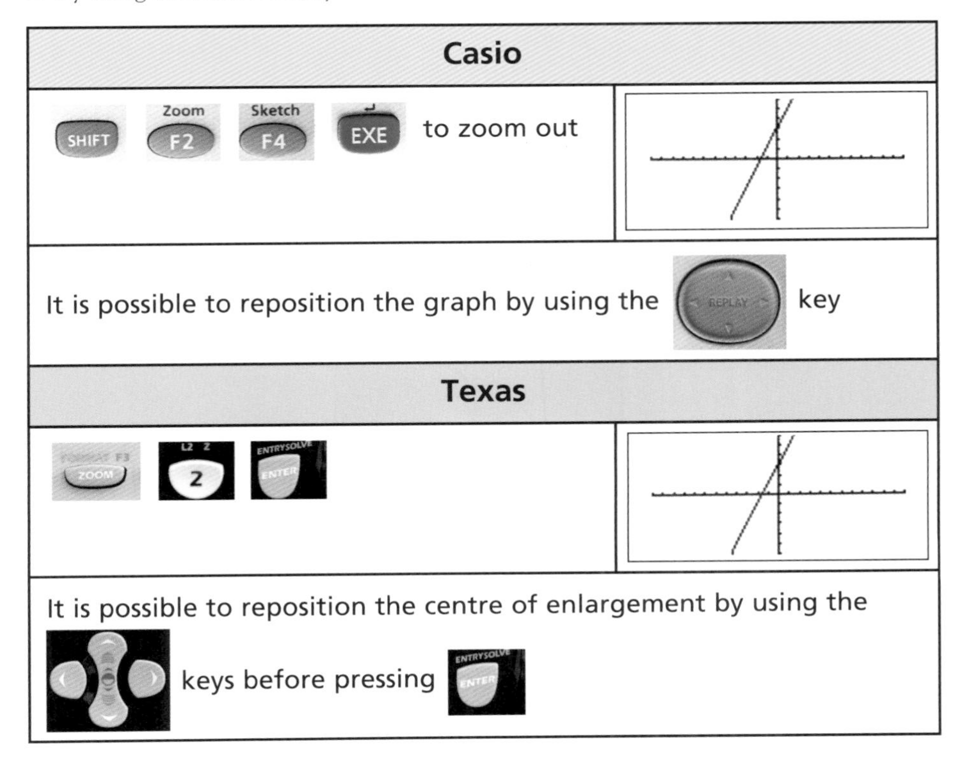

**Casio**

SHIFT Zoom F2 Sketch F4 EXE to zoom out

It is possible to reposition the graph by using the REPLAY key

**Texas**

ZOOM 2 ENTER

It is possible to reposition the centre of enlargement by using the keys before pressing ENTER

- By changing the scale manually

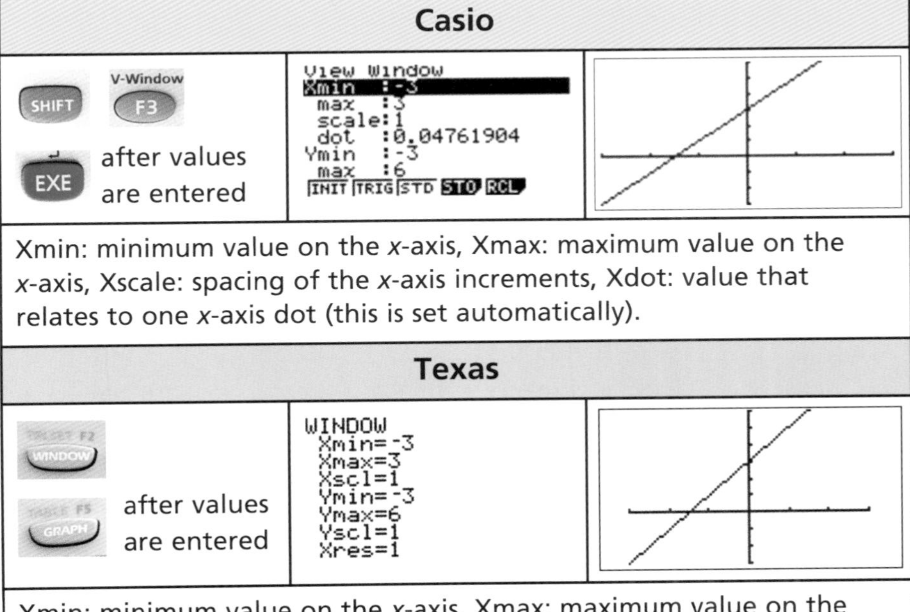

**Casio**

SHIFT V-Window F3

EXE after values are entered



Xmin: minimum value on the *x*-axis, Xmax: maximum value on the *x*-axis, Xscale: spacing of the *x*-axis increments, Xdot: value that relates to one *x*-axis dot (this is set automatically).

**Texas**

WINDOW

GRAPH after values are entered



Xmin: minimum value on the *x*-axis, Xmax: maximum value on the *x*-axis, Xscl: spacing of the *x*-axis increments.

## Exercise 3

In the following questions, the axes have been set to their default settings, i.e. Xmin = −10, Xmax = 10, Xscale = 1, Ymin = −10, Ymax = 10, Yscale = 1.

i)

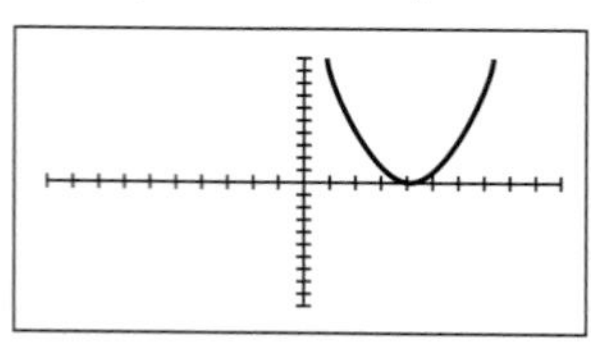

ii)

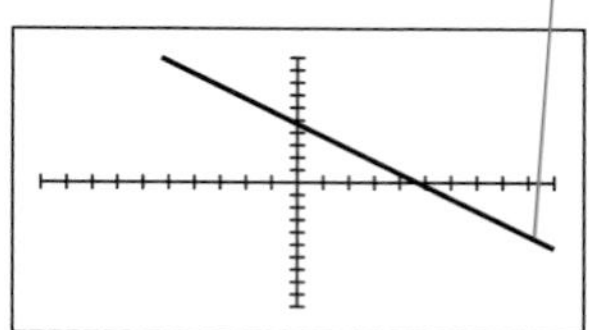

iii)

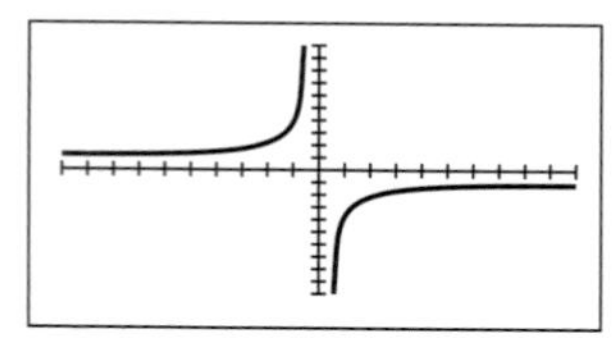

iv)

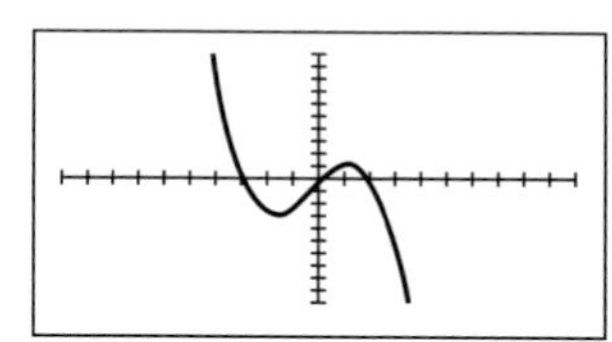

v)

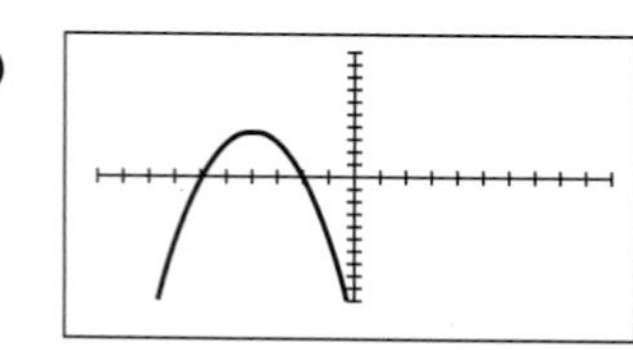

vi)

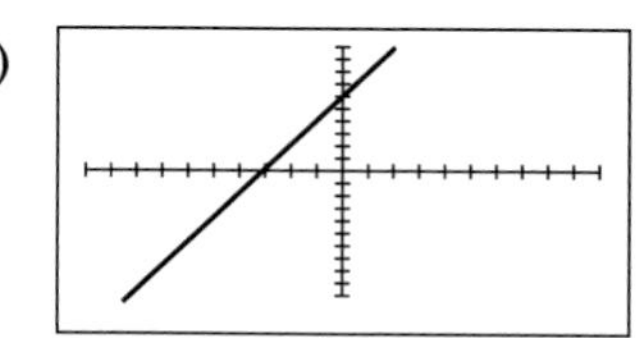

vii)

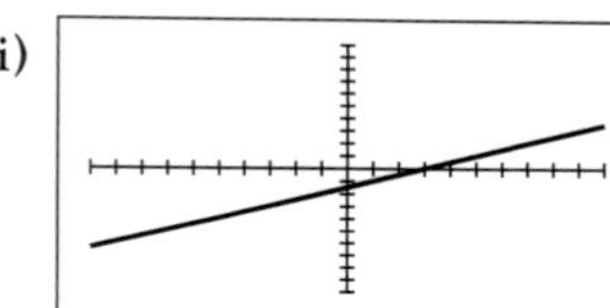

viii)

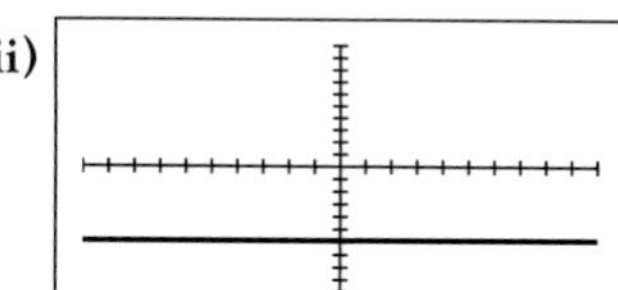

ix)

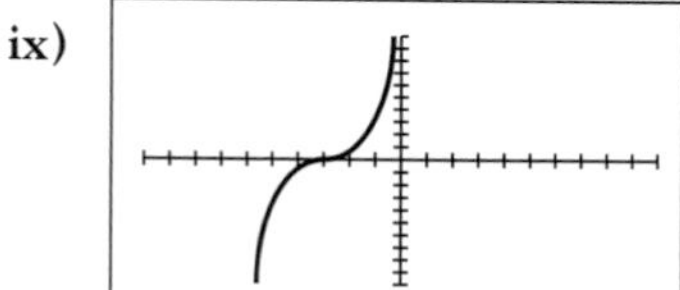

x)

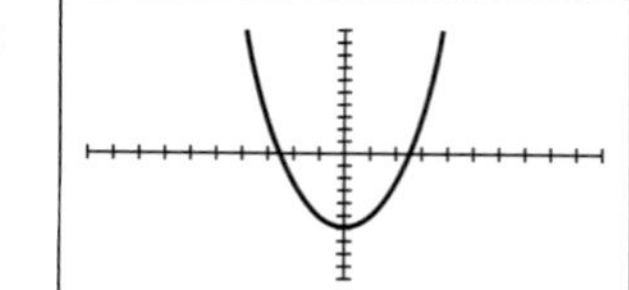

1 By using your GDC to graph the following functions, match each of them to the correct graph.

**a** $y = 2x + 6$

**b** $y = \frac{1}{2}x - 2$

**c** $y = -x + 5$

**d** $y = -\frac{5}{x}$

**e** $y = x^2 - 6$

**f** $y = (x - 4)^2$

**g** $y = -(x + 4)^2 + 4$

**h** $y = \frac{1}{2}(x + 3)^3$

**i** $y = -\frac{1}{3}x^3 + 2x - 1$

**j** $y = -6$

2 In each of the following, a function and its graph are given. Using your GDC, enter a function that produces a reflection of the original function in the $x$-axis.

**a** $y = x + 5$

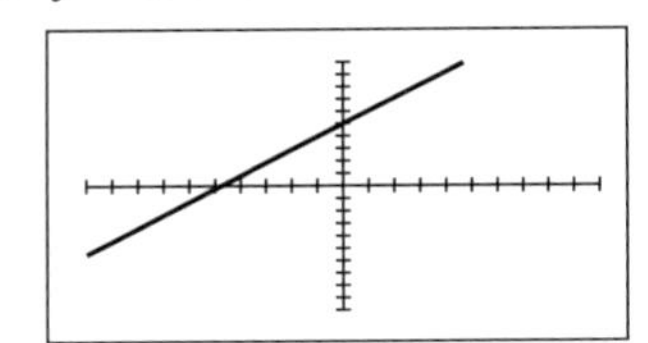

**b** $y = -2x + 4$

**c** $y = (x + 5)^2$

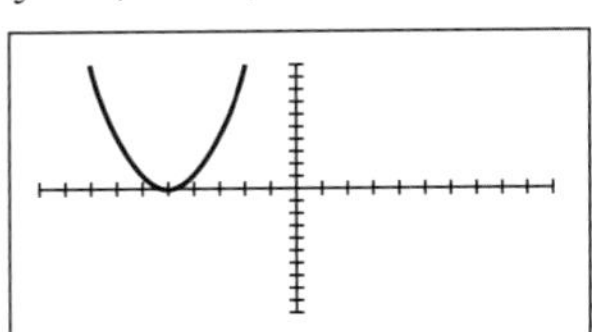

**d** $y = (x - 5)^2 + 3$

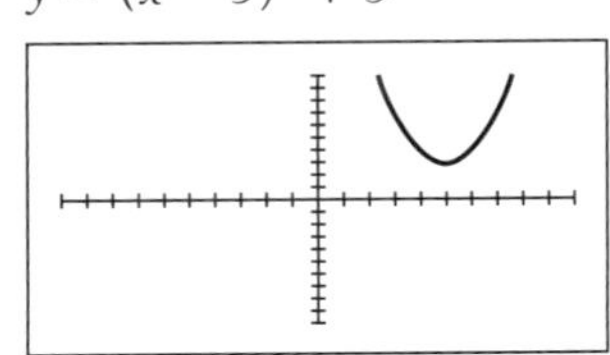

3 Using your GDC, enter a function that produces a reflection in the $y$-axis of each of the original functions in question 2.

4 By entering appropriate functions into your calculator:

**i)** make your GDC screen look like the ones shown

**ii)** write down the functions you used.

**a**

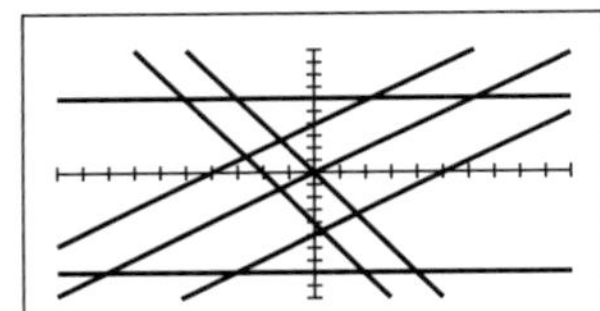

**b**

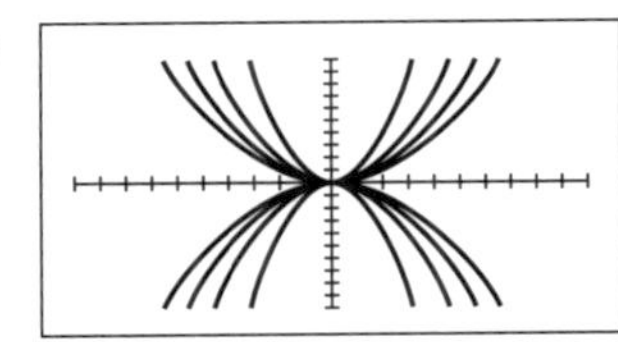

**c**

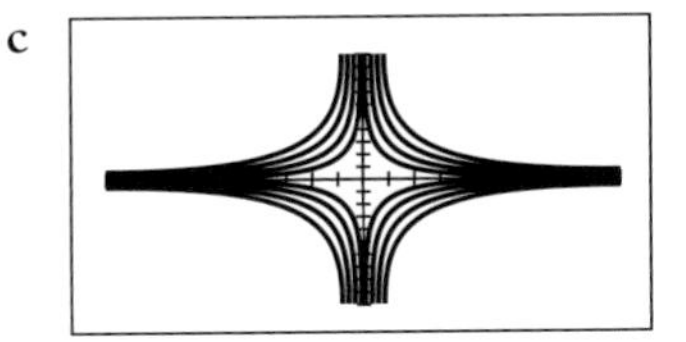

**d**

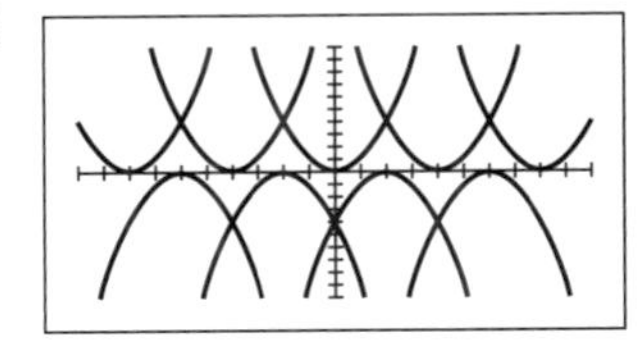

## Intersections

When graphing a function it is often necessary to find where it intersects one or both of the axes. If more than one function is graphed simultaneously, it may also be necessary to find where the graphs intersect each other. GDCs have a 'Trace' facility which gives an approximate coordinate of a cursor on the screen. More accurate methods are available and will be introduced in the topics as appropriate.

**Worked example**

Find where the graph of $y = \frac{1}{5}(x + 3)^3 + 2$ intersects both the $x$- and $y$-axes.

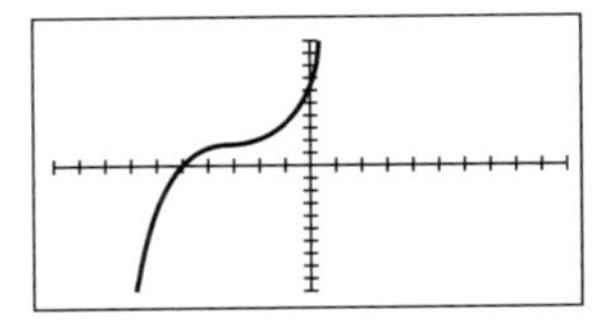

The graph shows that $y = \frac{1}{5}(x + 3)^3 + 2$ intersects each of the axes once.

To find the approximate coordinates of the points of intersection:

**Casio**

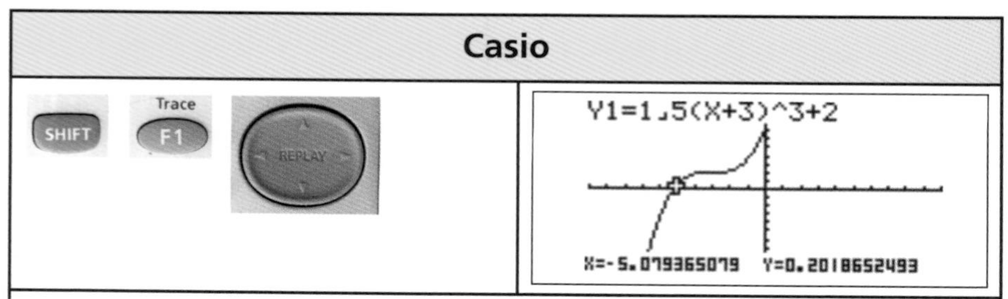


Because of the pixels on the screen, the 'Trace' facility will usually only give an approximate value. The $y$-coordinate of the point of intersection with the $x$-axis will always be zero. However, the calculator's closest result is $y = 0.202$. The $x$-value of –5.079 will also therefore be only an approximation. By moving the cursor to the point of intersection with the $y$-axis, values of $x = 0$ and $y = 7.4$ are obtained.

**Texas**

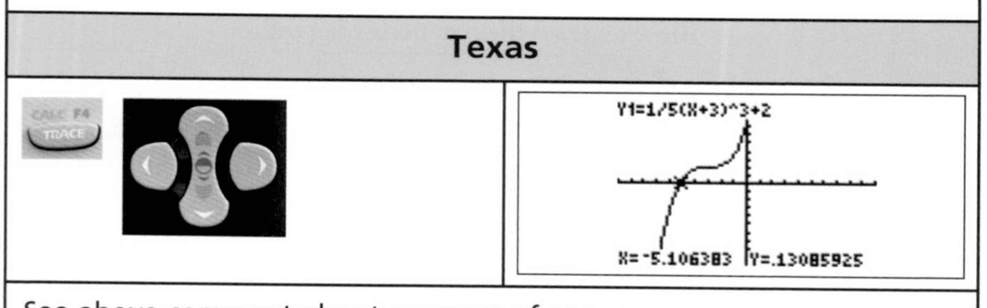


See above comment about accuracy of answers.

## Exercise 4

1 Find an approximate solution to where the following graphs intersect both the $x$- and $y$-axes using your GDC.

**a** $y = x^2 - 3$

**b** $y = (x + 3)^2 + 2$

**c** $y = \frac{1}{2}x^3 - 2x^2 + x + 1$

**d** $y = \frac{-5}{x + 2} + 6$

2 Find the coordinates of the point(s) of intersection of each of the following pairs of equations using your GDC.

**a** $y = x + 3$ and $y = -2x - 2$

**b** $y = -x + 1$ and $y = \frac{1}{2}(x^2 - 3)$

**c** $y = -x^2 + 1$ and $y = \frac{1}{2}(x^2 - 3)$

**d** $y = -\frac{1}{4}x^3 + 2x^2 - 3$ and $y = \frac{1}{2}x^2 - 2$

## Tables of results

A function such as $y = \frac{3}{x} + 2$ implies that there is a relationship between $y$ and $x$.

To plot the graph manually, the coordinates of several points on the graph need to be calculated and then plotted. GDCs have the facility to produce a table of values giving the coordinates of some of the points on the line.

**Worked example**

For the function $y = \frac{3}{x} + 2$, complete the following table of values using the table facility of your GDC.

| $x$ | −3 | −2 | −1 | 0 | 1 | 2 | 3 |
|---|---|---|---|---|---|---|---|
| $y$ | | | | | | | |

### Casio

Enter function $y = \frac{3}{x} + 2$

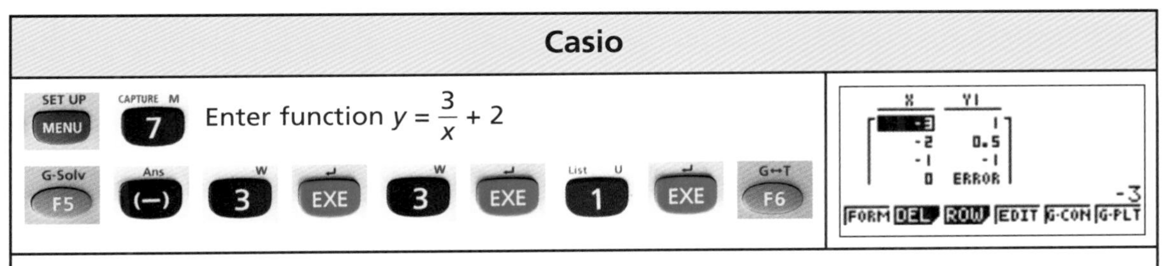

The screen

Table Setting
X
Start:-3
End :3
Step :1

shows that the $x$-values range from −3 to 3 in increments of 1.

Once the table is displayed, the remaining results can be viewed by using

### Texas

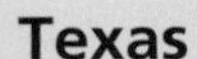

Enter function $y = \frac{3}{x} + 2$

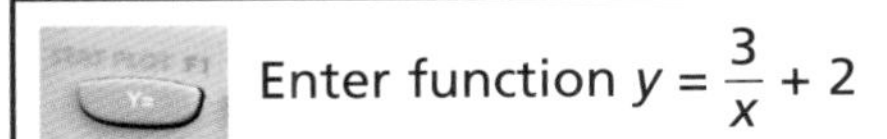

The screen

TABLE SETUP
TblStart=-3
ΔTbl=1
Indpnt: Auto Ask
Depend: Auto Ask

shows that the $x$-values start at −3 and increase in increments of 1.

Once the table is displayed, further results can be viewed by using

## Exercise 5

1 Copy and complete the tables of values for the following functions using the table facility of your GDC.

a $y = x^2 + x - 4$

| $x$ | −3 | −2 | −1 | 0 | 1 | 2 | 3 |
|---|---|---|---|---|---|---|---|
| $y$ | | | | | | | |

b $y = x^3 + x^2 - 10$

| $x$ | −3 | −2 | −1 | 0 | 1 | 2 | 3 |
|---|---|---|---|---|---|---|---|
| $y$ | | | | | | | |

c $y = \frac{4}{x}$

| $x$ | 0 | 0.5 | 1 | 1.5 | 2 | 2.5 | 3 |
|---|---|---|---|---|---|---|---|
| $y$ | | | | | | | |

d $y = \sqrt{(x+1)}$

| $x$ | −1 | −0.5 | 0 | 0.5 | 1 | 1.5 | 2 | 2.5 | 3 |
|---|---|---|---|---|---|---|---|---|---|
| $y$ | | | | | | | | | |

2 A car accelerates from rest. Its speed, $y\,\text{m}\,\text{s}^{-1}$, $x$ seconds after starting, is given by the equation $y = 1.8x$.

a Using the table facility of your GDC, calculate the speed of the car every 2 seconds for the first 20 seconds.

b How fast was the car travelling after 10 seconds?

3 A ball is thrown vertically upwards. Its height $y$ metres, $x$ seconds after launch, is given by the equation $y = 15x - 5x^2$.

a Using the table facility of your GDC, calculate the height of the ball each $\frac{1}{2}$ second during the first 4 seconds.

b What is the greatest height reached by the ball?

c How many seconds after its launch did the ball reach its highest point?

d After how many seconds did the ball hit the ground?

e In the context of this question, why can the values for $x = 3.5$ and $x = 4$ be ignored?

## Lists

Data is often collected and then analyzed so that observations and conclusions can be made. GDCs have the facility for storing data as lists. Once stored as a list, many different types of calculations can be carried out. This section will explain how to enter data as a list and then how to carry out some simple calculations.

**Worked example**

An athlete records her time (seconds) in ten races for running 100 m. These are shown below.

12.4 12.7 12.6 12.9 12.4 12.3 12.7 12.4 12.5 13.1

Calculate the mean, median and mode for this set of data using the list facility of your GDC.

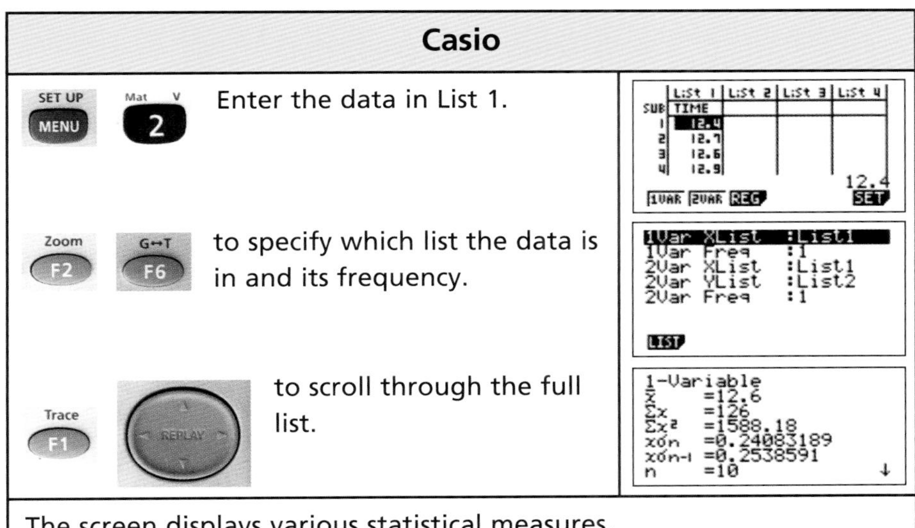


The screen displays various statistical measures.
$\overline{x}$ is the mean, $n$ is the number of data items, Med is the median, Mod is the modal value, Mod: F is the frequency of the modal values.

## Texas

 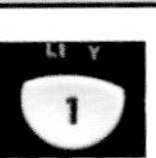 Enter the data in List 1.

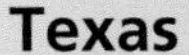

  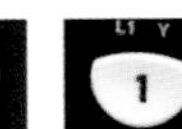   

to apply calculations to the data in List 1.

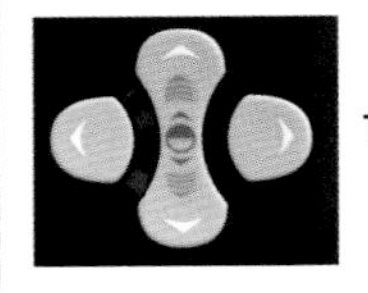 to scroll through the full list.

The screen displays various statistical measures.
$\overline{x}$ is the mean, $n$ is the number of data items, Med is the median. The T1-84 does not display the modal value.

If a lot of data is collected, it is often presented in a frequency table.

**Worked example**

The numbers of pupils in 30 maths classes are shown in the frequency table.

| Number of pupils | Frequency |
|---|---|
| 27 | 4 |
| 28 | 6 |
| 29 | 9 |
| 30 | 7 |
| 31 | 3 |
| 32 | 1 |

Calculate the mean, median and mode for this set of data using the list facility of your GDC.

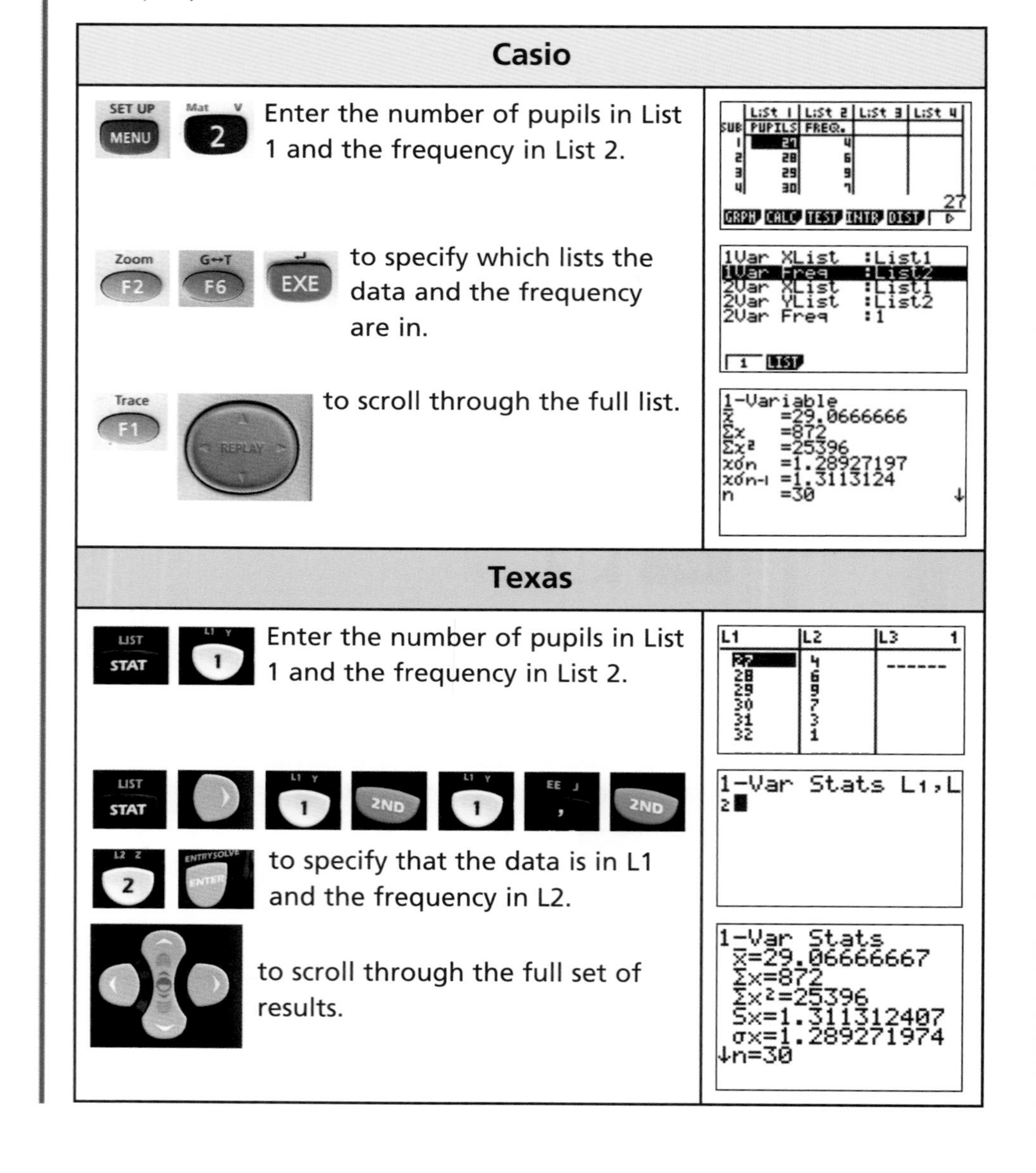

## Exercise 6

1 Find the mean, the median and, if possible, the mode of these sets of numbers using the list facility of your GDC.
 **a** 3, 6, 10, 8, 9, 10, 12, 4, 6, 10, 9, 4
 **b** 12.5, 13.6, 12.2, 14.4, 17.1, 14.8, 20.9, 12.2

2 During a board game, a player makes a note of the numbers he rolls on the dice. These are shown in the frequency table below.

| **Number on dice** | 1 | 2 | 3 | 4 | 5 | 6 |
|---|---|---|---|---|---|---|
| **Frequency** | 3 | 8 | 5 | 2 | 5 | 7 |

Find the mean, the median and, if possible, the modal dice roll using the list facility of your GDC.

3 A class of 30 pupils sat two maths tests. Their scores out of 10 are recorded in the frequency tables below.

| **Test A** | | | | | | | | | | |
|---|---|---|---|---|---|---|---|---|---|---|
| **Score** | 1 | 2 | 3 | 4 | 5 | 6 | 7 | 8 | 9 | 10 |
| **Frequency** | 3 | 2 | 4 | 3 | 1 | 8 | 3 | 1 | 3 | 2 |

| **Test B** | | | | | | | | | | |
|---|---|---|---|---|---|---|---|---|---|---|
| **Score** | 1 | 2 | 3 | 4 | 5 | 6 | 7 | 8 | 9 | 10 |
| **Frequency** | 4 | 1 | 0 | 0 | 0 | 24 | 0 | 0 | 0 | 1 |

 **a** Find the mean, the median and, if possible, the mode for each test using the list facility of your GDC.
 **b** Comment on any similarities or differences in your answers to part **a**.
 **c** Which test did the class find easiest? Give reasons for your answer.

# Applications, project ideas and theory of knowledge

1 The use of mathematical tables has declined since the development of scientific calculators. Find examples of both maritime (naval) and other mathematical books of tables and discuss their use.

2 You may be able to get access to an abacus and slide rule. How were they used?

3 Use of devices such as a slide rule and books of calculation tables took time to learn. At what point did using a slide rule justify the time taken to learn its use? Speculate on which professionals might have used such a tool.

4 Investigate what an 'Enigma Machine' was. Where, when and for what purpose was it designed?

5 What are punch cards and how were they used? Investigate simple early computer languages (like basic). This could be a starting point for a project.

6 'Modern graphic display calculators are more powerful than 20-year-old computers.' Discuss this statement.

7 Investigate and learn how to use logarithm and other mathematical tables.

8 'The problem with calculators is that nobody can estimate any more.' Discuss this statement. Design an experiment to test the ability of your classmates to estimate quickly and accurately. This could form the basis of a project.

9 What is meant by 'The law of diminishing returns'? Use your calculator to illustrate it.

10 Explain the difference between 'knowing how' and 'knowing that' with reference to the use of a calculator.

11 'Truth, belief and knowledge are interconnected.' Discuss.

Topic 1

# Number and algebra

## Syllabus content

**1.1** Natural numbers, $\mathbb{N}$; integers, $\mathbb{Z}$; rational numbers, $\mathbb{Q}$; and real numbers, $\mathbb{R}$.

**1.2** Approximation: decimal places; significant figures. Percentage errors.

Estimation.

**1.3** Expressing numbers in the form $a \times 10^k$ where $1 \le a < 10$ and $k$ is an integer. Operations with numbers in this form.

**1.4** SI (*Système International*) and other basic units of measurement: for example, kilogram (kg), metre (m), second (s), litre (l), metre per second ($\text{m s}^{-1}$), Celsius scale.

**1.5** Currency conversions.

**1.6** Use of a GDC to solve: pairs of linear equations in two variables; quadratic equations.

**1.7** Arithmetic sequences and series, and their applications.

Use of the formulae for the $n$th term and the sum of the first $n$ terms of the sequence.

**1.8** Geometric sequences and series.

Use of the formulae for the $n$th term and the sum of the first $n$ terms of the sequence.

**1.9** Financial applications of geometric sequences and series: compound interest; annual depreciation.

## Introduction

The first written records showing the origin and development of the use of money were found in the city of Eridu in Mesopotamia (modern Iraq). The records were on tablets like these ones found at Uruk.

The Sumerians, as the people of this region were known, used a system of recording value, known as 'Cuneiform', five thousand years ago. This writing is now believed to be simply accounts of grain surpluses. This may sound insignificant now, but the change from a hunter–gatherer society to a farming-based society led directly to the kind of sophisticated way of life we have today.

Our present number system has a long history, originating from the Indian Brahmi numerals around 300 BCE.

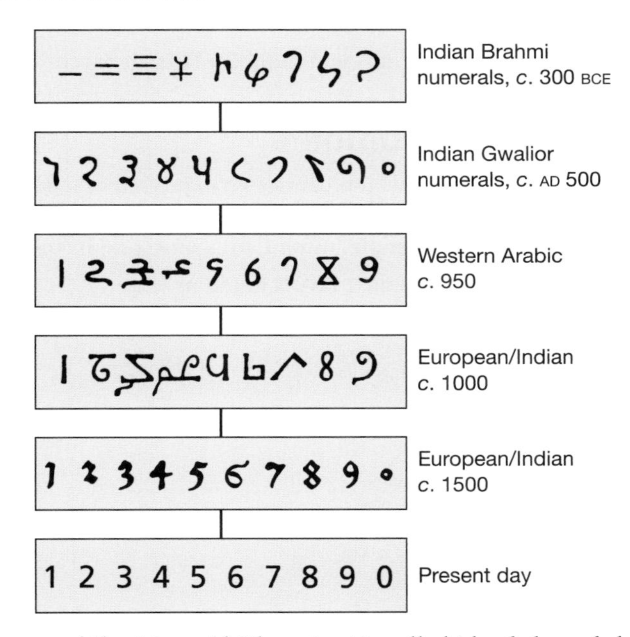


Abu Ja'far Muhammad Ibn Musa Al-Khwarizmi is called 'the father of algebra'. He was born in Baghdad in 790 AD and wrote the book *Hisab al-jabr w'al-muqabala* in 830 AD from which the word algebra (*al-jabr*) is taken. He worked at the university of Baghdad, then the greatest in the world.

The poet Omar Khayyam is known for his long poem the *Rubaiyat*. He was also a fine mathematician who worked on the Binomial theorem. He introduced the symbol 'shay' which became our '$x$'.

# 1.1 Sets of numbers

## Natural numbers

A child learns to count: 'one, two, three, four, …'. These are sometimes called the counting numbers or whole numbers.

The child will say 'I am three', or 'I live at number 73'.

If we include the number 0, then we have the set of numbers called the **natural numbers**. The set of natural numbers $\mathbb{N} = \{0, 1, 2, 3, 4, \ldots\}$.

## Integers

On a cold day, the temperature may be 4 °C at 10p.m. If the temperature drops by a further 6 °C, then the temperature is 'below zero'; it is −2 °C.

If you are overdrawn at the bank by £200, this could be shown as −£200.

The set of **integers** $\mathbb{Z} = \{\ldots -3, -2, -1, 0, 1, 2, 3, \ldots\}$.

$\mathbb{Z}$ is therefore an extension of $\mathbb{N}$. Every natural number is an integer.

## Rational numbers

A child may say 'I am three'; she may also say 'I am three and a half', or even 'three and a quarter'. $3\frac{1}{2}$ and $3\frac{1}{4}$ are **rational numbers**. All rational numbers can be written as a fraction whose denominator is not zero. All terminating and recurring decimals are rational numbers as they can be written as fractions too, e.g.

$0.2 = \frac{1}{5}$ $\quad 0.3 = \frac{3}{10}$ $\quad 7 = \frac{7}{1}$ $\quad 1.53 = \frac{153}{100}$ $\quad 0.\dot{2} = \frac{2}{9}$

The set of rational numbers $\mathbb{Q}$ is an extension of the set of integers.

## Real numbers

Numbers which cannot be expressed as a fraction are not rational numbers; they are **irrational numbers**.

For example, using Pythagoras' rule in the triangle shown below, the length of the hypotenuse AC is found as $\sqrt{2}$:

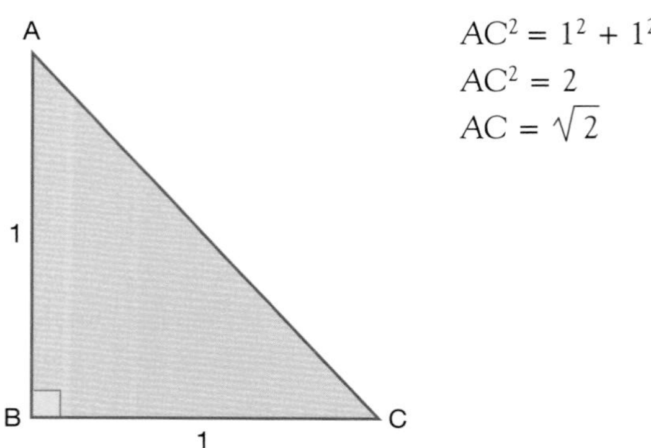


$$AC^2 = 1^2 + 1^2$$
$$AC^2 = 2$$
$$AC = \sqrt{2}$$

$\sqrt{2} = 1.41421356\ldots$ . The digits in this number do not recur or repeat. This is a property of all irrational numbers. Another example of an irrational number which you will come across is $\pi$ (pi) $= 3.141592654\ldots$ . The set of rational and irrational numbers together form the set of **real numbers** $\mathbb{R}$. There are also numbers, called imaginary numbers, which are not real, but all the numbers that you will come across in this textbook are real numbers.

## Exercise 1.1.1

1 State to which of the sets $\mathbb{N}$, $\mathbb{Z}$, $\mathbb{Q}$ and $\mathbb{R}$ these numbers belong.

a 3   b $-5$   c $\sqrt{3}$   d $11.\dot{3}$

In questions 2–6, state, giving reasons, whether each number is rational or irrational.

2 a 1.3   b $0.\dot{6}$   c $\sqrt{3}$

3 a $-2\frac{2}{5}$   b $\sqrt{25}$   c $\sqrt[3]{8}$

4 a $\sqrt{7}$ b 0.625 c $0.\dot{1}$

5 a $\sqrt{2} \times \sqrt{3}$ b $\sqrt{2} + \sqrt{3}$ c $(\sqrt{2} \times \sqrt{3})^2$

6 a $\dfrac{\sqrt{8}}{\sqrt{2}}$ b $\dfrac{2\sqrt{5}}{\sqrt{20}}$ c $4 + (\sqrt{9} - 4)$

In questions 7–10, state, giving reasons, whether the quantity required is rational or irrational.

7

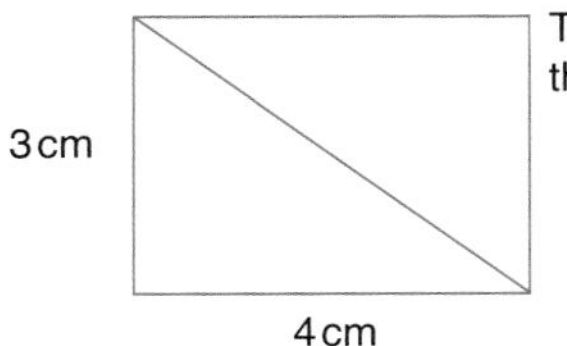


8

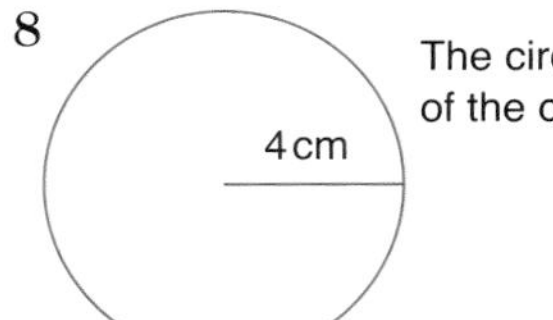


9

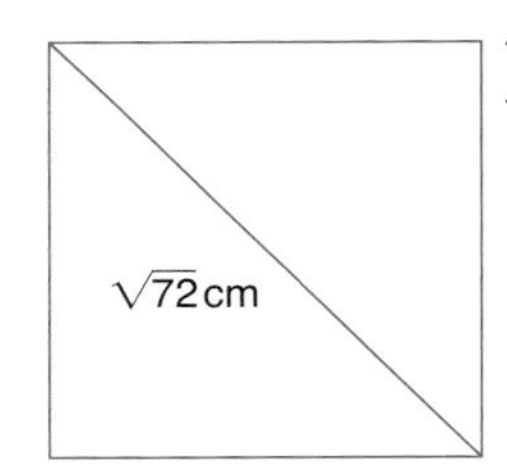


10

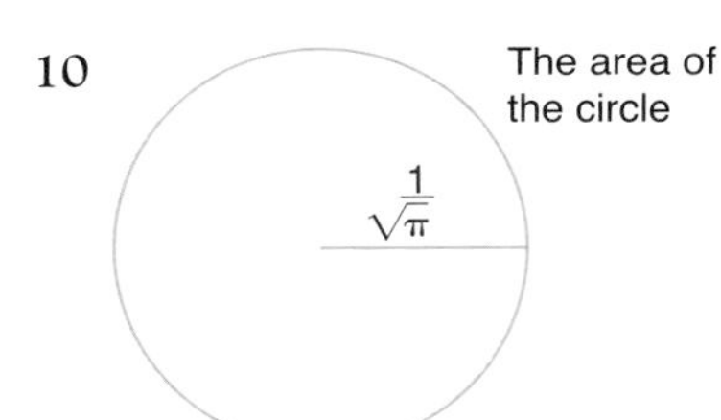


# 1.2 Approximation

In many instances exact numbers are not necessary or even desirable. In those circumstances, approximations are given. The approximations can take several forms. The common types of approximation are explained below.

## Rounding

If 28 617 people attend a gymnastics competition, this number can be reported to various levels of accuracy.

To the nearest 10 000 the number would be rounded up to 30 000.
To the nearest 1000 the number would be rounded up to 29 000.
To the nearest 100 the number would be rounded down to 28 600.

In this type of situation it is unlikely that the exact number would be reported.

## ■ Exercise 1.2.1

1 Round the following numbers to the nearest 1000.

a 68 786 b 74 245 c 89 000
d 4020 e 99 500 f 999 999

2 Round the following numbers to the nearest 100.

| | | |
|---|---|---|
| a 78540 | b 6858 | c 14099 |
| d 8084 | e 950 | f 2984 |

3 Round the following numbers to the nearest 10.

| | | |
|---|---|---|
| a 485 | b 692 | c 8847 |
| d 83 | e 4 | f 997 |

## Decimal places

A number can also be approximated to a given number of decimal places (d.p.). This refers to the number of digits written after a decimal point.

**Worked examples**

1 Give 7.864 to one decimal place.

The answer needs to be written with one digit after the decimal point. However, to do this, the second digit after the decimal point needs to be considered. If it is 5 or more then the first digit is rounded up. In this case it is 6, so the 8 is rounded up to 9, i.e.

7.864 is written as 7.9 to 1 d.p.

2 Give 5.574 to two decimal places.

The answer here is to be given with two digits after the decimal point. In this case the third digit after the decimal point needs to be considered. As the third digit after the decimal point is less than 5, the second digit is not rounded up, i.e.

5.574 is written as 5.57 to 2 d.p.

## Exercise 1.2.2

1 Give the following to one decimal place.

| | | |
|---|---|---|
| a 5.58 | b 0.73 | c 11.86 |
| d 157.39 | e 4.04 | f 15.045 |
| g 2.95 | h 0.98 | i 12.049 |

2 Give the following to two decimal places.

| | | |
|---|---|---|
| a 6.473 | b 9.587 | c 16.476 |
| d 0.088 | e 0.014 | f 9.3048 |
| g 99.996 | h 0.0048 | i 3.0037 |

## Significant figures

Numbers can also be approximated to a given number of significant figures (s.f.). In the number 43.25, the 4 is the most significant figure as it has a value of 40. In contrast, the 5 is the least significant as it has a value of only 5 hundredths. If you are not told otherwise, you are expected to round any answers that are not exact to three significant figures.

**Worked examples**

1 Give 43.25 to three significant figures.

Only the three most significant figures are written, but the fourth figure needs to be considered to see whether the third figure is to be rounded up or not. Since the fourth figure is 5, the third figure is rounded up, i.e.

43.25 is written as 43.3 to three significant figures.

2 Give 0.0043 to one significant figure.

In this example only two figures have any significance, the 4 and the 3. The 4 is the most significant and therefore is the only one of the two to be written in the answer, i.e.

0.0043 is written as 0.004 to one significant figure.

## Exercise 1.2.3

1 Give the following to the number of significant figures written in brackets.

| | | |
|---|---|---|
| **a** 48 599 (1 s.f.) | **b** 48 599 (3 s.f.) | **c** 6841 (1 s.f.) |
| **d** 7538 (2 s.f.) | **e** 483.7 (1 s.f.) | **f** 2.5728 (3 s.f.) |
| **g** 990 (1 s.f.) | **h** 2045 (2 s.f.) | **i** 14.952 (3 s.f.) |

2 Give the following to the number of significant figures written in brackets.

| | | |
|---|---|---|
| **a** 0.085 62 (1 s.f.) | **b** 0.5932 (1 s.f.) | **c** 0.942 (2 s.f.) |
| **d** 0.954 (1 s.f.) | **e** 0.954 (2 s.f.) | **f** 0.003 05 (1 s.f.) |
| **g** 0.003 05 (2 s.f.) | **h** 0.009 73 (2 s.f.) | **i** 0.009 73 (1 s.f.) |

3 Determine the following, giving your answer to three significant figures.

| | | |
|---|---|---|
| **a** $23.456 \times 17.89$ | **b** $0.4 \times 12.62$ | **c** $18 \times 9.24$ |
| **d** $76.24 \div 3.2$ | **e** $7.6^2$ | **f** $16.42^3$ |
| **g** $\frac{2.3 \times 3.37}{4}$ | **h** $\frac{8.31}{2.02}$ | **i** $9.2 \div 4^2$ |

### Estimating answers to calculations

Even though many calculations can be done quickly and effectively on a calculator, an estimate for an answer is often a useful check. This is done by rounding each of the numbers in a way that makes the calculation relatively straightforward.

**Worked examples**

1 Estimate the answer to $57 \times 246$.

Here are two possibilities:
**i)** $60 \times 200 = 12\,000$
**ii)** $50 \times 250 = 12\,500$.

2 Estimate the answer to $6386 \div 27$.

$6000 \div 30 = 200$.

## Exercise 1.2.4

1 Without using a calculator, estimate the answers to the following.

a $62 \times 19$ b $270 \times 12$ c $55 \times 60$
d $4950 \times 28$ e $0.8 \times 0.95$ f $0.184 \times 475$

2 Without using a calculator, estimate the answers to the following.

a $3946 \div 18$ b $8287 \div 42$ c $906 \div 27$
d $5520 \div 13$ e $48 \div 0.12$ f $610 \div 0.22$

3 Without using a calculator, estimate the answers to the following.

a $78.45 + 51.02$ b $168.3 - 87.09$ c $2.93 \div 3.14$
d $84.2 \div 19.5$ e $\frac{4.3 \times 752}{15.6}$ f $\frac{(9.8)^3}{(2.2)^2}$

4 Using estimation, identify which of the following are definitely incorrect. Explain your reasoning clearly.

a $95 \times 212 = 20\,140$ b $44 \times 17 = 748$
c $689 \times 413 = 28\,457$ d $142\,656 \div 8 = 17\,832$
e $77.9 \times 22.6 = 2512.54$ f $\frac{8.4 \times 46}{0.2} = 19\,366$

5 Estimate the shaded areas of the following shapes. Do *not* work out an exact answer.

a

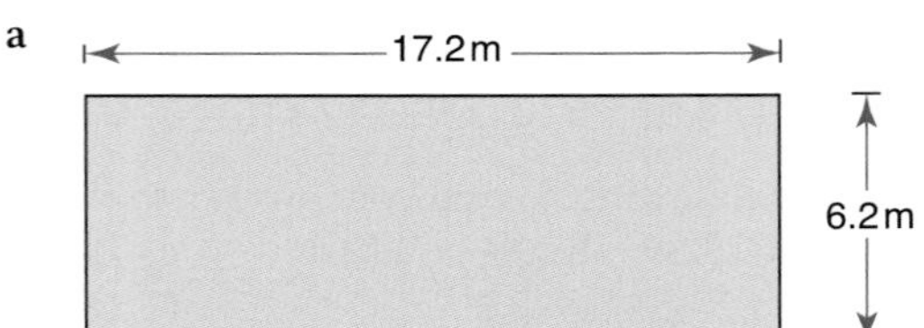


b

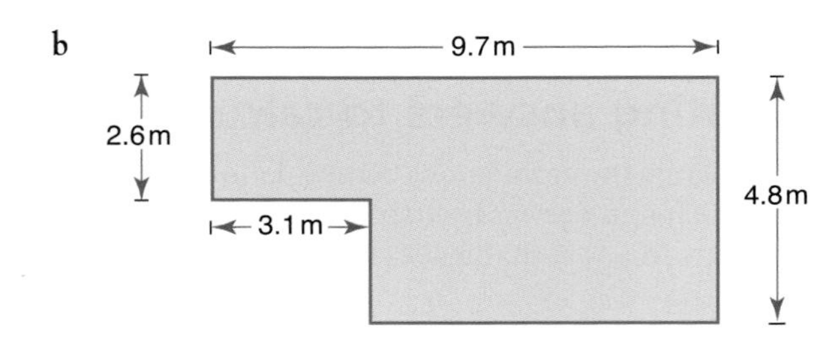


c

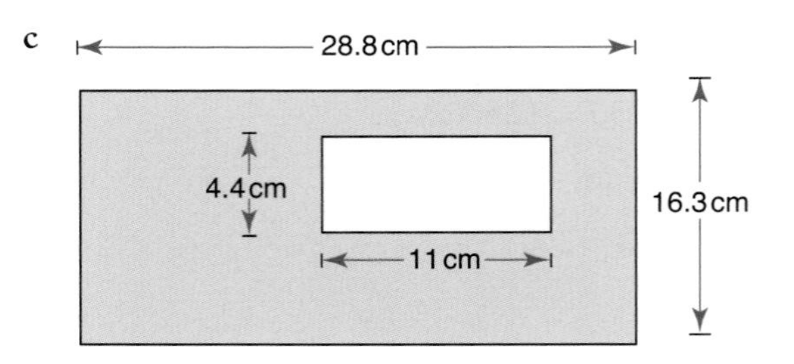

## Exercise 1.3.1

1 Which of the following are not in the form $a \times 10^k$, where $1 \leq a < 10$ and $k \in \mathbb{Z}$?

a $6.2 \times 10^5$ b $7.834 \times 10^{16}$ c $8.0 \times 10^5$
d $0.46 \times 10^7$ e $82.3 \times 10^6$ f $6.75 \times 10^1$

2 Write down the following numbers in the form $a \times 10^k$, where $1 \leq a < 10$ and $k \in \mathbb{Z}$.

a 600 000 b 48 000 000 c 784 000 000 000
d 534 000 e 7 million f 8.5 million

3 Write down the following in the form $a \times 10^k$, where $1 \leq a < 10$ and $k \in \mathbb{Z}$.

a $68 \times 10^5$ b $720 \times 10^6$ c $8 \times 10^5$
d $0.75 \times 10^8$ e $0.4 \times 10^{10}$ f $50 \times 10^6$

4 Multiply the following and write down your answers in the form $a \times 10^k$, where $1 \leq a < 10$ and $k \in \mathbb{Z}$.

a $200 \times 3000$ b $6000 \times 4000$
c 7 million $\times$ 20 d 500 $\times$ 6 million
e 3 million $\times$ 4 million f $4500 \times 4000$

5 Light from the Sun takes approximately 8 minutes to reach Earth. If light travels at a speed of $3 \times 10^8\,\mathrm{m\,s^{-1}}$, calculate to three significant figures (s.f.) the distance from the Sun to the Earth.

6 Find the value of the following and write down your answers in the form $a \times 10^k$, where $1 \leq a < 10$ and $k \in \mathbb{Z}$.

a $(4.4 \times 10^3) \times (2 \times 10^5)$ b $(6.8 \times 10^7) \times (3 \times 10^3)$
c $(4 \times 10^5) \times (8.3 \times 10^5)$ d $(5 \times 10^9) \times (8.4 \times 10^{12})$
e $(8.5 \times 10^6) \times (6 \times 10^{15})$ f $(5.0 \times 10^{12})^2$

7 Find the value of the following and write down your answers in the form $a \times 10^k$, where $1 \leq a < 10$ and $k \in \mathbb{Z}$.

a $(3.8 \times 10^8) \div (1.9 \times 10^6)$ b $(6.75 \times 10^9) \div (2.25 \times 10^4)$
c $(9.6 \times 10^{11}) \div (2.4 \times 10^5)$ d $\dfrac{1.8 \times 10^{12}}{9.0 \times 10^7}$
e $\dfrac{2.3 \times 10^{11}}{9.2 \times 10^4}$ f $\dfrac{2.4 \times 10^8}{6.0 \times 10^3}$

8 Find the value of the following and write down your answers in the form $a \times 10^k$, where $1 \leq a < 10$ and $k \in \mathbb{Z}$.

a $(3.8 \times 10^5) + (4.6 \times 10^4)$ b $(7.9 \times 10^9) + (5.8 \times 10^8)$
c $(6.3 \times 10^7) + (8.8 \times 10^5)$ d $(3.15 \times 10^9) + (7.0 \times 10^6)$
e $(5.3 \times 10^8) - (8.0 \times 10^7)$ f $(6.5 \times 10^7) - (4.9 \times 10^6)$
g $(8.93 \times 10^{10}) - (7.8 \times 10^9)$ h $(4.07 \times 10^7) - (5.1 \times 10^6)$

9 The following list shows the distance of the planets of the Solar System from the Sun.

| | |
|---|---|
| Jupiter | 778 million kilometres |
| Mercury | 58 million kilometres |
| Mars | 228 million kilometres |
| Uranus | 2870 million kilometres |
| Venus | 108 million kilometres |
| Neptune | 4500 million kilometres |
| Earth | 150 million kilometres |
| Saturn | 1430 million kilometres |

Write down each of the distances in the form $a \times 10^k$, where $1 \leq a < 10$ and $k \in \mathbb{Z}$ and then arrange them in order of magnitude, starting with the distance of the planet closest to the Sun.

## A negative index

A negative index is used when writing a number between 0 and 1 in the form $a \times 10^k$, where $1 \leq a < 10$ and $k \in \mathbb{Z}$.

e.g.

$$100 = 1 \times 10^2$$
$$10 = 1 \times 10^1$$
$$1 = 1 \times 10^0$$
$$0.1 = 1 \times 10^{-1}$$
$$0.01 = 1 \times 10^{-2}$$
$$0.001 = 1 \times 10^{-3}$$
$$0.0001 = 1 \times 10^{-4}$$

Note that $a$ must still lie within the range $1 \leq a < 10$.

**Worked examples**

1 Write down 0.0032 in the form $a \times 10^k$, where $1 \leq a < 10$ and $k \in \mathbb{Z}$.

$3.2 \times 10^{-3}$

2 Write down these numbers in order of magnitude, starting with the largest.

$3.6 \times 10^{-3}$ $5.2 \times 10^{-5}$ $1 \times 10^{-2}$ $8.35 \times 10^{-2}$ $6.08 \times 10^{-8}$

$8.35 \times 10^{-2}$ $1 \times 10^{-2}$ $3.6 \times 10^{-3}$ $5.2 \times 10^{-5}$ $6.08 \times 10^{-8}$

## Exercise 1.3.2

1 Write down the following numbers in the form $a \times 10^k$, where $1 \leq a < 10$ and $k \in \mathbb{Z}$.

a 0.0006 b 0.000 053 c 0.000 864
d 0.000 000 088 e 0.000 000 7 f 0.000 414 5

2 Write down the following numbers in the form $a \times 10^k$, where $1 \leq a < 10$ and $k \in \mathbb{Z}$.

a $68 \times 10^{-5}$ b $750 \times 10^{-9}$ c $42 \times 10^{-11}$
d $0.08 \times 10^{-7}$ e $0.057 \times 10^{-9}$ f $0.4 \times 10^{-10}$

3 Deduce the value of $k$ in each of the following cases.

a $0.000\,25 = 2.5 \times 10^k$ b $0.00357 = 3.57 \times 10^k$

c $0.000\,000\,06 = 6 \times 10^k$ d $0.004^2 = 1.6 \times 10^k$

e $0.000\,65^2 = 4.225 \times 10^k$

4 Write down these numbers in order of magnitude, starting with the largest.

$3.2 \times 10^{-4}$ $6.8 \times 10^5$ $5.57 \times 10^{-9}$ $6.2 \times 10^3$

$5.8 \times 10^{-7}$ $6.741 \times 10^{-4}$ $8.414 \times 10^2$

## 1.4 SI units of measurement

A soldier in Julius Caesar's army could comfortably march 20 miles in one day, wearing full kit, and then help to build a defensive blockade.

The mile was a unit of length based on 1000 strides of a Roman legionary. The measurement was sufficiently accurate for its purpose but only an approximate distance.

Most measures started as rough estimates. The yard (3 feet or 36 inches) was said to be the distance from the king's nose (reputed to be Henry I of England) to the tip of his extended finger. As it became necessary to have standardization in measurement, the measures themselves became more exact.

In 1791 during the French Revolution, a new unit of measurement, the metre, was defined in France. Originally it was defined as 'one ten-millionth of the length of the quadrant of the Earth's meridian through Paris'. This use of this unit of measurement became law in France in 1795.

However, this measurement was not considered sufficiently accurate and further definitions were required.

In 1927 a metre was defined as the distance between two marks on a given platinum–iridium bar. This bar is kept in Paris.

In 1960 the definition was based on the emission of a krypton-86 lamp.

At the 1983 General Conference on Weights and Measures, the metre was redefined as the length of the path travelled by light in a vacuum in $\frac{1}{299\,792\,488}$ second. This definition, although not very neat, can be considered one of the few 'accurate' measures. Most measures are only to a degree of accuracy.

SI is an abbreviation of *Système International d'Unités*. Its seven base units are listed below.

You will come across SI units of measurement in the Chemistry and Physics Diploma courses.

| Quantity | Unit | Symbol |
|---|---|---|
| Distance | metre | m |
| Mass | kilogram | kg |
| Time | second | s |
| Electrical current | ampere | A |
| Temperature | kelvin | K |
| Substance | mole | mol |
| Intensity of light | candela | cd |

The SI has other derived units. The following questions highlight some of the more common derived units and their relationship to the base units.

## Exercise 1.4.1

1 Copy and complete the sentences below.
   a There are ______ centimetres in one metre.
   b A centimetre is ______ part of a metre.
   c There are ______ metres in one kilometre.
   d A metre is ______ part of a kilometre.
   e There are ______ grams in one kilogram.
   f A gram is ______ part of a kilogram.
   g A kilogram is ______ part of a tonne.
   h There are ______ millilitres in one litre.
   i One thousandth of a litre is ______ .
   j There are ______ grams in one tonne.

2 Which of the units below would be used to measure each of the following?

| millimetre | centimetre | metre | kilometre |
|---|---|---|---|
| milligram | gram | kilogram | tonne |
| millilitre | litre | | |

   a Your mass (weight)
   b The length of your foot
   c Your height
   d The amount of water in a glass
   e The mass of a ship
   f The height of a bus
   g The capacity of a swimming pool
   h The length of a road
   i The capacity of the fuel tank of a truck
   j The size of your waist

3 Draw five lines of different lengths.
   a Estimate the length of each line in millimetres.
   b Measure the length of each line to the nearest millimetre.

4 Write down an estimate for each of the following using the correct unit.
   a Your height
   b Your weight (mass)
   c The capacity of a cup
   d The distance to the nearest town
   e The mass of an orange
   f The quantity of blood in the human body
   g The depth of the Pacific Ocean
   h The distance to the moon
   i The mass of a car
   j The capacity of a swimming pool

## Converting from one unit to another

### Length

1 km is 1000 m, so:

to change from kilometres to metres, multiply by 1000
to change from metres to kilometres, divide by 1000.

**Worked examples**

1 Convert 5.84 km to metres.

1 km = 1000 m so multiply by 1000
$5.84 \times 1000 = 5840$ m

2 Convert 3640 mm to metres.

1 m = 1000 mm so divide by 1000
$3640 \div 1000 = 3.64$ m

## Exercise 1.4.2

1 Convert these to millimetres.

| | | | |
|---|---|---|---|
| a 4 cm | b 6.2 cm | c 28 cm | d 1.2 m |
| e 0.88 m | f 3.65 m | g 0.008 m | h 0.23 cm |

2 Convert these to metres.

| | | | |
|---|---|---|---|
| a 260 cm | b 8900 cm | c 2.3 km | d 0.75 km |
| e 250 cm | f 0.4 km | g 3.8 km | h 25 km |

3 Convert these to kilometres.

| | | | |
|---|---|---|---|
| a 2000 m | b 26 500 m | c 200 m | d 750 m |
| e 100 m | f 5000 m | g 15 000 m | h 75 600 m |

### Mass

1 tonne is 1000 kg, so:

to change from tonnes to kilograms multiply by 1000
to change from kilograms to tonnes divide by 1000.

**Worked examples**

1 Convert 0.872 tonne to kilograms.

1 tonne is 1000 kg so multiply by 1000.
$0.872 \times 1000 = 872$ kg

2 Convert 4200 kg to tonnes.

1 tonne = 1000 kg so divide by 1000.
$4200 \div 1000 = 4.2$ tonnes

### Capacity

1 litre is 1000 ml, so:

to change from litres to millilitres multiply by 1000
to change from millilitres to litres divide by 1000.

**Worked examples**

1 Convert 2.4 ℓ to millilitres.

1 ℓ is 1000 ml so multiply by 1000.
$2.4 \times 1000 = 2400$ ml

2 Convert 4500 ml to litres.

1 ℓ = 1000 ml so divide by 1000
$4500 \div 1000 = 4.5$ ℓ

## Exercise 1.4.3

1 Convert these to kilograms.
 a 2 tonne  b 7.2 tonne  c 2800 g  d 750 g
 e 0.45 tonne  f 0.003 tonne  g 6500 g  h 7 000 000 g

2 Convert these to millilitres.
 a 2.6 ℓ  b 0.7 ℓ  c 0.04 ℓ  d 0.008 ℓ

3 Convert these to litres.
 a 1500 ml  b 5280 ml  c 750 ml  d 25 ml

4 The masses of four containers loaded on a ship are 28 tonnes, 45 tonnes, 16.8 tonnes and 48 500 kg.
 a What is the total mass in tonnes?
 b What is the total mass in kilograms? Write your answer in the form $a \times 10^k$, where $1 \leq a < 10$ and $k \in \mathbb{Z}$.

5 Three test tubes contain 0.08 ℓ, 0.42 ℓ and 220 ml.
 a What is the total in millilitres?
 b How many litres of water need to be added to make the solution up to 1.25 ℓ?

## Temperature scales

You will have come across two temperature scales in science lessons.

Celsius °C  kelvin K

The kelvin is the official SI unit of temperature. It is identical to the Celsius scale (in that a 1 °C change is equivalent to a 1 K change), except that it starts at 0 K, which is equivalent to −273 °C. This temperature is known as absolute zero.

Absolute zero = 0 K = −273 °C

The table shows the conversion between Celsius and kelvin scales of temperature.

| Scale | Freezing point of water | Boiling point of water |
|---|---|---|
| Celsius | 0 | 100 |
| Kelvin | 273 | 373 |

## 1.5 Currency conversions

The kingdom of the Lydian King Croesus (Lydia is now part of Turkey) is credited as being the first to mint coins in about 560 BC. The expression 'As rich as Croesus' came to describe people of vast wealth.

For a substance to be used as money it must be a 'scarce good'. It may be red ochre, diamonds or, in some circumstances, cigarettes. Scarce goods like gold are called 'commodity money'. Bank notes, which came later, are called 'representative money' as the paper itself has no value, but it can be exchanged for other goods. One pound sterling is so called because it could, on demand, be exchanged for one pound weight of sterling silver.

What are the advantages and disadvantages of paper money compared to commodity money?

The Eurozone is a term that describes those European countries that replaced their previous currencies with one common currency, i.e. the euro.

In 2002, just after its launch, its exchange rate against the US dollar (USD or \$) was below parity, i.e one euro was worth less than one dollar.

In June 2008 before the banking crisis, one euro was worth approximately one dollar and sixty cents. In April 2012, one euro was worth approximately one dollar and thirty cents. The value of a bank note relative to other currencies can change, sometimes very rapidly.

When changing currencies, banks take a commission (a fee). This meant that when Europe had a large number of different currencies, traders lost money paying these commissions and currency fluctuations meant long-term planning was difficult for exporters.

Commission can either be a fixed sum or a percentage of the money exchanged. In addition, when you exchange money, there are two rates; one for when selling and another for buying. For example, a bank might buy £1 sterling (GBP) for \$1.30 and sell for \$1.35. So if you changed £1000 into dollars you would receive \$1300 but if you changed your dollars back into pounds you would receive $1300 \div 1.35 =$ £963, a cost of £37. Had they charged a percentage commission of 3%, it would have cost £30 to make the original exchange. So you would get £970 × 1.30 = \$1261.

The table below shows the rate at which countries entering the euro in 2002 changed their old currency for one euro.

| Country | Currency | Exchange for 1 euro |
|---|---|---|
| France | Franc | 6.56 francs |
| Germany | Deutsche mark | 1.96 Deutsche marks |
| Italy | Lira | 1940 lire |
| Spain | Peseta | 166 pesetas |
| Holland | Guilder | 2.20 guilders |
| Austria | Schilling | 13.76 schillings |
| Belgium | Franc | 40 francs |
| Finland | Markka | 5.95 markkas |
| Greece | Drachma | 341 drachmas |
| Ireland | Punt | 0.79 punts |
| Portugal | Escudo | 200 escudos |
| Luxembourg | Franc | 40 francs |

The euro became the single currency for the twelve countries above on 1st January 2002. It was the biggest change in currencies Europe had ever seen. By 2008 Andorra, Cyprus, Malta, Monaco, Montenegro, San Marino, Slovenia and Vatican City had joined the euro. European countries still to join include Great Britain, Sweden and Denmark. Why do you think these countries have not joined the euro?

Note: There is no universal agreement about whether the € sign should go before or after the number. This depends to a great extent on the conventions that were in place in each country with its previous currency.

**Worked example**

Using the exchange rate in the table above, convert 15 000 pesetas into guilders.

From the table: 166 pesetas = 1 euro = 2.20 guilders

166 pesetas = 2.20 guilders

$$1 \text{ peseta} = \frac{2.20}{166} \text{ guilders}$$

$$15\,000 \text{ pesetas} = \frac{2.20}{166} \times 15\,000 \text{ guilders}$$

Therefore, 15 000 pesetas = 198.80 guilders.

## Exercise 1.5.1

Using the exchange rates in the table above, convert each of the following.

1 100 Deutsche marks into French francs
2 500 guilder into drachmas
3 20 000 lira into schillings
4 7500 pesetas into escudos
5 1000 Belgian francs into French francs
6 1 million punt into markkas
7 200 000 lira into pesetas
8 3000 Deutsche marks and 1000 schillings into punts
9 5000 Deutsche marks into lira
10 2500 guilder into shillings

You will come across exchange rates in the Economics Diploma course.

This gives you some idea of the many changes occurring every day in Europe before 2002.

The pound sterling was a 'reserve currency'. Many economists think that the euro could become a 'reserve currency'. What does the term mean?

Currencies change their value with respect to each other because of the change in economic circumstances in each country.

For example, in 2002 €1 could be exchanged for $1.04. In 2008 1 euro could be exchanged for $1.54.

The exchange rate today can be found from what is called the currency exchange market.

In 2008, visitors to the USA from Germany would have found the country cheap to visit. Americans visiting Europe would have had the opposite experience, since their dollars did not buy as many euros as in previous years. In theory the external exchange rate does not affect the internal worth of the currency. In practice, because countries need to import and export goods and services, the strength or weakness of a currency will directly affect people. For example if your currency weakens compared to the dollar, then the cost of oil and gas, which are priced in dollars, will increase. Sometimes a currency becomes so weak that it is impossible to import goods. In 2008, the currency of Zimbabwe was not accepted for overseas trade payments. The economy suffered badly and many Zimbabweans could not find work.

**Worked example**

In 2001 Kurt and his family travelled to the USA. The flights cost \$1650, hotels cost \$2200 and other expenses were \$4000. At that time, the exchange rate was €1 = \$1.05.

In 2008 they took the same vacation but the costs in dollars had increased by 10%. The exchange rate was now 1 euro = 1.55 dollars.

a What was the total cost in euros in 2001?
b What was the total cost in euros in 2008?

a Total cost in dollars = \$7850
\$1.05 = €1 ⇒ \$1 = €0.95
Therefore total cost in euro was 7850 × 0.95 = €7476.

b A 10% increase is equivalent to a multiplier of 1.10.
Total cost in dollars = 7850 × 1.10 = \$8635
\$1.55 = €1 ⇒ \$1 = €0.65
Therefore total cost in euros was 8635 × 0.65 = €5613.

## Exercise 1.5.2

1 The rate of exchange of the South African rand is £1 = 12.8 rand. £1 = 1.66 US dollars.
  a How many pounds could be exchanged for 100 rand?
  b How many rand could be exchanged for \$1000?

2 An apartment in Barcelona is offered for sale for €350 000. The exchange rate is £1 = €1.24.
  a What was the cost of the apartment in pounds?
  b The value of the pound against the euro increases by 10%.What is the new cost of the apartment in pounds?

3 The Japanese yen trades at 190¥ to £1. The dollar trades at \$1.90 to the pound.
  a What is the yen–dollar exchange rate?
  b What is the dollar–yen exchange rate?

4 The Russian rouble trades at 24 roubles to \$1. The Israeli shekel trades at 3.5 shekels to \$1.
  a What is the rouble–shekel exchange rate?
  b What is the shekel–rouble exchange rate?

5 Gold is priced at \$930 per ounce. 16 ounces = 1 pound weight.
  a What is the cost of 1 ton (2240 pounds) of gold?
  1 dollar = 0.62 euros and 1 dollar = 41 Indian rupees.
  b What does 1 ton of gold cost in rupees?
  c What does 1 ton of gold cost in euros?
  d What weight (in ounces) of gold could be bought for 1 million euros?

# 1.6 Graphical solution of equations

A linear equation, when plotted, produces a straight line.

The following are all examples of linear equations:

$y = x + 1 \qquad y = 2x - 1 \qquad y = 3x \qquad y = -x - 2 \qquad y = 4$

They all have a similar format, i.e. $y = mx + c$.

In the equation $y = x + 1$, $m = 1$ and $c = 1$
$y = 2x - 1$, $m = 2$ and $c = -1$
$y = 3x$, $m = 3$ and $c = 0$
$y = -x - 2$, $m = -1$ and $c = -2$
$y = 4$, $m = 0$ and $c = 4$

Their graphs are shown below.

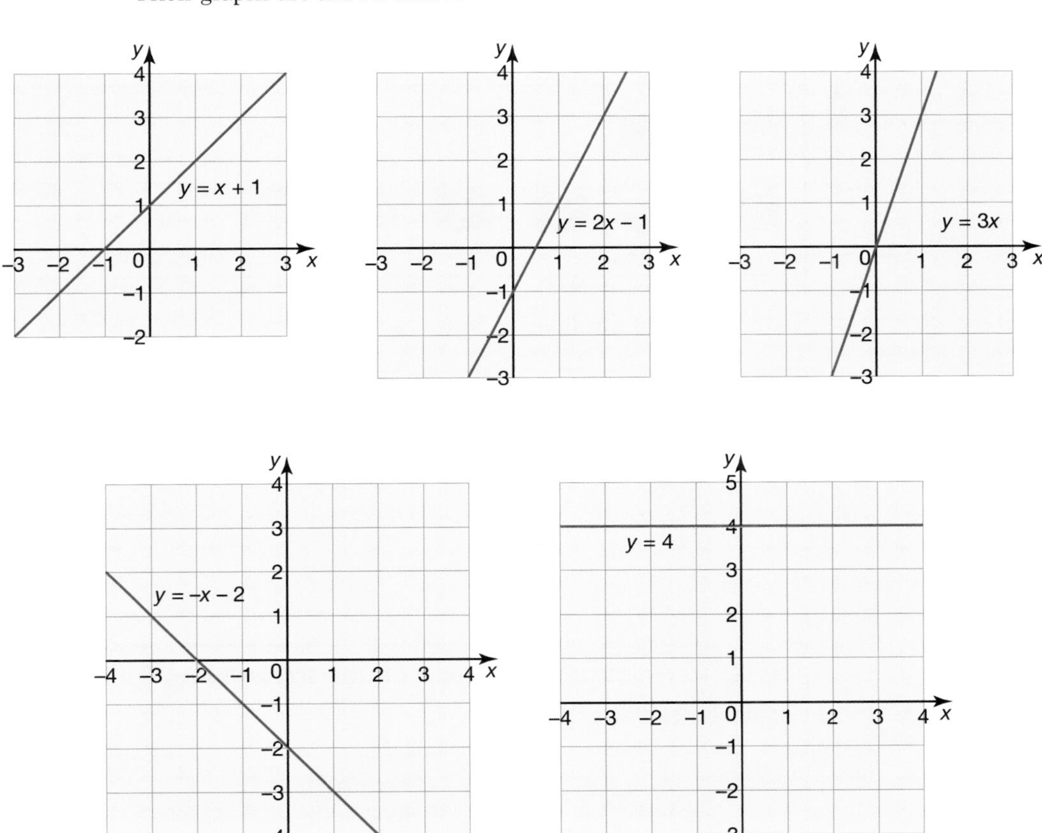


In Section 5.1 plotting and solving linear equations are studied in more detail. In this section we look exclusively at how to use your GDC and graphing software to draw linear graphs and to solve simple problems involving them.

## Using a GDC or graphing software to plot a linear equation

In the Introduction to the graphic display calculator you saw how to plot a single linear equation using your GDC. For example, to graph the linear equation $y = 2x + 3$:

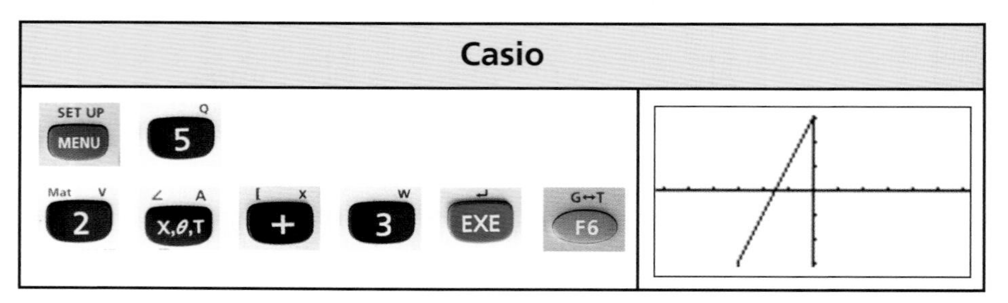


With graphing software too, the process is relatively straightforward:

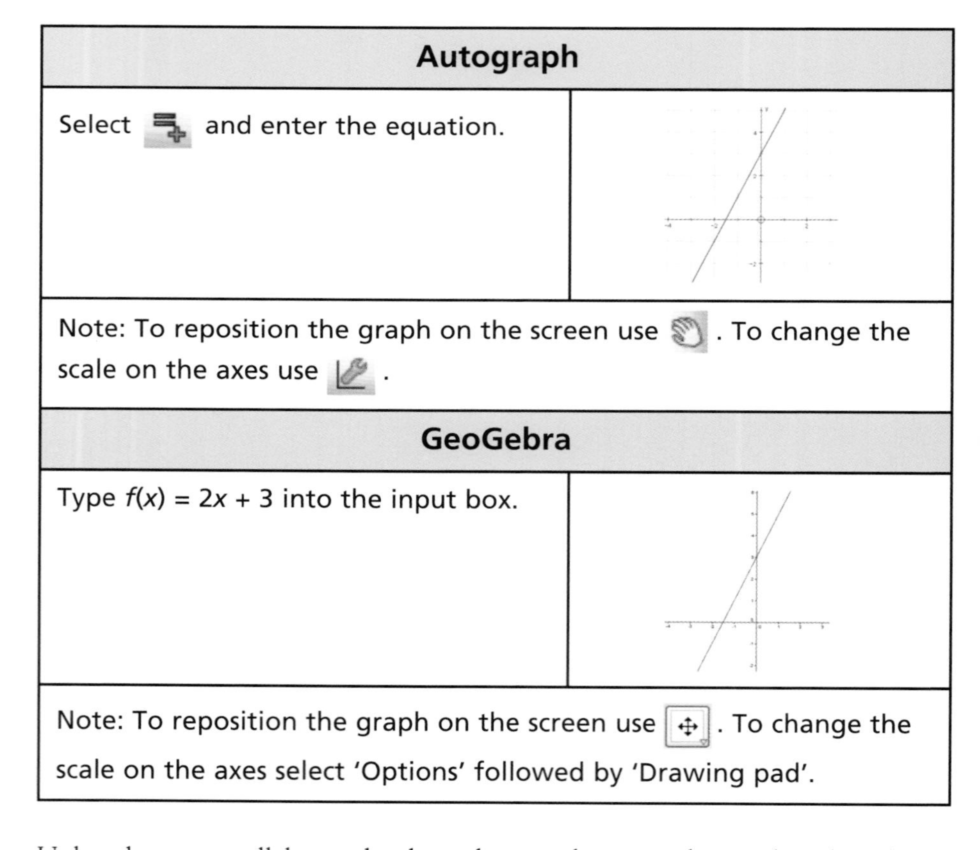


Unless they are parallel to each other, when two linear graphs are plotted on the same axes, they will intersect at one point. Solving the equations simultaneously will give the coordinates of the point of intersection. Your GDC and graphing software will also be able to work out the coordinates of the point of intersection.

**Worked example**

Find the point of intersection of these linear equations.

$$y = 2x - 1 \text{ and } y = \tfrac{1}{2}x + 2$$

Using a GDC:

## Casio

  and enter $y = 2x - 1$, 

Enter $y = \frac{1}{2}x + 2$, EXE

 to graph the equations.

  followed by  to select 'intersect' in the 'graph solve' menu. The results are displayed at the bottom of the screen.

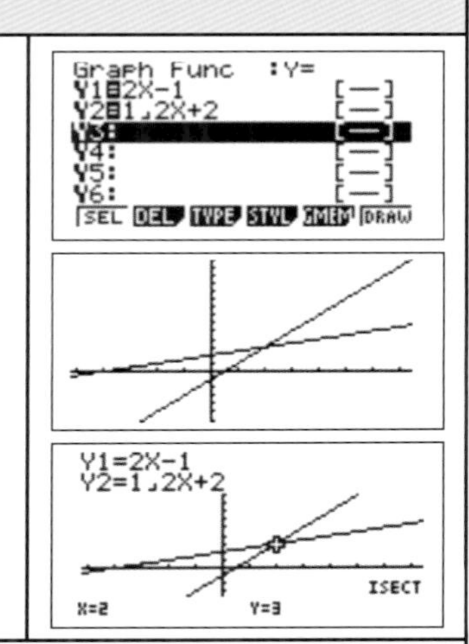


Note: Equations of lines have to be entered in the form $y = \ldots$, e.g. the equation $2x - 3y = 9$ would need to be rearranged to make $y$ the subject, i.e. $y = \frac{2x - 9}{3}$ or $y = \frac{2}{3}x - 3$.

## Texas

 and enter $y = 2x - 1$, 

Then $y = \frac{1}{2}x + 2$, 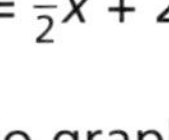

GRAPH to graph the equations.

  followed by  to select 'intersect' in the 'graph calc' menu.

Once the two lines are selected using , the results are displayed at the bottom of the screen.

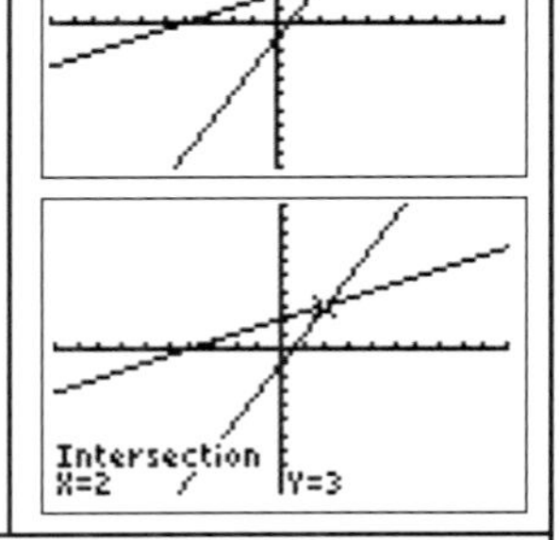


Note: See the note for the Casio above.

Using graphing software:

**Autograph**

Select [icon] and enter each equation in turn.

Use [icon] and select both lines.

Click on 'Object' followed by the 'solve f(x) = g(x)' sub-menu.

Click [icon] to display the 'results box'.

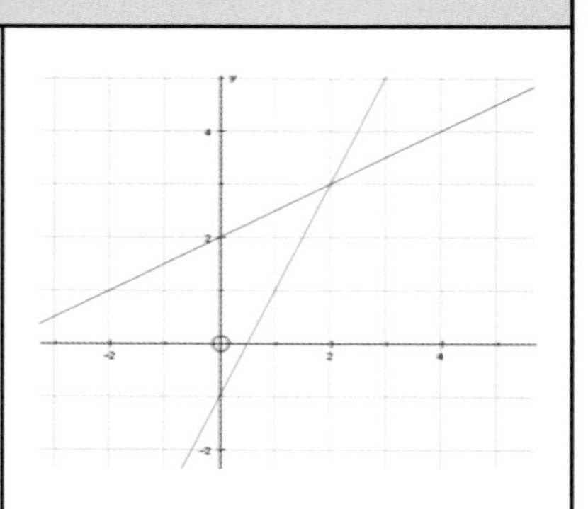

Note: To select the second line, keep the shift key pressed.

**GeoGebra**

Enter $f(x) = 2x - 1$ and $g(x) = \frac{1}{2}x + 2$ in the input field.

Enter 'Intersect [$f(x)$, $g(x)$]' in the input field.

The point of intersection is displayed and its coordinate written in the algebra window.

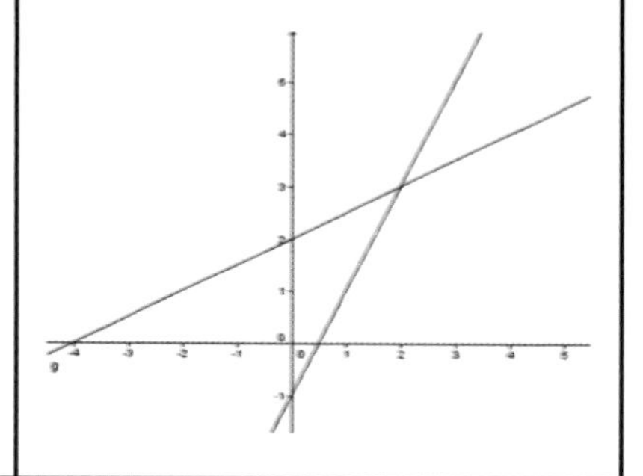

## Exercise 1.6.1

1 Using either a GDC or graphing software, find the coordinates of the points of intersection of the following pairs of linear graphs.

- **a** $y = 5 - x$ and $y = x - 1$
- **b** $y = 7 - x$ and $y = x - 3$
- **c** $y = -2x + 5$ and $y = x - 1$
- **d** $x + 3y = -1$ and $y = \frac{1}{2}x + 3$
- **e** $x - y = 6$ and $x + y = 2$
- **f** $3x - 2y = 13$ and $2x + y = 4$
- **g** $4x - 5y = 1$ and $2x + y = -3$
- **h** $x = y$ and $x + y + 6 = 0$
- **i** $2x + y = 4$ and $4x + 2y = 8$
- **j** $y - 3x = 1$ and $y = 3x - 3$

2 By referring to the lines, explain your answers to parts **i** and **j** above.

## Arithmetic sequences

In an **arithmetic sequence** there is a common difference ($d$) between successive terms, e.g.

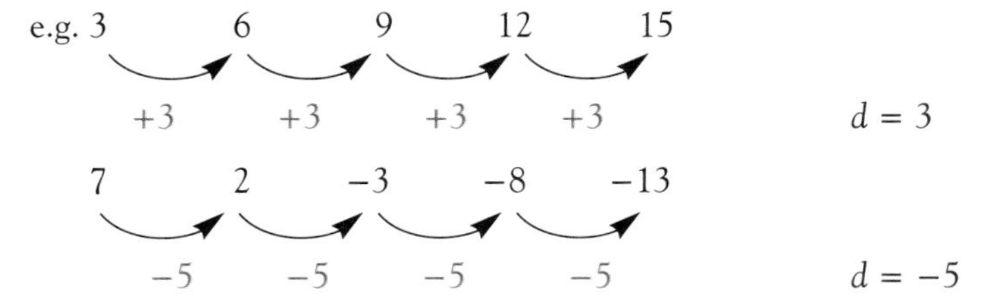


### Formulae for the terms of an arithmetic sequence

There are two main ways to describe a sequence.

1 A term-to-term rule, known as a **recurrence relation**.
In the following sequence,

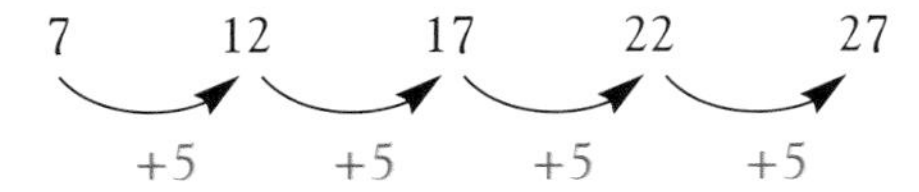


the recurrence relation is +5, i.e. $u_2 = u_1 + 5$, $u_3 = u_2 + 5$, etc.

The general form is therefore written as $u_{n+1} = u_n + 5$, $u_1 = 7$, where $u_n$ is the $n$th term and $u_{n+1}$ is the term after the $n$th term.

Note: It is important to give the value of one of the terms, e.g. $u_1$, so that the exact sequence can be generated.

2 A formula for the $n$th term of a sequence.
This type of rule links each term to its position in the sequence, e.g.

| Position | 1 | 2 | 3 | 4 | 5 | $n$ |
|---|---|---|---|---|---|---|
| Term | 7 | 12 | 17 | 22 | 27 | |

We can deduce from the figures above that each term can be calculated by multiplying its position number by 5 and adding 2. Algebraically this can be written as the formula for the $n$th term:

$$u_n = 5n + 2$$

A GDC can be used to generate the terms of a sequence from the recurrence relation. For example, to calculate $u_2$, $u_3$ and $u_4$ for the sequence $u_{n+1} = 2u_n - 5$, $u_1 = 1$, the following can be done:

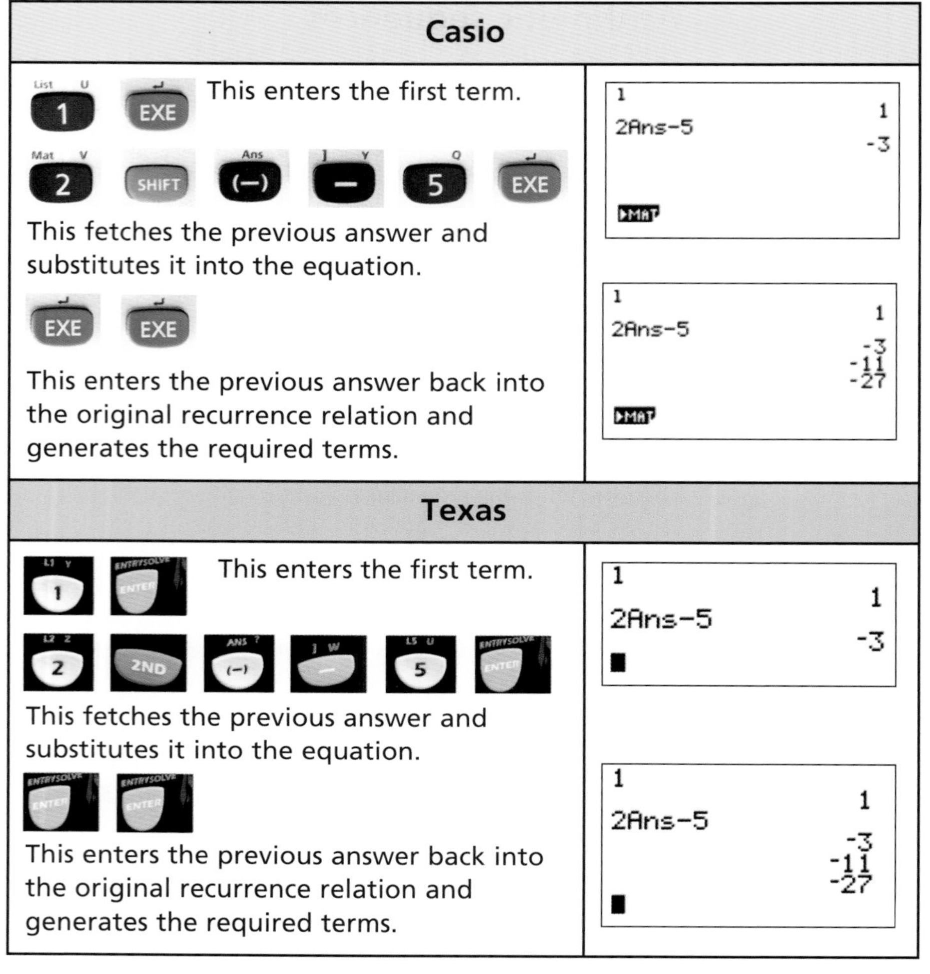


Therefore $u_2 = -3$, $u_3 = -11$ and $u_4 = -27$.

Note: This sequence is not arithmetic as the difference between successive terms is not constant.

With an arithmetic sequence, the rule for the $n$th term can be easily deduced by looking at the common difference, e.g.

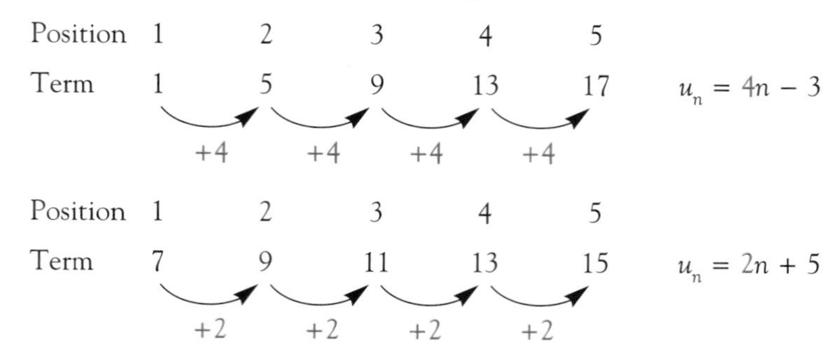

## Exercise 1.9.1

All rates of interest are annual rates.

1 Find the simple interest paid in each of the following cases.

| | Capital | Rate | Time period |
|---|---|---|---|
| a | NZ$300 | 6% | 4 years |
| b | £750 | 8% | 7 years |
| c | 425¥ | 6% | 4 years |
| d | 2800 baht | 4.5% | 2 years |
| e | HK$880 | 6% | 7 years |

2 How long will it take for the following amounts of interest to be earned?

| | C | *R* | *I* |
|---|---|---|---|
| a | 500 baht | 6% | 150 baht |
| b | 5800¥ | 4% | 950¥ |
| c | AU$4000 | 7.5% | AU$1500 |
| d | £2800 | 8.5% | £1904 |
| e | €900 | 4.5% | €243 |
| f | 400 Ft | 9% | 252 Ft |

3 Calculate the rate of interest per year which will earn the given amount of interest in the given time period.

| | Capital | Time period | Interest |
|---|---|---|---|
| a | €400 | 4 years | €1120 |
| b | US$800 | 7 years | US$224 |
| c | 2000 baht | 3 years | 210 baht |
| d | £1500 | 6 years | £675 |
| e | €850 | 5 years | €340 |
| f | AU$1250 | 2 years | AU$275 |

4 Calculate the capital required to earn the interest stated in the number of years and with the rates given.

| | Interest | Time period | Rate |
|---|---|---|---|
| a | 80 Ft | 4 years | 5% |
| b | NZ$36 | 3 years | 6% |
| c | €340 | 5 years | 8% |
| d | 540 baht | 6 years | 7.5% |
| e | €540 | 3 years | 4.5% |
| f | US$348 | 4 years | 7.25% |

5 What rate of interest is paid on a deposit of £2000 that earns £400 interest in 5 years?

6 How long will it take a capital of €350 to earn €56 interest at 8% per year?

7 A capital of 480 Ft earns 108 Ft interest in 5 years. What rate of interest was being paid?

8 A capital of €750 becomes a total of €1320 in 8 years. What rate of interest was being paid?

9 AU$1500 is invested for 6 years at 3.5% per year. What is the interest earned?

10 500 baht is invested for 11 years and becomes 830 baht in total. What rate of interest was being paid?

## Compound interest

In this section we move on from simple interest to look at compound interest. For the first time period, there is no difference between the two types: a percentage of the capital is paid as interest. However, simple interest is calculated using the original amount whereas compound interest is paid on the total amount, which includes the interest paid in the first time period. Simple interest has few applications in real life: when we talk about interest in real life, it will usually be compound interest. If you have a savings account the money in it will earn compound interest and when people take out a loan they will pay compound interest on the money they have borrowed.

For example, a builder is going to build six houses on a plot of land in Spain. He borrows 500 000 euros at 10% p.a. and will pay off the loan in full after 3 years, when he expects to have finished building the houses and to have sold them.
At the end of the first year he will owe:

€500 000 + 10% of €500 000 i.e. €500 000 × 1.10 = €550 000

At the end of the second year he will owe:

€550 000 + 10% of €550 000 i.e. €550 000 × 1.10 = €605 000

At the end of the third year he will owe:

€605 000 + 10% of €605 000 i.e. €605 000 × 1.10 = €665 500

The amount of interest he has to pay is:

€665 500 – €500 000 = €165 500

The simple interest is €50 000 per year, i.e. a total of €150 000.
The difference of €15 500 is the compound interest.
The time taken for a debt to grow at compound interest can be calculated as shown in the example below.

**Worked example**

How long will it take for a debt to double with a compound interest rate of 27% p.a.?
An interest rate of 27% implies a multiplier of 1.27.

| **Time (years)** | 0 | 1 | 2 | 3 |
|---|---|---|---|---|
| **Debt** | $C$ | $1.27C$ | $1.27^2C = 1.61C$ | $1.27^3C = 2.05C$ |

× 1.27 × 1.27 × 1.27

The debt will have more than doubled after 3 years.

Using the example of the builder's loan above, if C represents the capital he borrows, then after 1 year his debt will be given by the formula:

$D = C\left(1 + \frac{r}{100}\right)$, where $r$ is the rate of interest.

This is a geometric sequence.

After 2 years: $D = C\left(1 + \frac{r}{100}\right)\left(1 + \frac{r}{100}\right)$

After 3 years: $D = C\left(1 + \frac{r}{100}\right)\left(1 + \frac{r}{100}\right)\left(1 + \frac{r}{100}\right)$

After $n$ years: $D = C\left(1 + \frac{r}{100}\right)^n$

This formula for the debt includes the original capital loan. By subtracting C, the compound interest is calculated.

$$I = C\left(1 + \frac{r}{100}\right)^n - C$$

Compound interest is an example of a geometric sequence. You studied geometric sequences in more detail in Topic 1.8.

The $n$th term of a geometric sequence is given by:

$$u_n = u_1 r^{n-1}$$

Compare this with the formula for the amount of money earning compound interest at $r$%:

$$A = C\left(1 + \frac{r}{100}\right)^n$$

Note: The differences between this and the formula for calculating compound interest. $u_n$ is analogous to the amount in an account, A, but $n$ is used differently in the two formulae. The initial amount in an account, C, is when $n = 0$ whereas the first term of a geometric sequence, $u_1$, is when $n = 1$. $r$ represents the common difference in the formula for the $n$th term of a geometric sequence. The common difference in a sequence of the amount of money in an account earning compound interest is $\left(1 + \frac{r}{100}\right)$, where $r$ is the rate of interest.

The interest is usually calculated annually, but there can be other time periods. Compound interest can be charged or credited yearly, half-yearly, quarterly, monthly or daily. (In theory, any time period can be chosen.)

**Worked examples**

1 Alex deposits €1500 in his savings account. The interest rate offered by the savings account is 6% compound interest each year for a 10-year period. Assuming Alex leaves the money in the account, calculate how much interest he has gained after 10 years.

$$I = 1500\left(1 + \frac{6}{100}\right)^{10} - 1500$$

$$I = 2686.27 - 1500 = 1186.27$$

The amount of interest gained is €1186.27.

2 Adrienne deposits £2000 in her savings account. The interest rate offered by the bank for this account is 8% compound interest per year. Calculate the number of years Adrienne needs to leave the money in her account for it to double in value.

An interest rate of 8% implies a common ratio of 1.08.
This can be found by generating the geometric sequence using the recurrence rule $u_{n+1} = u_n \times 1.08$ on your calculator as shown on page 48. Think of the initial amount as $u_0$.

$u_1 = 2000 \times 1.08 = 2160$
$u_2 = 2160 \times 1.08 = 2332.80$
$u_3 = 2332.80 \times 1.08 = 2519.42$
...
$u_9 = 3998.01$
$u_{10} = 4317.85$

Adrienne needs to leave the money in the account for 10 years in order for it to double in value.

3 Use your GDC to find the compound interest paid on a loan of \$600 for 3 years at an annual percentage rate (A.P.R.) of 5%.

The total payment is \$694.58 so the interest due is \$694.58 – \$600 = \$94.58.

4 Use a GDC to calculate the compound interest when \$3000 is invested for 18 months at an APR of 8.5%. The interest is calculated every 6 months.

Note: The interest for each time period of 6 months is $\frac{8.5}{2}$%. There will therefore be three time periods of 6 months each.

$3000 \times 1.0425^3 = £3398.99$

The final sum is \$3399, so the interest is \$3399 – \$3000 = \$399.

## Exercise 1.9.2

1 A shipping company borrows \$70 million at 5% p.a. compound interest to build a new cruise ship. If it repays the debt after 3 years, how much interest will the company pay?

2 A woman borrows €100 000 for home improvements. The interest rate is 15% p.a. and she repays it in full after 3 years. Calculate the amount of interest she pays.

3 A man owes \$5000 on his credit cards. The APR is 20%. If he doesn't repay any of the debt, calculate how much he will owe after 4 years.

4 A school increases its intake by 10% each year. If it starts with 1000 students, how many will it have at the beginning of the fourth year of expansion?

5 8 million tonnes of fish were caught in the North Sea in 2005. If the catch is reduced by 20% each year for 4 years, what weight is caught at the end of this time?

6 How many years will it take for a debt to double at 42% p.a. compound interest?

4 **Part A**

Daniel wants to invest $25 000 for a total of three years. There are three investment options.

**Option One** pays simple interest at annual rate of interest of 6%.

**Option Two** pays compound interest at a nominal annual rate of interest of 5%, compounded **annually**.

**Option Three** pays compound interest at a nominal annual rate of interest of 4.8%, compounded **monthly**.

a Calculate the value of his investment at the end of the third year for each investment option, **correct to two decimal places.** [8]

b Determine Daniel's best investment option. [1]

**Part B**

An arithmetic sequence is defined as

$$U_n = 135 + 7n, \quad n = 1, 2, 3, \ldots$$

a Calculate $u_1$, the first term in the sequence. [2]

b Show that the common difference is 7. [2]

$S_n$ is the sum of the first $n$ terms of the sequence.

c Find an expression for $S_n$. Give your answer in the form $S_n = An^2 + Bn$, where $A$ and $B$ are constants. [3]

The first term, $v_1$, of a geometric sequence is 20 and its fourth term $v_4$ is 67.5.

d Show that the common ratio, $r$, of the geometric sequence is 1.5. [2]

$T_n$ is the sum of the first $n$ terms of the geometric sequence.

e Calculate $T_7$, the sum of the first seven terms of the geometric sequence. [2]

f Use your graphic display calculator to find the smallest value for $n$ for which $T_n > S_n$. [2]

**Paper 2, May 10, Q5**

# Applications, project ideas and theory of knowledge

1 Is there an underlying reason why the letters ℕ, ℤ, ℚ, ℝ were chosen to represent sets of numbers?

2 The series 1, 2, 3, 4, 5, … pairs with the series of square numbers 1, 4, 9, 16, 25, …. Discuss whether there are more members of the first series than the second series.

3 What is the difference between:

a) legal proof and mathematical proof?

b) 'real' and 'imaginary' in normal life compared with mathematical solutions?

4 Does the use of S.I. notation help to make mathematics a 'universal language'? Investigate systems of measurement that preceded the S.I. system.

5 What is the 'Golden ratio' and why is it said to be 'beautiful'? A possible topic for a project would be the Fibonacci series. But be warned, whole lives have been spent on this study.

6 Is it possible to get an exact numerical answer to a problem? Is an error necessarily a mistake?

7 Some series are neither arithmetic nor geometric series. Investigate some different series.

8 Investigate the way that a knight moves on a chess board. This could form the basis of a project.

9 $n^2 - n + 41$ is always a prime number. Investigate this statement and try to prove or disprove it. Find other similar quadratic expressions as a basis for a project.

10 Does zero exist or is it merely a mathematician's 'idea'?

11 'The first space programme back in the 1960s cost billions of dollars, and that was when a billion dollars was a lot of money'. What did the speaker mean? Do billions and trillions have any real meaning in 'normal' life?

12 Approximation is used in different ways in science: Physics, Meteorology and Biology are examples. Scales of number and approximation could form the basis of a project.

13 Newton, Gauss and Einstein were geniuses. Could their ability be measured? Are the creators of Facebook and Google geniuses? If so, can their ability be measured?

14 Currencies are traded like shares in companies. Should this trading be banned since it can result in mass unemployment?

Topic 2

# Descriptive statistics

## Syllabus content

**2.1** Classification of data as discrete or continuous.

**2.2** Simple discrete data: frequency tables.

**2.3** Grouped discrete or continuous data: frequency tables; mid-interval values; upper and lower boundaries.

Frequency histograms.

**2.4** Cumulative frequency tables for grouped discrete data and for grouped continuous data; cumulative frequency curves; median and quartiles.

Box and whisker diagrams.

**2.5** Measures of central tendency.

For simple discrete data: mean; median; mode.

For grouped discrete and continuous data: estimate of a mean; modal class.

**2.6** Measures of dispersion: range; interquartile range; standard deviation.

## Introduction

The word *statistics* comes from the Latin *status* meaning state. So statistics was related to information useful to the state.

Statistics is often not considered to be a branch of mathematics, and many universities have a separate statistics department.

Statistics developed out of studies of probability.

Societies such as the London Statistical Society, established in 1834, brought the study of statistics to new heights, but only the advent of computers has brought the ability to handle and analyze very large amounts of data.

## 2.1 Discrete and continuous data

You will come across discrete and continuous data in the Biology and Psychology Diploma courses.

**Discrete data** can only take specific values, for example the number of tickets sold for a concert can only be positive integer values.

**Continuous data**, on the other hand, can take any value within a certain range, for example the time taken to run 100 m will typically fall in the range 10–20 seconds. Within that range, however, the time stated will depend on the accuracy required. So a time stated as 13.8 s could have been 13.76 s, 13.764 s or 13.7644 s, etc.

**GeoGebra**

Type
Histogram[{5.75, 6.25, 6.75, 7.25, 7.75, 8.25, 8.75, 9.25}, {2, 3, 3, 6, 4, 1, 1}]

The first set of numbers represent the upper and lower bounds of each value, whilst the second set represent the frequency.

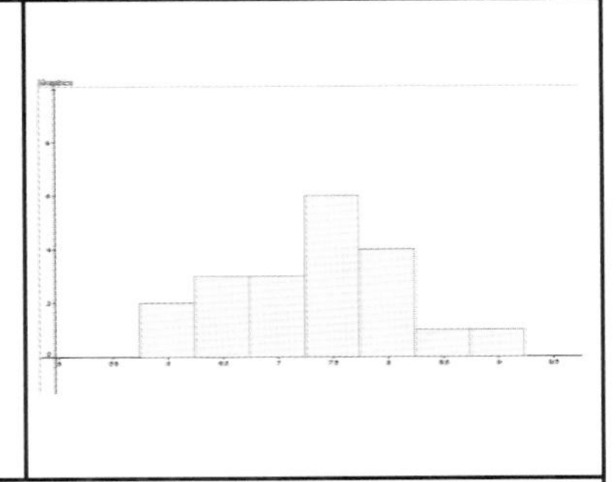

Note: A similar graph is produced using the command BarChart[{6, 6.5, 7, 7.5, 8, 8.5, 9}, {2, 3, 3, 6, 4, 1, 1}].

## Exercise 2.2.1

1 The figures in the list below give the total number of chocolate sweets in each of 20 packets of sweets.

35, 36, 38, 37, 35, 36, 38, 36, 37, 35, 36, 36, 38, 36, 35, 38, 37, 38, 36, 38

- **a** Draw a frequency table of the data.
- **b** Draw a frequency histogram of the data.

2 Record the shoe sizes of everybody in your class.

- **a** Draw a frequency table of the results.
- **b** Draw a frequency histogram of the data.
- **c** What conclusions can you draw from your results?

# 2.3 Grouped discrete or continuous data

You will come across grouped data in the Geography Diploma course.

If there is a large range in the data, it is sometimes easier and more useful to group the data in a grouped frequency table.

The discrete data below shows the scores for the first round of a golf competition.

| | | | | | | | | | | | | | | |
|---|---|---|---|---|---|---|---|---|---|---|---|---|---|---|
| 71 | 75 | 82 | 96 | 83 | 75 | 76 | 82 | 103 | 85 | 79 | 77 | 83 | 85 | 88 |
| 104 | 76 | 77 | 79 | 83 | 84 | 86 | 88 | 102 | 95 | 96 | 99 | 102 | 75 | 72 |

One possible way of grouping this data in a grouped frequency table is shown opposite.

Note: The groups are arranged so that no score can appear in two groups.

| Score | Frequency |
|---|---|
| 71–75 | 5 |
| 76–80 | 6 |
| 81–85 | 8 |
| 86–90 | 3 |
| 91–95 | 1 |
| 96–100 | 3 |
| 101–105 | 4 |

## Exercise 2.3.1

1 The following data gives the percentage scores obtained by students from two classes, 12X and 12Y, in a mathematics exam.

12X

42 73 93 85 68 58 33 70 71 85 90 99 41 70 65
80 73 89 88 93 49 50 57 64 78 79 94 80 50 76 99

12Y

70 65 50 89 96 45 32 64 55 39 45 58 50 82 84
91 92 88 71 52 33 44 45 53 74 91 46 48 59 57 95

a Draw a grouped tally and frequency table for each of the classes.
b Comment on any similarities or differences between the results.

2 The number of apples collected from 50 trees is recorded below.

35 78 15 65 69 32 12 9 89 110 112 148 98
67 45 25 18 23 56 71 62 46 128 7 133 96
24 38 73 82 142 15 98 6 123 49 85 63 19
111 52 84 63 78 12 55 138 102 53 80

Choose suitable groups for this data and use them to draw a grouped frequency table.

With grouped continuous data, the groups are presented in a different way.
The results below are the times given (in h:min:s) for the first 50 people completing a marathon.

2:07:11 2:08:15 2:09:36 2:09:45 2:10:45
2:10:46 2:11:42 2:11:57 2:12:02 2:12:11
2:13:12 2:13:26 2:14:26 2:15:34 2:15:43
2:16:25 2:16:27 2:17:09 2:18:29 2:19:26
2:19:27 2:19:31 2:20:00 2:20:23 2:20:29
2:21:47 2:21:52 2:22:32 2:22:48 2:23:08
2:23:17 2:23:28 2:23:46 2:23:48 2:23:57
2:24:04 2:24:12 2:24:15 2:24:24 2:24:29
2:24:45 2:25:18 2:25:34 2:25:56 2:26:10
2:26:22 2:26:51 2:27:14 2:27:23 2:27:37

The data can be arranged into a grouped frequency table as follows.

| Group | Frequency |
|---|---|
| 2:05:00 ≤ $t$ < 2:10:00 | 4 |
| 2:10:00 ≤ $t$ < 2:15:00 | 9 |
| 2:15:00 ≤ $t$ < 2:20:00 | 9 |
| 2:20:00 ≤ $t$ < 2:25:00 | 19 |
| 2:25:00 ≤ $t$ < 2:30:00 | 9 |

Note that, as with discrete data, the groups do not overlap. However, as the data is continuous, the groups are written using inequalities. The first group includes all times from 2 h 5 min *up to but not including* 2 h 10 min.

With continuous data, the upper and lower bound of each group are the numbers written as the limits of the group. In the example above, for the group 2:05:00 $\leq t <$ 2:10:00, the lower bound is 2:05:00; the upper bound is considered to be 2:10:00 despite it not actually being included in the inequality.

## Frequency histograms for grouped data

A **frequency histogram** displays the frequency of either continuous or grouped discrete data in the form of bars. There are several important features of a frequency histogram for grouped data.

- The bars touch.
- The horizontal axis is labelled with a scale.
- The bars can be of varying width. (Note: In this course, the width of all bars will be constant).
- The frequency of the data is represented by the area of the bar and not the height. (Note: In the case of bars of equal width, the area is directly proportional to the height of the bar and so the height is usually used as the measure of frequency.)

**Worked example**

The table shows the marks out of 100 in a mathematics exam for a class of 32 students.

Draw a histogram representing this data.

| Test marks | Frequency |
|---|---|
| 1–10 | 0 |
| 11–20 | 0 |
| 21–30 | 1 |
| 31–40 | 2 |
| 41–50 | 5 |
| 51–60 | 8 |
| 61–70 | 7 |
| 71–80 | 6 |
| 81–90 | 2 |
| 91–100 | 1 |

All the class intervals are the same. As a result the bars of the histogram will all be of equal width, and the frequency can be plotted on the vertical axis. The histogram is shown. Note that the upper and lower bounds are used to draw the bars.

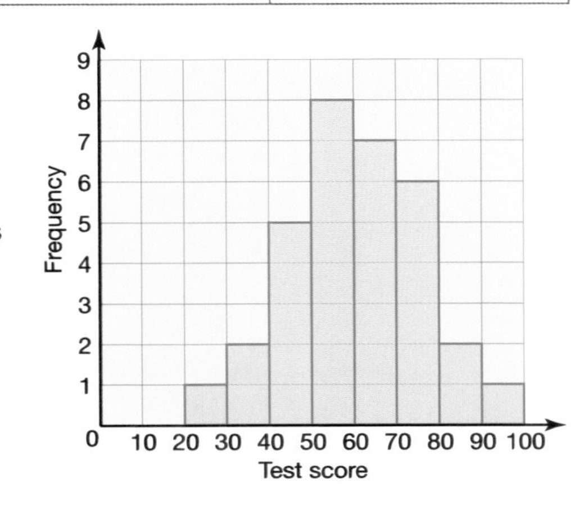

## Exercise 2.3.2

1 The table shows the distances travelled to school by a class of 30 students.

| Distance (km) | Frequency |
|---|---|
| $0 \le d < 1$ | 8 |
| $1 \le d < 2$ | 5 |
| $2 \le d < 3$ | 6 |
| $3 \le d < 4$ | 3 |
| $4 \le d < 5$ | 4 |
| $5 \le d < 6$ | 2 |
| $6 \le d < 7$ | 1 |
| $7 \le d < 8$ | 1 |

Draw this information on a histogram.

2 The heights of students in a class were measured. The results are shown in the table.

| Height (cm) | Frequency |
|---|---|
| 145– | 1 |
| 150– | 2 |
| 155– | 4 |
| 160– | 7 |
| 165– | 6 |
| 170– | 3 |
| 175– | 2 |
| 180–185 | 1 |

Draw a histogram to represent this data.

Note: In the context this question, 145– means the height ($h$) falls within the range $145 \le h < 150$.

# 2.4 Cumulative frequency

A cumulative frequency graph is particularly useful when trying to calculate the median (the middle value) of a large set of grouped discrete data or continuous data, or when trying to establish how consistent a set of results are. Calculating the cumulative frequency is done by adding up the frequencies as we go along.

**Worked example**

The duration of two different brands of battery, A and B, is tested. Fifty batteries of each type are randomly selected and tested in the same way. The duration of each battery is then recorded. The results of the tests are shown in the table below.

| Type A: duration (h) | Frequency | Cumulative frequency |
|---|---|---|
| $0 \leq t < 5$ | 3 | 3 |
| $5 \leq t < 10$ | 5 | 8 |
| $10 \leq t < 15$ | 8 | 16 |
| $15 \leq t < 20$ | 10 | 26 |
| $20 \leq t < 25$ | 12 | 38 |
| $25 \leq t < 30$ | 7 | 45 |
| $30 \leq t < 35$ | 5 | 50 |

| Type B: duration (h) | Frequency | Cumulative frequency |
|---|---|---|
| $0 \leq t < 5$ | 1 | 1 |
| $5 \leq t < 10$ | 1 | 2 |
| $10 \leq t < 15$ | 10 | 12 |
| $15 \leq t < 20$ | 23 | 35 |
| $20 \leq t < 25$ | 9 | 44 |
| $25 \leq t < 30$ | 4 | 48 |
| $30 \leq t < 35$ | 2 | 50 |

**a** Plot a cumulative frequency curve for each brand of battery.
**b** Estimate the median duration for each brand.

**a** The points are plotted at the upper boundary of each class interval rather than at the middle of the interval. So, for type A, points are plotted at (5, 3), (10, 8), etc. The points are joined with a smooth curve which is extended to include (0, 0).

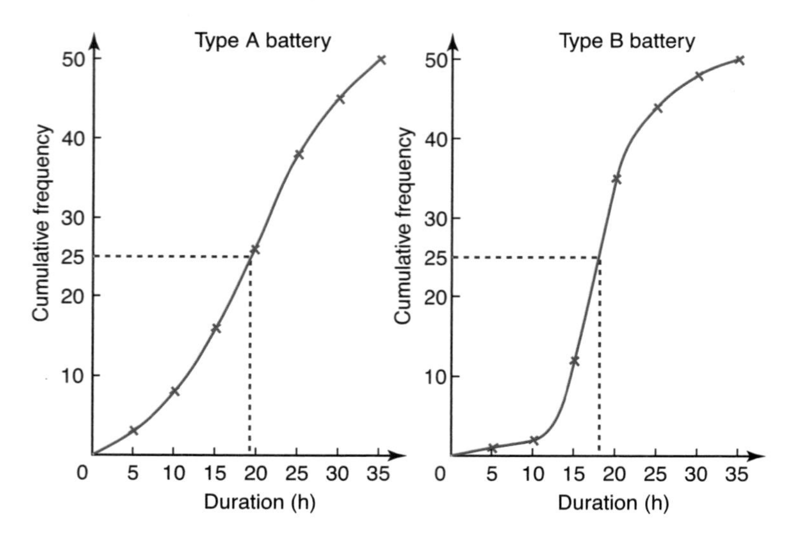


Both cumulative frequency curves are plotted above.

**b** The median value is the value which occurs half-way up the cumulative frequency axis. This is shown with broken lines on the graphs. Therefore:
median for type A batteries ≈ 19 h
median for type B batteries ≈ 18 h
This tells us that, on average, batteries of type A last longer (19 hours) than batteries of type B (18 hours).

Graphing software can produce a cumulative frequency curve and help with calculating the median value.

The example on the next page takes the data for battery A above.

## Autograph

Select  to produce a statistics page.

to enter the grouped data properties as shown. Click 'OK'.

Click on the cumulative frequency graph icon .

Ensure 'cumulative frequency' and 'curve fit' are selected.

Click 'OK'.

To change the scale on the axes use  .

To calculate the median, click on the 'cumulative

frequency diagram measurement' icon and select 'Median'. Click 'OK'.

A horizontal line is drawn at the middle value on the cumulative frequency axis. A vertical line can be dragged at its base until it intersects the horizontal line on the curve.

The median result is shown at the base of the screen.

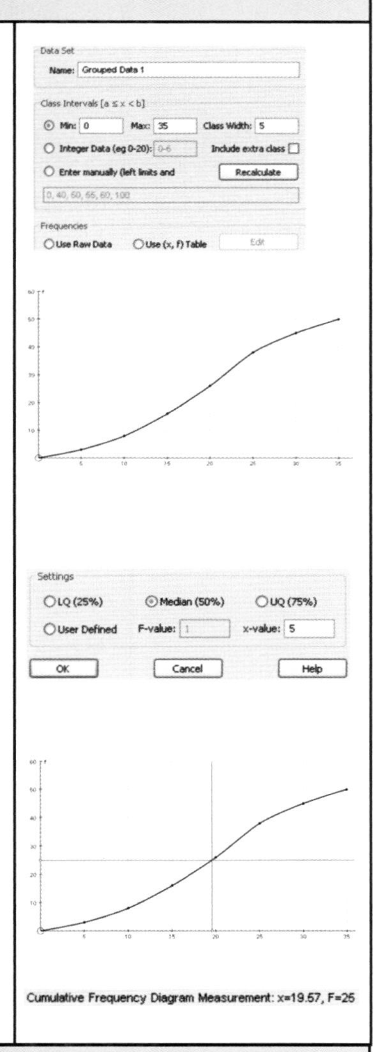


## GeoGebra

At present GeoGebra does not have a cumulative frequency graphing facility.

## Quartiles

The cumulative frequency axis can also be represented in terms of percentiles. A percentile divides the cumulative frequency scale into hundredths. The maximum value of cumulative frequency is found at the 100th percentile. Similarly the median, being the middle value, is also known as the 50th percentile.

In addition to these, the cumulative frequency scale can be divided into quarters. The 25th percentile is known as the lower quartile ($Q_1$), whilst the 75th percentile is the upper quartile ($Q_3$).

Graphing software can also be used to calculate the upper and lower quartiles.

### Autograph

Enter the data for battery A and produce the cumulative frequency graph as shown on the previous page.

To calculate each of the quartiles, click on the 'cumulative frequency diagram measurement' icon and select 'LQ(25%)'. Click 'OK'.

A horizontal line is drawn at the lower quartile value on the cumulative frequency axis. A vertical line can be dragged at its base until it intersects the horizontal line on the curve.

The lower quartile result is shown at the base of the screen.

The above procedure can be repeated for the upper quartile.

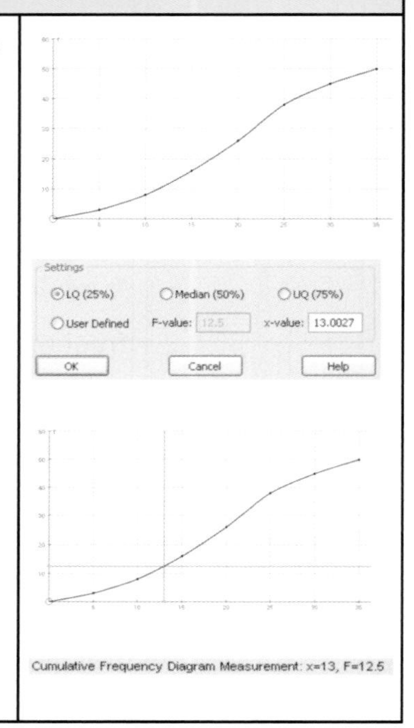

## Exercise 2.4.1

1 Sixty athletes enter a cross-country race. Their finishing times are recorded and are shown in the table below.

| **Finishing time (h)** | 0– | 0.5– | 1.0– | 1.5– | 2.0– | 2.5– | 3.0–3.5 |
|---|---|---|---|---|---|---|---|
| **Frequency** | 0 | 0 | 6 | 34 | 16 | 3 | 1 |
| **Cumulative freq.** | | | | | | | |

a Copy the table and calculate the values for the cumulative frequency.
b Draw a cumulative frequency curve of the results.
c Show how your graph could be used to find the approximate median finishing time.
d What does the median value tell us?

2 Three mathematics groups take the same test in preparation for their final exam. Their raw scores are shown below.

Group A 12, 21, 24, 30, 33, 36, 42, 45, 53, 53, 57, 59, 61, 62, 74, 88, 92, 93
Group B 48, 53, 54, 59, 61, 62, 67, 78, 85, 96, 98, 99
Group C 10, 22, 36, 42, 44, 68, 72, 74, 75, 83, 86, 89, 93, 96, 97, 99, 99

a Using the class intervals $0 \leq x < 20$, $20 \leq x < 40$ etc., draw a grouped frequency table and cumulative frequency table for each group.
b Draw a cumulative frequency curve for each group.
c Show how your graph could be used to find the median score for each group.
d What does the median value tell us?
e From your graph, estimate the upper and lower quartile for each group.

3 The table below shows the heights of students in a class over a three-year period.

| **Height (m)** | **Frequency 2010** | **Frequency 2011** | **Frequency 2012** |
|---|---|---|---|
| $150 \leq h < 155$ | 6 | 2 | 2 |
| $155 \leq h < 160$ | 8 | 9 | 6 |
| $160 \leq h < 165$ | 11 | 10 | 9 |
| $165 \leq h < 170$ | 4 | 4 | 8 |
| $170 \leq h < 175$ | 1 | 3 | 2 |
| $175 \leq h < 180$ | 0 | 2 | 2 |
| $180 \leq h < 185$ | 0 | 0 | 1 |

a Construct a cumulative frequency table for each year.
b Draw the cumulative frequency curve for each year.
c Show how your graph could be used to find the median height for each year.
d What does the median value tell us?
e From your graph, estimate the upper and lower quartile height for each year.
f Comment on your results in part **e** above.

## Box and whisker diagrams

So far we have seen how cumulative frequency curves enable us to make estimes of the medium and quartiles of grouped data.

**Box and whisker diagrams** (also known as box and whisker plots or **box plots**) provide another visual way of representing data. The diagram below demonstrates what a 'typical' box and whisker diagram looks like and also highlights its main features.

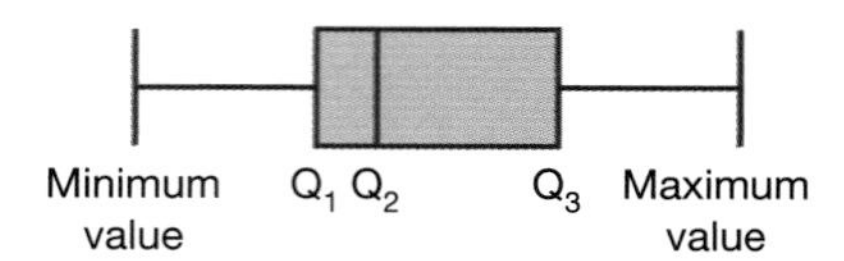


The box and whisker diagram shows all the main features of the data, i.e. the minimum and maximum values, the upper and lower quartiles and the median. The box represents the middle 50% of the data (the interquartile range) and the whiskers represent the whole of the data (the range).

For discrete data, the median position is given by the formula $\frac{n+1}{2}$, where $n$ represents the number of values. Similarly the position of the lower quartile can be calculated using the formula $\frac{n+1}{4}$ and the upper quartile by the formula $\frac{3(n+1)}{4}$.

**Worked example**

The shoe sizes of 15 boys and 15 girls from the same class are recorded in the frequency table below.

| **Shoe size** | 5 | $5\frac{1}{2}$ | 6 | $6\frac{1}{2}$ | 7 | $7\frac{1}{2}$ | 8 | $8\frac{1}{2}$ | 9 | $9\frac{1}{2}$ |
|---|---|---|---|---|---|---|---|---|---|---|
| **Frequency (boys)** | 0 | 0 | 1 | 2 | 1 | 2 | 3 | 4 | 1 | 1 |
| **Frequency (girls)** | 1 | 3 | 4 | 4 | 1 | 1 | 1 | 0 | 0 | 0 |

a Calculate the lower quartile, median and upper quartile shoe sizes for the boys and girls in the class.
b Compare this data using two box and whisker diagrams (one for boys and one for girls).
c What conclusions can be drawn from the box and whisker diagrams?

a Lower quartile boy $= \frac{15+1}{4} = 4\text{th}$ i.e. $q_1 = 7$

Median boy $= \frac{15+1}{2} = 8\text{th}$ i.e. $q_2 = 8$

Upper quartile boy $= \frac{3(15+1)}{4} = 12\text{th}$ i.e. $q_3 = 8\frac{1}{2}$

Lower quartile girl is the 4th i.e. $q_1 = 5\frac{1}{2}$
Median girl is the 8th i.e. $q_2 = 6$
Upper quartile girl is the 12th i.e. $q_3 = 6\frac{1}{2}$

**b** For box and whisker diagrams it is necessary to know the maximum and minimum values.
Minimum boy shoe size is 6, maximum boy shoe size is $9\frac{1}{2}$.
Minimum girl shoe size is 5, maximum girl shoe size is 8.

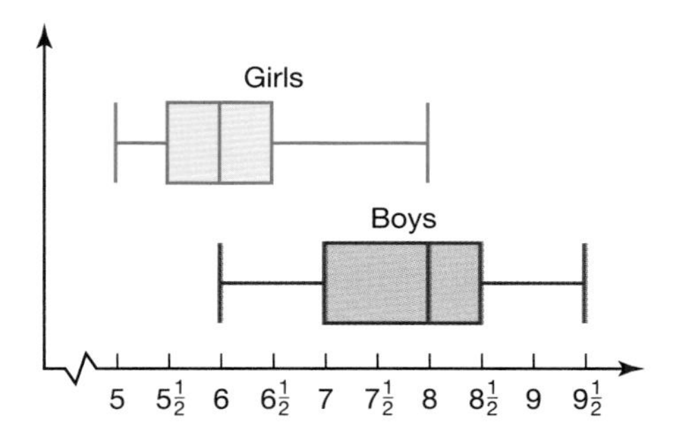


Note: There is no scale on the *y*-axis as it is not relevant in a box and whisker diagram. Consequently, the height of a box does not have a particular value.

**c**
- The overall range of data is greater for boys than it is for girls.
- The middle 50% of girls have a narrower spread of shoe size than the middle 50% of boys.

A GDC can also produce a box and whisker diagram. The instructions that follow are for the boys' data above.

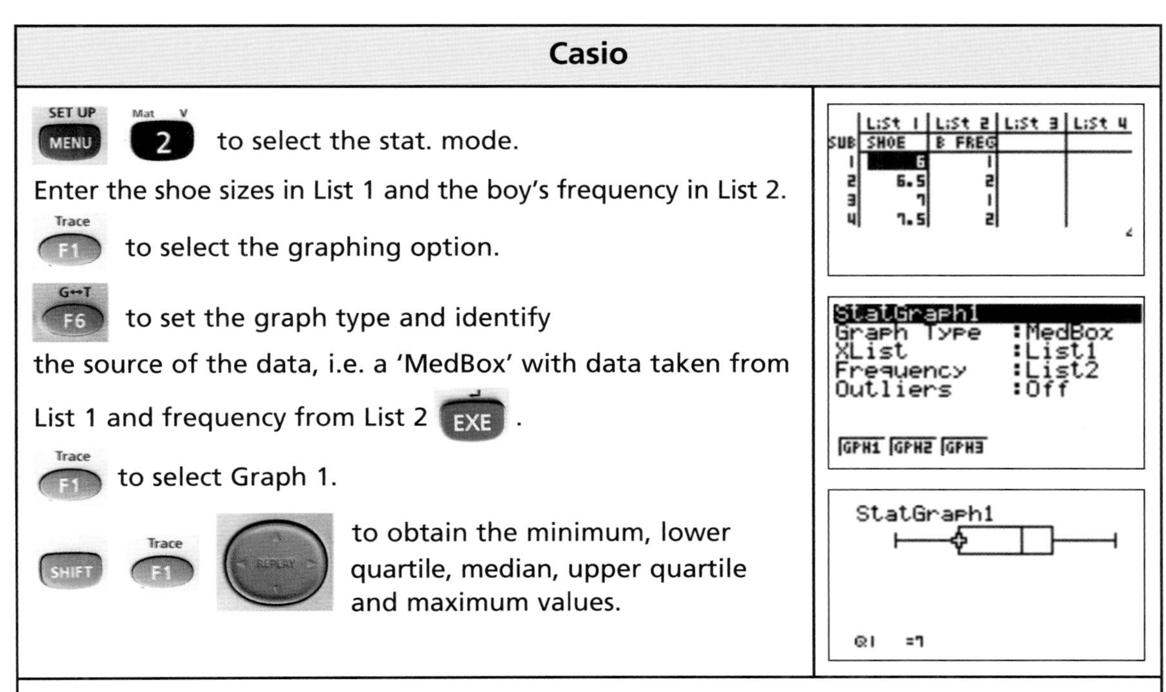

## Texas

to enter the data into lists.

Enter the boy's shoe size into List 1 and the frequency in List 2.

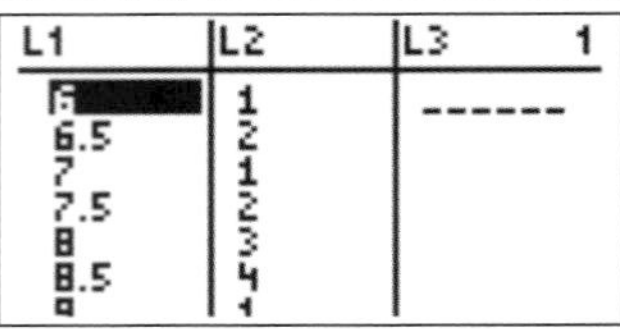

to select the type of graph and its properties, i.e. turn Plot 1 on, choose the box and whisker plot graph and ensure that data is taken from L1 and its frequency from L2 as shown.

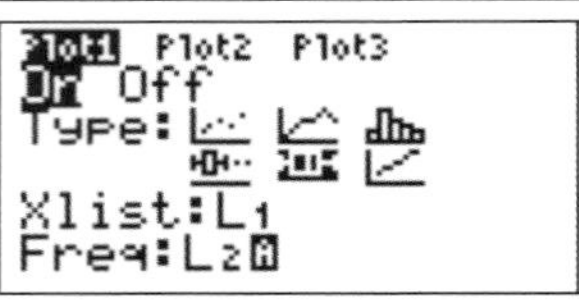

to determine the scale of the axes.

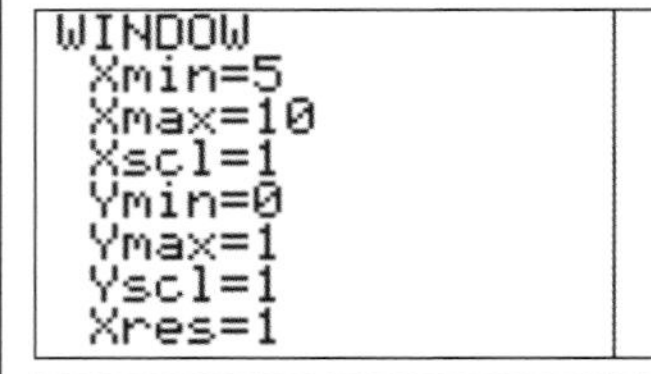

to graph the box and whisker diagram.

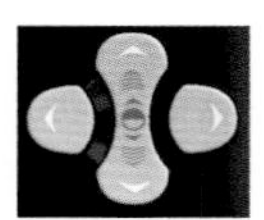
to obtain the minimum, lower quartile, median, upper quartile and maximum values.

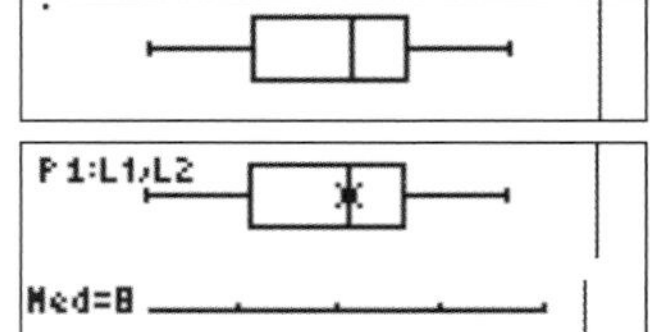


Note: Although a *y*-axis scale needs to be entered, it has no effect on the shape of the box and whisker diagram.

## Exercise 2.4.2

Using a GDC or otherwise, answer the following questions.

1 A football team records, over 20 matches, the number of goals it scored and the number of goals it let in in each match. The results are shown in the table below.

| **Number of goals** | 0 | 1 | 2 | 3 | 4 | 5 |
|---|---|---|---|---|---|---|
| **Frequency of goals scored** | 6 | 9 | 3 | 1 | 0 | 1 |
| **Frequency of goals let in** | 3 | 3 | 8 | 3 | 3 | 0 |

a For goals scored and goals let in, find:
  i) the mean
  ii) the median
  iii) the lower quartile
  iv) the upper quartile.

b Using the same scale, draw box and whisker diagrams to represent the data.

c Write a brief report about what the box and whisker diagrams tell you about the team's results.

2 Two competing holiday resorts record the number of hours of sunshine they have each day during the month of August. The results are shown below.

| Hours of sunshine | 4 | 5 | 6 | 7 | 8 | 9 | 10 | 11 | 12 |
|---|---|---|---|---|---|---|---|---|---|
| Resort A | 1 | 2 | 3 | 5 | 5 | 4 | 4 | 4 | 3 |
| Resort B | 0 | 0 | 0 | 4 | 12 | 10 | 5 | 0 | 0 |

a For each resort find the number of hours of sunshine represented by:
i) the mean
ii) the median
iii) the lower quartile
iv) the upper quartile.
b Using the same scale, draw box and whisker diagrams to represent the data.
c Based on the data and referring to your box and whisker diagrams, explain which resort you would choose to go to for a beach holiday in August.

3 A teacher decides to tackle the problem of students arriving late to his class. To do this, he records how late they are to the nearest minute. His results are shown in the table below.

| Number of minutes late | 0 | 1 | 2 | 3 | 4 | 5 | 6 | 7 | 8 | 9 | 10 |
|---|---|---|---|---|---|---|---|---|---|---|---|
| Number of students | 6 | 4 | 4 | 5 | 7 | 3 | 1 | 0 | 0 | 0 | 0 |

After two weeks of trying to improve the situation, he records a new set of results. These are shown below.

| Number of minutes late | 0 | 1 | 2 | 3 | 4 | 5 | 6 | 7 | 8 | 9 | 10 |
|---|---|---|---|---|---|---|---|---|---|---|---|
| Number of students | 14 | 7 | 4 | 1 | 1 | 1 | 0 | 0 | 1 | 0 | 1 |

The teacher decides to analyze these sets of data using box and whisker diagrams. By carrying out any necessary calculations and drawing the relevant box and whisker diagrams, decide whether his strategy has improved student punctuality. Give detailed reasons for your answer.

## 2.5 Measures of central tendency

'Average' is a word which, in general use, is taken to mean somewhere in the middle. For example, a woman may describe herself as being of average height. A student may think that he or she is of average ability in maths. Mathematics is more precise and uses three main methods to measure average.

- The **mode** is the value occurring most often.
- The **median** is the middle value when all the data is arranged in order of size.
- The **mean** is found by adding together all the values of the data and then dividing the total by the number of data values.

**Worked example**

The numbers below represent the number of goals scored by a team in the first 15 matches of the season. Find the mean, median and mode of the goals.

1 0 2 4 1 2 1 1 2 5 5 0 1 2 3

$$\text{Mean} = \frac{1+0+2+4+1+2+1+1+2+5+5+0+1+2+3}{15} = 2$$

Arranging all the data in order and then picking out the middle number gives the median: 0 0 1 1 1 1 1 ②2 2 2 3 4 5 5

The mode is the number that appears most often. Therefore the mode is 1.

Note: If there is an even number of data values, then there will not be one middle number, but a middle pair. The median is calculated by working out the mean of the middle pair.

Your GDC is also capable of calculating the mean and median of this set of data.

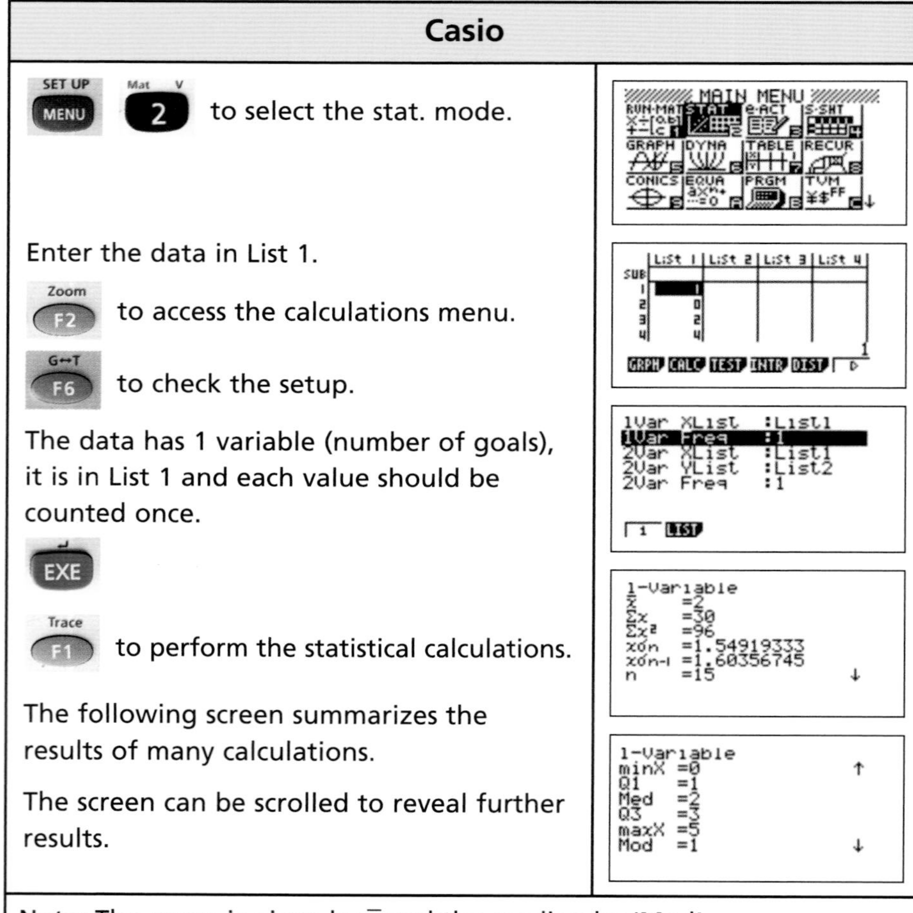


Note: The mean is given by $\bar{x}$ and the median by 'Med'.
The lower quartile, $q_1$, and upper quartile, $q_3$, are also displayed on this screen, as Q1 and Q3 respectively.

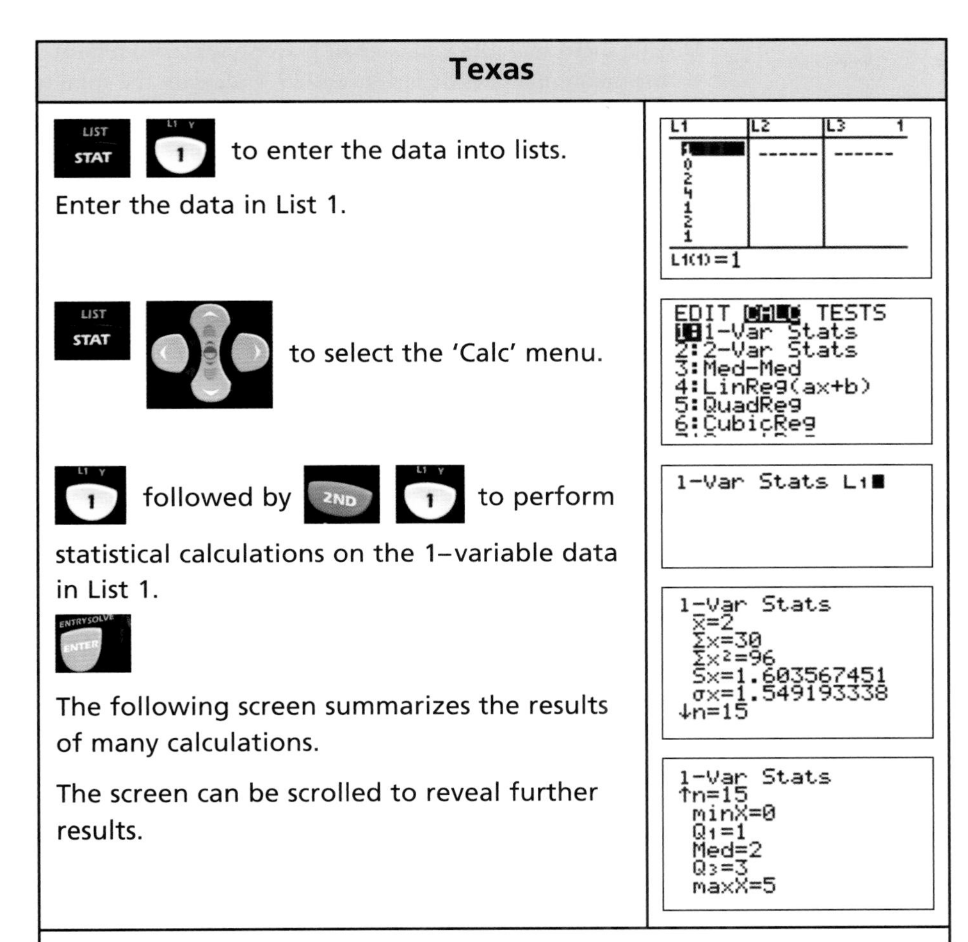


Note: The mean is given by $\bar{x}$ and the median by 'Med'.
The lower quartile, $q_1$, and upper quartile, $q_3$, are also displayed on this screen, as $Q_1$ and $Q_3$ respectively.

## Exercise 2.5.1

1 Find the mean, median and mode for each set of data.

  a The number of goals scored by a hockey team in each of 15 matches
  1 0 2 4 0 1 1 1 2 5 3 0 1 2 2

  b The total scores when two dice are rolled
  7 4 5 7 3 2 8 6 8 7 6 5 11 9 7 3 8 7 6 5

  c The number of students present in a class over a three-week period
  28 24 25 28 23 28 27 26 27 25 28 28 28 26 25

  d An athlete's training times (seconds) for the 100 metres
  14.0 14.3 14.1 14.3 14.2 14.0 13.9 13.8
  13.9 13.8 13.8 13.7 13.8 13.8 13.8

2 The mean mass of the 11 players in a football team is 80.3 kg. The mean mass of the team plus a substitute is 81.2 kg. Calculate the mass of the substitute.

3 After eight matches a basketball player had scored a mean of 27 points. After three more matches his mean was 29. Calculate the total number of points he scored in the last three games.

## Large amounts of data

When there are only three values in a set of data, the median value is given by the second value, i.e. 1 (2) 3.

When there are four values in a set of data, the median value is given by the mean of the second and third values, i.e 1 (2 3) 4.

When there are five values in a set of data, the median value is given by the third value.

If this pattern is continued, it can be deduced that for $n$ sets of data, the median value is given by the value at $\frac{n+1}{2}$. This is useful when finding the median of large sets of data.

**Worked example**

The shoe sizes of 49 people are recorded in the table below. Calculate the median, mean and modal shoe size.

| **Shoe size** | 3 | $3\frac{1}{2}$ | 4 | $4\frac{1}{2}$ | 5 | $5\frac{1}{2}$ | 6 | $6\frac{1}{2}$ | 7 |
|---|---|---|---|---|---|---|---|---|---|
| **Frequency** | 2 | 4 | 5 | 9 | 8 | 6 | 6 | 5 | 4 |

As there are 49 data values, the median value is the 25th value. We can use the cumulative frequency to identify which class this falls within.

| **Shoe size** | 3 | $3\frac{1}{2}$ | 4 | $4\frac{1}{2}$ | 5 | $5\frac{1}{2}$ | 6 | $6\frac{1}{2}$ | 7 |
|---|---|---|---|---|---|---|---|---|---|
| **Cumulative frequency** | 2 | 6 | 11 | 20 | 28 | 34 | 40 | 45 | 49 |

The median occurs within shoe size 5. So the median shoe size is 5.
To calculate the mean of a large data set, we use the formula

$\bar{x} = \frac{\sum fx}{n}$ where $n = \sum f$ and $\sum fx$ means the sum of the product of $f$ and $x$.

| **Shoe size, *x*** | 3 | $3\frac{1}{2}$ | 4 | $4\frac{1}{2}$ | 5 | $5\frac{1}{2}$ | 6 | $6\frac{1}{2}$ | 7 |
|---|---|---|---|---|---|---|---|---|---|
| **Frequency, *f*** | 2 | 4 | 5 | 9 | 8 | 6 | 6 | 5 | 4 |
| ***fx*** | 6 | 14 | 20 | 40.5 | 40 | 33 | 36 | 32.5 | 28 |

Mean shoe size $= \frac{250}{49} = 5.10$

Note: The mean value is not necessarily a data value which appears in the set or a real shoe size.

The modal shoe size is $4\frac{1}{2}$.

**Worked example**

A certain type of matchbox claims to contain 50 matches in each box. A sample of 60 boxes produced the following results.

| Number of matches, $x$ | Frequency, $f$ | $fx$ | $(x - \bar{x})$ | $(x - \bar{x})^2$ | $f(x - \bar{x})x^2$ |
|---|---|---|---|---|---|
| 48 | 3 | 144 | −2.05 | 4.2025 | 12.6075 |
| 49 | 11 | 539 | −1.05 | 1.1025 | 12.1275 |
| 50 | 28 | 1400 | −0.05 | 0.0025 | 0.07 |
| 51 | 16 | 816 | 0.95 | 0.9025 | 14.44 |
| 52 | 2 | 104 | 1.95 | 3.8025 | 7.605 |
| Total | 60 | 3003 | | | 48.85 |

Calculate:
**a** the mean
**b** the standard deviation.

**a** $\bar{x} = \dfrac{\Sigma fx}{\Sigma f} = \dfrac{3003}{60} = 50.05$

**b** $s_n = \sqrt{\dfrac{\Sigma f\,(x - \bar{x})^2}{\Sigma f}} = \sqrt{\dfrac{46.85}{60}} = 0.884$

## Exercise 2.6.2

Using a GDC or otherwise, calculate:
**i)** the mean
**ii)** the range
**iii)** the interquartile range
**iv)** the standard deviation
for the data given in questions 1–4.

**1** **a** 6, 2, 3, 8, 7, 5, 9, 9, 2, 4
**b** 72, 84, 83, 81, 69, 77, 85, 79
**c** 1.6, 2.9, 3.7, 5.5, 4.2, 3.9, 2.8, 4.5, 4.2, 5.1, 3.9

**2** The number of goals a hockey team scores during each match in a season

| **Number of goals** | 0 | 1 | 2 | 3 | 4 | 5 | 6 |
|---|---|---|---|---|---|---|---|
| **Frequency** | 3 | 8 | 11 | 7 | 4 | 2 | 1 |

**3** The number of shots in a round for each player in a golf tournament

| **Number of shots** | 66 | 67 | 68 | 69 | 70 | 71 | 72 | 73 | 74 |
|---|---|---|---|---|---|---|---|---|---|
| **Frequency** | 1 | 2 | 1 | 4 | 11 | 17 | 43 | 18 | 3 |

4 The number of letters posted to 50 houses in a street

| Number of letters | 0 | 1 | 2 | 3 | 4 | 5 | 6 |
|---|---|---|---|---|---|---|---|
| Frequency | 5 | 8 | 12 | 8 | 7 | 8 | 2 |

5 The results for a series of experiments are given below.

| Experiment | 1 | 2 | 3 | 4 | 5 | 6 | 7 | 8 | 9 | 10 |
|---|---|---|---|---|---|---|---|---|---|---|
| Result | 6.2 | 6.1 | 6.3 | 6.3 | 6.7 | 6.1 | 6.2 | 6.3 | 6.1 | 5.9 |
| Experiment | 11 | 12 | 13 | 14 | 15 | 16 | 17 | 18 | 19 | 20 |
| Result | 6.0 | 6.2 | 6.1 | 6.3 | 6.0 | 6.1 | 6.2 | 6.3 | 6.1 | |

The result for experiment 20 is obscured. However it is known that the mean ($\bar{x}$) for all 20 experiments is 6.2.

Calculate:

a the value of the 20th result

b the standard deviation of all the results.

## Student assessment 1

1 State which of the following types of data are discrete and which are continuous.

a The number of goals scored in a hockey match

b The price of a kilogram of carrots

c The speed of a car

d The number of cars passing the school gate each hour

e The time taken to travel to school each morning

f The wingspan of butterflies

g The height of buildings

2 Four hundred students sit their mathematics IGCSE exam. Their marks (as percentages) are shown in the table below.

| Mark (%) | Frequency | Cumulative frequency |
|---|---|---|
| 31–40 | 21 | |
| 41–50 | 55 | |
| 51–60 | 125 | |
| 61–70 | 74 | |
| 71–80 | 52 | |
| 81–90 | 45 | |
| 91–100 | 28 | |

a Copy and complete the above table by calculating the cumulative frequency.

b Draw a cumulative frequency curve of the results.

c Using the graph, estimate a value for:

i) the median exam mark

ii) the upper and lower quartiles

iii) the interquartile range.

3 Eight hundred students sit an exam. Their marks (as percentages) are shown in the table opposite.

| Mark (%) | Frequency | Cumulative frequency |
|---|---|---|
| 1–10 | 10 | |
| 11–20 | 30 | |
| 21–30 | 40 | |
| 31–40 | 50 | |
| 41–50 | 70 | |
| 51–60 | 100 | |
| 61–70 | 240 | |
| 71–80 | 160 | |
| 81–90 | 70 | |
| 91–100 | 30 | |

a Copy and complete the above table by calculating the cumulative frequency.
b Draw a cumulative frequency curve of the results.
c An A grade is awarded to any student achieving at or above the upper quartile. Using your graph, deduce the minimum mark required for an A grade.
d Any student below the lower quartile is considered to have failed the exam. Using your graph, identify the minimum mark needed so as not to fail the exam.
e How many students failed the exam?
f How many students achieved an A grade?

4 A businesswoman travels to work in her car each morning in one of two ways; either using the country lanes or using the motorway. She records the time taken to travel to work each day. The results are shown in the table below.

| Time (mins) | Motorway frequency | Country lanes frequency |
|---|---|---|
| $10 \le t < 15$ | 3 | 0 |
| $15 \le t < 20$ | 5 | 0 |
| $20 \le t < 25$ | 7 | 9 |
| $25 \le t < 30$ | 2 | 10 |
| $30 \le t < 35$ | 1 | 1 |
| $35 \le t < 40$ | 1 | 0 |
| $40 \le t < 45$ | 1 | 0 |

a Complete a cumulative frequency table for each of the sets of results shown above.
b Using your cumulative frequency tables, plot two cumulative frequency curves – one for the time taken to travel to work using the motorway, the other for the time taken to travel to work using country lanes.
c Use your graph to find the following for each method of travel:
   i) the median travelling time
   ii) the upper and lower quartile travelling times
   iii) the interquartile range for the travelling times.
d With reference to your graph or calculations, describe which is the most reliable way for the businesswoman to get to work.
e If she had to get to work one morning within 25 minutes of leaving home, which way would you recommend she goes? Explain your answer fully.

5 Two classes take a maths test. One class contains students of similar ability; the other is a mixed ability class. The results of the tests for each class are presented using the box and whisker diagrams below.

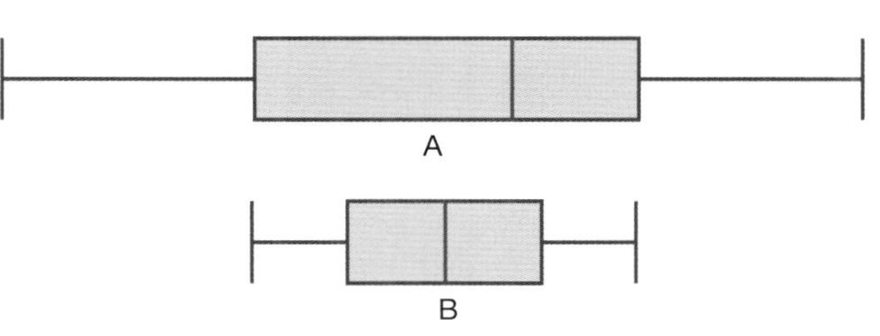


Describe clearly, giving your reasons, which of the two box and whisker diagrams is **likely** to belong to the mixed ability maths class and which is likely to belong to the other maths class.

## Student assessment 2

1 Find the mean, median and mode of the following sets of data.
  a 4 5 5 6 7
  b 63 72 72 84 86
  c 3 8 12 18 18 24
  d 4 9 3 8 7 11 3 5 3 8

2 The mean mass of the 15 players in a rugby team is 85 kg. The mean mass of the team plus a substitute is 83.5 kg. Calculate the mass of the substitute.

3 Thirty children were asked about the number of pets they had. The results are shown in the table below.

| Number of pets | 0 | 1 | 2 | 3 | 4 | 5 | 6 |
|---|---|---|---|---|---|---|---|
| Frequency | 5 | 5 | 3 | 7 | 3 | 1 | 6 |

  a Calculate the mean number of pets per child.
  b Calculate the median number of pets per child.
  c Calculate the modal number of pets.

4 Thirty families live in a street. The number of children in each family is given in the table below.

| Number of children | 0 | 1 | 2 | 3 | 4 | 5 | 6 |
|---|---|---|---|---|---|---|---|
| Frequency | 3 | 5 | 8 | 9 | 3 | 0 | 2 |

  a Calculate the mean number of children per family.
  b Calculate the median number of children per family.
  c Calculate the modal number of children.

5 The number of people attending a disco at a club's over 30s evenings are shown below.

| | | | | | | | |
|---|---|---|---|---|---|---|---|
| 89 | 94 | 32 | 45 | 57 | 68 | 127 | 138 |
| 23 | 77 | 99 | 47 | 44 | 100 | 106 | 132 |
| 28 | 56 | 59 | 49 | 96 | 103 | 90 | 84 |
| 136 | 38 | 72 | 47 | 58 | 110 | | |

  a Using groups 0–19, 20–39, 40–59, etc., draw a grouped frequency table of the above data.
  b Using your grouped data, calculate an estimate for the mean number of people going to the disco each night.

6 The number of people attending thirty screenings of a film at a local cinema is given below.

| | | | | | | | |
|---|---|---|---|---|---|---|---|
| 21 | 30 | 66 | 71 | 10 | 37 | 24 | 21 |
| 62 | 50 | 27 | 31 | 65 | 12 | 38 | 34 |
| 53 | 34 | 19 | 43 | 70 | 34 | 27 | 28 |
| 52 | 57 | 45 | 25 | 30 | 39 | | |

  a Using groups 10–19, 20–29, 30–39, etc., draw a grouped frequency table of the above data.
  b Using your grouped data, calculate an estimate for the mean number of people attending each screening.

7 Find the standard deviation of the following set of numbers.

8, 8, 10, 10, 10, 12, 14, 15, 17, 20

8 A hockey team scores the following number of goals in their matches over a season.

| Goals scored | 0 | 1 | 2 | 3 | 4 | 5 |
|---|---|---|---|---|---|---|
| Frequency | 4 | 12 | 8 | 11 | 4 | 1 |

Calculate:
  a the mean
  b the range
  c the standard deviation.

# Examination questions

1 120 Mathematics students in a school sat an examination. Their scores (given as a percentage) were summarized on a cumulative frequency diagram. This diagram is given below.

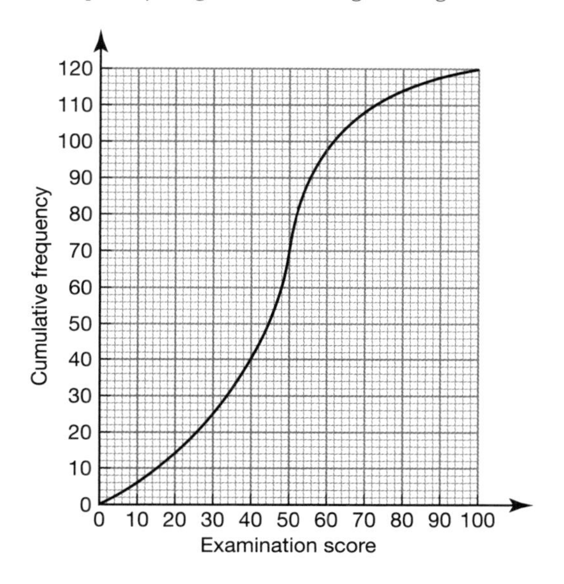


a Copy and complete the grouped frequency table for the students. [3]

| Examination score $x$ (%) | Frequency |
|---|---|
| $0 \leq x \leq 20$ | 14 |
| $20 < x \leq 40$ | 26 |
| $40 < x \leq 60$ | |
| $60 < x \leq 80$ | |
| $80 < x \leq 100$ | |

b Write down the mid-interval value of the $40 \leq x < 60$ interval. [1]

c Calculate an estimate of the mean examination score of the students. [2]

**Paper 1, May 10, Q9**

2 56 students were given a test out of 40 marks. The teacher used the following box and whisker plot to represent the marks of the students.

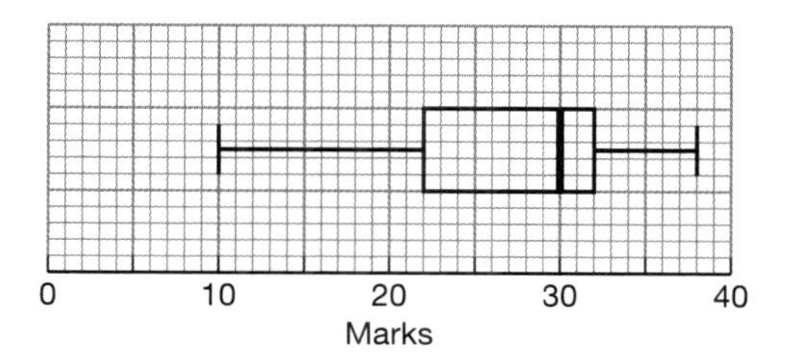


a Write down
i) the median mark;
ii) the 75th percentile mark;
iii) the range of marks. [4]

b Estimate the number of students who achieved a mark greater than 32. [2]

**Paper 1, Nov 10, Q5**

3 The weights of 90 students in a school were recorded. The information is displayed in the following table.

| Weight (kg) | Number of students |
|---|---|
| $40 \leq w < 50$ | 7 |
| $50 \leq w < 60$ | 28 |
| $60 \leq w < 70$ | 35 |
| $70 \leq w < 80$ | 20 |

a Write down the mid-interval value for the interval $50 \leq w < 60$. [1]

b Use your graphic display calculator to find an estimate for
i) the mean weight;
ii) the standard deviation. [3]

c Find the weight that is 3 standard deviations below the mean. [2]

**Paper 1, Nov 10, Q2**

# Topic 2 Applications, project ideas and theory of knowledge

1 Have statistics published about the African continent had a detrimental effect upon an ordinary person's view of that continent?

2 How far is it likely that the taking of opinion polls before an election itself influences the result of the election? If it does, should the taking of such polls be regulated?

3 Can any sample be an accurate reflection of a larger group? Can there be a perfect question in a questionnaire which is independent of both the interviewer and interviewee?

4 Early political opinion polls in the USA presidential election, which were conducted by telephone, turned out to be wildly inaccurate. Why were they so inaccurate?

# 3.2 Sets and logical reasoning

## Symbols used in logic

There are some symbols that you will need to become familiar with when we study logic in more detail.

The following symbols refer to the relationship between two propositions $p$ and $q$.

| Symbol | Meaning |
|---|---|
| $\wedge$ | $p$ and $q$ (conjunction) |
| $\vee$ | $p$ or $q$ or both (inclusive disjunction) |
| $\veebar$ | $p$ or $q$ but not both (exclusive disjunction) |
| $\Rightarrow$ | If $p$ then $q$ (implication) |
| $\Leftrightarrow$ | If $p \Rightarrow q$ and $q \Rightarrow p$ the statements are equivalent, i.e. $p \Leftrightarrow q$ (equivalence) |
| $\neg$ | If $p$ is true, $q$ cannot be true. $p \neg q$ (negation) |

**Proposition:** A proposition is a stated fact. It may also be called a **statement**. It can be true or false. For example,

'Nigeria is in Africa' is a true proposition.
'Japan is in Europe' is a false proposition.
These are examples of simple propositions.

**Compound statement:** Two or more simple propositions can be combined to form a compound proposition or compound statement.

**Conjunction:** Two simple propositions are combined with the word **and**, e.g.

$p$: Japan is in Asia.
$q$: The capital of Japan is Tokyo.

These can be combined to form: Japan is in Asia **and** the capital of Japan is Tokyo.

This is written $p \wedge q$, where $\wedge$ represents the word **and**.

**Negation:** The negation of any simple proposition can be formed by putting '**not**' into the statement, e.g.

$p$: Ghana is in Africa.
$q$: Ghana is not in Africa.

Therefore $q = \neg p$ (i.e. $p$ is the negation of $q$).

If $p$ is true then $q$ cannot also be true.

**Implication:** For two simple propositions $p$ and $q$, $p \Rightarrow q$ means **if** $p$ is true **then** $q$ is also true, e.g.

$p$: It is raining.
$q$: I am carrying an umbrella.

Then $p \Rightarrow q$ states: If it is raining then I am carrying an umbrella.

**Converse:** This is the reverse of a proposition. In the example above the converse of $p \Rightarrow q$ is $q \Rightarrow p$. Note, however, although $p \Rightarrow q$ is true, i.e. If it is raining then I must be carrying an umbrella, its converse $q \Rightarrow p$ is not necessarily true, i.e. it is not necessarily the case that: If I am carrying an umbrella then it is raining.

**Equivalent propositions:** If two propositions are true and converse, then they are said to be equivalent. For, example if we have two propositions

$p$: Pedro lives in Madrid.
$q$: Pedro lives in the capital city of Spain.

these propositions can be combined as a compound statement:

If Pedro lives in Madrid, then Pedro lives in the capital city of Spain.

i.e. $p$ implies $q$ $(p \Rightarrow q)$

This statement can be manipulated to form its converse:

If Pedro lives in the capital of Spain, then Pedro lives in Madrid.

i.e. $q$ implies $p$ $(q \Rightarrow p)$

The two combined statements are both true and converse so they are said to be logically equivalent ($q \Leftrightarrow p$). Logical equivalence will be discussed further in Section 3.4.

**Disjunction:** For two propositions, $p$ and $q$, $p \vee q$ means **either** $p$ **or** $q$ is true **or both** are true, e.g.

$p$: It is sunny.
$q$: I am wearing flip-flops.

Then $p \vee q$ states either it is sunny or I am wearing flip-flops or it is both sunny and I am wearing flip-flops.

**Exclusive disjunction:** For two propositions, $p$ and $q$, $p \veebar q$ means **either** $p$ **or** $q$ is true but **not both** are true, e.g.

$p$: It is sunny.
$q$: I am wearing flip-flops.

Then $p \veebar q$ states either it is sunny or I am wearing flip-flops only.

**Valid arguments:** An argument is valid if the conclusion follows from the premises (the statements). A premise is always assumed to be true, even though it might not be, e.g.

| | |
|---|---|
| London is in France. | the first premise |
| France is in Africa. | the second premise |
| Therefore London is in Africa. | the conclusion |

## Exercise 3.2.1

1 Which of the following are propositions?

**a** Are you from Portugal?
**b** Capetown is in South Africa.
**c** Catalan is a Spanish language.
**d** Be careful with that.
**e** $x = 3$
**f** $x \neq 3$
**g** I play football.
**h** Go outside and play.
**i** Apples are good to eat.
**j** J is a letter of the alphabet.

2 Form compound statements using the word 'and' from the two propositions given and state whether the compound statement is true or false.

**a** $t$: Teresa is a girl. $a$: Abena is a girl.
**b** $p$: $x < 8$ $q$: $x > -1$
**c** $a$: A pentagon has 5 sides. $b$: A triangle has 4 sides.
**d** $l$: London is in England. $e$: England is in Europe.
**e** $k$: $x < y$ $l$: $y < z$
**f** $m$: 5 is a prime number $n$: 4 is an even number
**g** $s$: A square is a rectangle. $t$: A triangle is a rectangle.
**h** $p$: Paris is the capital of France. $g$: Ghana is in Asia.
**i** $a$: 37 is a prime number $b$: 51 is a prime number
**j** $p$: parallelograms are rectangles $t$: trapeziums are rectangles

# 3.3 Truth tables

In probability experiments, a coin when tossed can land on heads or tails. These are complementary events, i.e. P(H) + P(T) = 1.

In logic, if a statement is not uncertain, then it is either true (T) or false (F). If there are two statements, then either both are true, both are false or one is true and one is false.

A truth table is a clear way of showing the possibilities of statements.

Let proposition $p$ be 'Coin A lands heads' and proposition $q$ be 'Coin B lands heads'. The truth table below shows the different possibilities when the two coins are tossed. Alongside is a two-way table also showing the different outcomes. Note the similarity between the two tables.

| $p$ | $q$ |
|---|---|
| T | T |
| T | F |
| F | T |
| F | F |

Truth table

| Coin A | Coin B |
|---|---|
| H | H |
| H | T |
| T | H |
| T | T |

Two-way table

## Conjunction, disjunction and negation

Extra columns can be added to a truth table.

$p \wedge q$ (conjunction) means that both $p$ and $q$ must be true for the statement to be true.

| $p$ | $q$ | $p \wedge q$ |
|---|---|---|
| T | T | T |
| T | F | F |
| F | T | F |
| F | F | F |

$p \vee q$ (inclusive disjunction) means that either $p$ or $q$, or both, must be true for the statement to be true.

| $p$ | $q$ | $p \vee q$ |
|---|---|---|
| T | T | T |
| T | F | T |
| F | T | T |
| F | F | F |

$p \veebar q$ (exclusive disjunction) means that either $p$ or $q$, but not both, must be true for the statement to be true.

| $p$ | $q$ | $p \veebar q$ |
|---|---|---|
| T | T | F |
| T | F | T |
| F | T | T |
| F | F | F |

$\neg p$ represents a negation, i.e. $p$ must not be true for the statement to be true.

| $p$ | $q$ | $\neg p$ |
|---|---|---|
| T | T | F |
| T | F | F |
| F | T | T |
| F | F | T |

## Exercise 3.3.1

1 Copy and complete the truth table for three propositions $p$, $q$ and $r$. It may help to think of spinning three coins and drawing a table of possible outcomes.

| $p$ | $q$ | $r$ |
|---|---|---|
| T | T | T |
| | | |
| | | |
| | | |
| | | |
| | | |
| | | |
| F | F | F |

2 Copy and complete the truth table below for the three statements $p$, $q$ and $r$.

| $p$ | $q$ | $r$ | $\neg p$ | $p \vee q$ | $\neg p \vee r$ | $(p \vee q) \wedge (\neg p \vee r)$ |
|---|---|---|---|---|---|---|
| T | T | T | F | T | T | T |
| T | T | F | | | | |
| T | | | | | | |
| T | | | | | | |
| F | | | | | | |
| F | | | | | | |
| F | | | | | | |
| F | | | | | | |

# Logical contradiction and tautology

## Logical contradiction

A contradiction or contradictory proposition is never true. For example, let $p$ be the proposition that Rome is in Italy.

$p$: Rome is in Italy.

Therefore $\neg p$, the negation of $p$, is the proposition: Rome is not in Italy.

If we write $p \wedge \neg p$ we are saying Rome is in Italy and Rome is not in Italy. This cannot be true at the same time. This is an example of a logical contradiction.

A truth table is shown below for the above statement.

| $p$ | $\neg p$ | $p \wedge \neg p$ |
|---|---|---|
| T | F | F |
| F | T | F |

Both entries in the final column are F. In other words a logical contradiction *must* be false.

**Worked example**

Show that the compound proposition below is a contradiction.

$(p \vee q) \wedge [(\neg p) \wedge (\neg q)]$

Construct a truth table:

| $p$ | $q$ | $\neg p$ | $\neg q$ | $p \vee q$ | $(\neg p) \wedge (\neg q)$ | $(p \vee q) \wedge [(\neg p) \wedge (\neg q)]$ |
|---|---|---|---|---|---|---|
| T | T | F | F | T | F | F |
| T | F | F | T | T | F | F |
| F | T | T | T | T | F | F |
| F | F | T | T | F | T | F |

Because the entries in the last column are all false, the statement is a logical contradiction.

## Tautology

The manager of the band *Muse* said to me recently: 'If *Muse's* album "Resistance" is a success, they will be a bigger band than U2.' He paused 'Or they will not'.

This is an example of a tautology: 'either it does or it doesn't'. It is always true.

A compound proposition is a tautology if it always true regardless of the truth values of its variables.

Consider the proposition: All students study maths or all students do not study maths. This is a tautology, as can be shown in a truth table by considering the result of $p \vee \neg p$.

| $p$ | $\neg p$ | $p \vee \neg p$ |
|---|---|---|
| T | F | T |
| F | T | T |

Since the entries in the final column $p \vee \neg p$ are all true, this is a tautology.

Worked example

Show that $(p \vee q) \vee [(\neg p) \wedge (\neg q)]$ is a tautology by copying and completing the truth table below.

| $p$ | $q$ | $\neg p$ | $\neg q$ | $p \vee q$ | $(\neg p) \wedge (\neg q)$ | $(p \vee q) \vee [(\neg p) \wedge (\neg q)]$ |
|---|---|---|---|---|---|---|
| T | T | F | F | T | F | T |
| T | F | F | T | T | F | T |
| F | T | T | F | T | F | T |
| F | F | T | T | F | T | T |

As the entries in the final column $(p \vee q) \vee [(\neg p) \wedge (\neg q)]$ are all true, the statement is a tautology.

## Exercise 3.3.2

1 Describe each of the following as a tautology, a contradiction or neither. Use a truth table if necessary.
   - a $p \wedge \neg q$
   - b $q \wedge \neg q$
   - c $p \vee \neg q$
   - d $q \vee \neg q$
   - e $[p \vee (\neg q)] \wedge [q \vee (\neg q)]$

2 By drawing a truth table in each case, deduce whether each of the following propositions is a tautology, contradiction or neither.
   - a $\neg p \wedge \neg q$
   - b $\neg(\neg p) \vee p$
   - c $q \wedge \neg r$
   - d $(p \wedge q) \wedge r$
   - e $(p \wedge q) \vee r$

# 3.4 Implication; converse; inverse; contrapositive and logical equivalence

## Implication

'If' is a word introducing a conditional clause.

Later in your life someone might say to you, 'If you get a degree, then I will buy you a car'.

Let us look at this in a truth table.

$p$: You get a degree.
$q$: I will buy you a car.

| $p$ | $q$ | $p \Rightarrow q$ |
|---|---|---|
| T | T | T |
| T | F | F |
| F | T | T |
| F | F | T |

The first row is simple:

You get a degree, I buy you a car, and therefore I have kept my promise.

The second row too is straightforward:

You get a degree, I don't buy you a car, and therefore I have broken my promise.

The last two rows seem more complicated, but think of them like this. If you do not get a degree, then I have kept my side of the bargain whether I buy you a car or not.

Therefore, the only way that this type of statement is false is if a 'promise' is broken.

Logically $p \Rightarrow q$ is true if:

$p$ is false
or $q$ is true
or $p$ is false *and* $q$ is true

Similarly $p \Rightarrow q$ is only false if $p$ is true and $q$ is false.

**Worked examples**

In the following statements, assume that the first phrase is $p$ and the second phrase $q$.

Determine whether the statement $p \Rightarrow q$ is logically true or false.

1 'If $5 \times 4 = 20$, then the Earth moves round the Sun.'
As both $p$ and $q$ are true, then $p \Rightarrow q$ is true, i.e. the statement $p \Rightarrow q$ is logically correct.

2 'If the Sun goes round the Earth, then I am an alien.'
Since $p$ is false, then $p \Rightarrow q$ is true whether I am an alien or not. Therefore the statement is logically true.

This means that witty replies like:

'If I could run faster, I could be a professional footballer'
'Yes and if you had wheels you'd be a professional skater' are logically true, since the premise $p$, 'if you had wheels', is false and therefore what follows is irrelevant.

## Exercise 3.4.1

1 In the following statements, assume that the first phrase is $p$ and the second phrase is $q$. Determine whether the statement $p \Rightarrow q$ is logically true or false.
  a If $2 + 2 = 5$ then $2 + 3 = 5$.
  b If the moon is round, then the Earth is flat.
  c If the Earth is flat, then the moon is flat.
  d If the Earth is round, then the moon is round.
  e If the Earth is round, then I am the man on the moon.

2 Descartes' phrase 'Cogito, ergo sum' translates as 'I think, therefore I am'.
  a Rewrite the sentence using one or more of the following: 'if', 'whenever' , 'it follows that', 'it is necessary' , 'unless' , 'only'.
  b Copy and complete the following sentence: 'Cogito ergo sum' only breaks down logically if Descartes thinks, but . . .'

## Logical equivalence

There are many different ways that we can form compound statements from the propositions $p$ and $q$ using connectives. Some of the different compound propositions have the same truth values. These propositions are said to be equivalent. The symbol for equivalence is $\Leftrightarrow$.

Two propositions are logically equivalent when they have identical truth values.

**Worked example**

Use a truth table to show that $\neg(p \wedge q)$ and $\neg p \vee \neg q$ are logically equivalent.

| $p$ | $q$ | $p \wedge q$ | $\neg(p \wedge q)$ | $\neg p$ | $\neg q$ | $\neg p \vee \neg q$ |
|---|---|---|---|---|---|---|
| T | T | T | F | F | F | F |
| T | F | F | T | F | T | T |
| F | T | F | T | T | F | T |
| F | F | F | T | T | T | T |

Since the truth values for $\neg(p \wedge q)$ and $\neg p \vee \neg q$ (columns 4 and 7) are identical, the two statements are logically equivalent.

## Converse

The statement 'All squares are rectangles' can be rewritten using the word 'if' as:

'If an object is a square, then it is a rectangle'. $p \Rightarrow q$. (true in this case)

The converse is:

$q \Rightarrow p$. 'If an object is a rectangle, then it is a square.' (false in this case)

## Inverse

The inverse of the statement 'If an object is a square, then it is a rectangle' ($p \Rightarrow q$) is:

$\neg p \Rightarrow \neg q$. 'If an object is not a square, then it is not a rectangle.' (false in this case)

## Contrapositive

The contrapositive of the statement 'If an object is a square, then it is a rectangle' ($p \Rightarrow q$) is:

$\neg q \Rightarrow \neg p$. 'If an object is not a rectangle, then it is not a square.' (true in this case)

Note:
A statement is logically equivalent to its contrapositive.
A statement is not logically equivalent to its converse or inverse.
The converse of a statement is logically equivalent to the inverse.

So if a statement is true, then its contrapositive is also true.
If a statement is false, then its contrapositive is also false.

And if the converse of a statement is true, then the inverse is also true.
If the converse of a statement is false, then the inverse is also false.

To summarize:

| | |
|---|---|
| given a conditional statement: | $p \Rightarrow q$ |
| the converse is: | $q \Rightarrow p$ |
| the inverse is: | $\neg p \Rightarrow \neg q$ |
| the contrapositive is: | $\neg q \Rightarrow \neg p$ |

**Worked example**

Statement: 'All even numbers are divisible by 2.'

**a** Rewrite the statement as a conditional statement.
**b** State the converse, inverse and contrapositive of the conditional statement. State whether each new statement is true or false.

**a** Conditional: 'If a number is even, then it is divisible by 2.' (true)
**b** Converse: 'If a number is divisible by 2, then it is an even number.' (true)
Inverse: 'If a number is not even, then it is not divisible by 2.' (true)
Contrapositive: 'If a number is not divisible by 2, then it is not an even number. (true)

Note: The contrapositive both switches the order and negates. It combines the converse and the inverse.

On a truth table it can be shown that a conditional statement and its contrapositive are logically equivalent

| $p$ | $q$ | $\neg p$ | $\neg q$ | Implication $p \Rightarrow q$ | Contrapositive $\neg q \Rightarrow \neg p$ |
|---|---|---|---|---|---|
| T | T | F | F | T | T |
| T | F | F | T | F | F |
| F | T | T | F | T | T |
| F | F | T | T | T | T |

Note: If we have a tautology, we must have logical equivalence. For example, 'If you cannot find the keys you have lost, then you are looking in the wrong place.' Obviously if you are looking in the right place then you can find your keys. (So the contrapositive is equivalent to the proposition.)

## Exercise 3.4.2

1 Write each of the following as a conditional statement and then write its converse, inverse or contrapositive, as indicated in brackets.

*Example* Being interested in the Romans means that you will enjoy Italy. (converse)

Solution: Conditional statement. If you are interested in the Romans, then you will enjoy Italy.

Converse. If you enjoy Italy, then you are interested in the Romans.

- a You do not have your mobile phone, so you cannot send a text. (inverse)
- b A small car will go a long way on 20 euros worth of petrol. (contrapositive)
- c Speaking in French means that you will enjoy France more. (converse)
- d When it rains I do not play tennis. (inverse)
- e We stop playing golf when there is a threat of lightning. (inverse)
- f The tennis serve is easy if you practise it. (contrapositive)
- g A six-sided polygon is a hexagon. (contrapositive)
- h You are less than 160 cm tall, so you are smaller than me. (inverse)
- i The bus was full, so I was late. (contrapositive)
- j The road was greasy, so the car skidded. (converse)

2 Rewrite these statements using the conditional 'if'. Then state the converse, inverse and contrapositive. State whether each new statement is true or false.

- a Any odd number is a prime number.
- b A polygon with six sides is called an octagon.
- c An acute-angled triangle has three acute angles.
- d Similar triangles are congruent.
- e Congruent triangles are similar.
- f A cuboid has six faces.
- g A solid with eight faces is a regular octahedron.
- h All prime numbers are even numbers.

# 3.5 Set theory

The modern study of set theory began with Georg Cantor and Richard Dedekind in an 1874 paper titled 'On a characteristic property of all real algebraic numbers'. It is most unusual to be able to put an exact date to the beginning of an area of mathematics.

The language of set theory is the most common foundation to all mathematics and is used in the definitions of nearly all mathematical objects.

A set is a well-defined group of objects or symbols. The objects or symbols are called the **elements** of the set. If an element $e$ belongs to a set $S$, this is represented as $e \in S$. If $e$ does not belong to set $S$ this is represented as $e \notin S$.

**Worked examples**

1 A particular set consists of the following elements: {South Africa, Namibia, Egypt, Angola, ...}.

- a Describe the set.
- b Add another two elements to the set.
- c Is the set finite or infinite?

a The elements of the set are countries of Africa.
b e.g. Zimbabwe, Ghana
c Finite. There is a finite number of countries in Africa.

2 Consider the set
{1, 4, 9, 16, 25, ...}.
a Describe the set.
b Write another two elements of the set.
c Is the set finite or infinite?

a The elements of the set are square numbers.
b e.g. 36, 49
c Infinite. There is an infinite number of square numbers.

## Exercise 3.5.1

1 For each of the following sets:
i) describe the set in words
ii) write down another two elements of the set.
a {Asia, Africa, Europe, ...}
b {2, 4, 6, 8, ...}
c {Sunday, Monday, Tuesday, ...}
d {January, March, July, ...}
e {1, 3, 6, 10, ...}
f {Mehmet, Michael, Mustapha, Matthew, ...}
g {11, 13, 17, 19, ...}
h {a, e, i, ...}
i {Earth, Mars, Venus, ...}
j $A = \{x \mid 3 \leq x \leq 12\}$
k $S = \{y \mid -5 \leq y \leq 5\}$

2 The number of elements in a set A is written as $n(A)$.
Give the value of $n(A)$ for the finite sets in question 1 above.

### Subsets

If all the elements of one set X are also elements of another set Y, then X is said to be a **subset** of Y.

This is written as $X \subseteq Y$.

If a set A is empty (i.e. it has no elements in it), then this is called the **empty set** and it is represented by the symbol $\varnothing$. Therefore $A = \varnothing$.

The empty set is a subset of all sets. For example, three girls, Winnie, Natalie and Emma, form a set A.

A = {Winnie, Natalie, Emma}
All the possible subsets of A are given below:
B = {Winnie, Natalie, Emma}
C = {Winnie, Natalie}
D = {Winnie, Emma}
E = {Natalie, Emma}

$F$ = {Winnie}
G = {Natalie}
$H$ = {Emma}
$I = \varnothing$

Note that the sets $B$ and $I$ above are considered as subsets of A, i.e. $A \subseteq A$ and $\varnothing \subseteq A$.

However, sets C, $D$, $E$, $F$, G and $H$ are considered **proper subsets** of A. This distinction in the type of subset is shown in the notation below. For proper subsets, we write:

$C \subset A$ and $D \subset A$ etc. instead of $C \subseteq A$ and $D \subseteq A$.

Similarly $G \nsubseteq H$ implies that G is not a subset of $H$
$G \not\subset H$ implies that G is not a proper subset of $H$.

**Worked example**

A = {1, 2, 3, 4, 5, 6, 7, 8, 9, 10}

**a** List the subset $B$ of even numbers.
**b** List the subset C of prime numbers.

**a** $B$ = {2, 4, 6, 8, 10}
**b** C = {2, 3, 5, 7}

## Exercise 3.5.2

1 $P$ is the set of whole numbers less than 30.
 a List the subset $Q$ of even numbers.
 b List the subset $R$ of odd numbers.
 c List the subset $S$ of prime numbers.
 d List the subset $T$ of square numbers.
 e List the subset $U$ of triangular numbers.

2 A is the set of whole numbers between 50 and 70.
 a List the subset $B$ of multiples of 5.
 b List the subset C of multiples of 3.
 c List the subset $D$ of square numbers.

3 $J$ = {p, q, r}
 a List all the subsets of $J$.
 b List all the proper subsets of $J$.

4 State whether each of the following statements is true or false.
 a {Algeria, Mozambique} $\subseteq$ {countries in Africa}
 b {mango, banana} $\subseteq$ {fruit}
 c {1, 2, 3, 4} $\subseteq$ {1, 2, 3, 4}
 d {1, 2, 3, 4} $\subset$ {1, 2, 3, 4}
 e {volleyball, basketball} $\nsubseteq$ {team sport}
 f {4, 6, 8, 10} $\not\subset$ {4, 6, 8, 10}
 g {potatoes, carrots} $\subseteq$ {vegetables}
 h {12, 13, 14, 15} $\not\subset$ {whole numbers}

## The universal set

The **universal set** ($U$) for any particular problem is the set which contains all the possible elements for that problem.

The **complement** of a set $A$ is the set of elements which are in $U$ but not in $A$. The set is identified as $A'$. Notice that $U' = \varnothing$ and $\varnothing' = U$.

**Worked examples**

1 If $U = \{1, 2, 3, 4, 5, 6, 7, 8, 9, 10\}$ and $A = \{1, 2, 3, 4, 5\}$, what set is represented by $A'$?

$A'$ consists of those elements in $U$ which are not in $A$.

Therefore $A' = \{6, 7, 8, 9, 10\}$.

2 If $U$ is the set of all three-dimensional shapes and $P$ is the set of prisms, what set is represented by $P'$?

$P'$ is the set of all three-dimensional shapes except prisms.

## Intersections and unions

The **intersection** of two sets is the set of all the elements that belong to both sets. The symbol $\cap$ is used to represent the intersection of two sets.

If $P = \{1, 2, 3, 4, 5, 6, 7, 8, 9, 10\}$ and $Q = \{2, 4, 6, 8, 10, 12, 14, 16, 18, 20\}$ then $P \cap Q = \{2, 4, 6, 8, 10\}$ as these are the numbers that belong to both sets.

The **union** of two sets is the set of all elements that belong to either or both sets and is represented by the symbol $\cup$.

Therefore in the example above,
$P \cup Q = \{1, 2, 3, 4, 5, 6, 7, 8, 9, 10, 12, 14, 16, 18, 20\}$.

Unions and intersections of sets can be shown diagrammatically using **Venn diagrams**.

## Venn diagrams

Venn diagrams are the principal way of showing sets diagrammatically. They are named after the mathematician John Venn (1834–1923). The method consists primarily of entering the elements of a set into a circle or circles.

Some examples of the uses of Venn diagrams are shown below.

$A = \{2, 4, 6, 8, 10\}$ can be represented as:

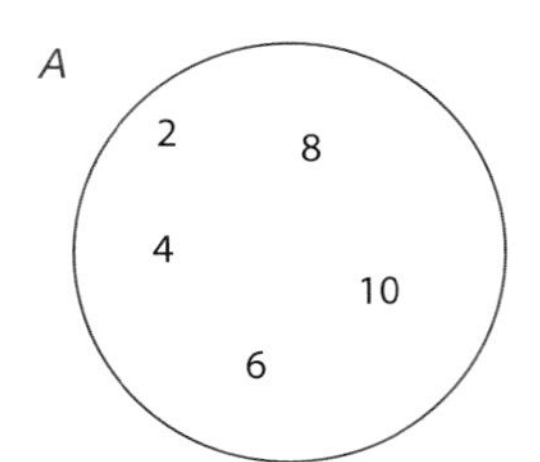

Elements which are in more than one set can also be represented using a Venn diagram.

$P = \{3, 6, 9, 12, 15, 18\}$ and $Q = \{2, 4, 6, 8, 10, 12\}$ can be represented as:

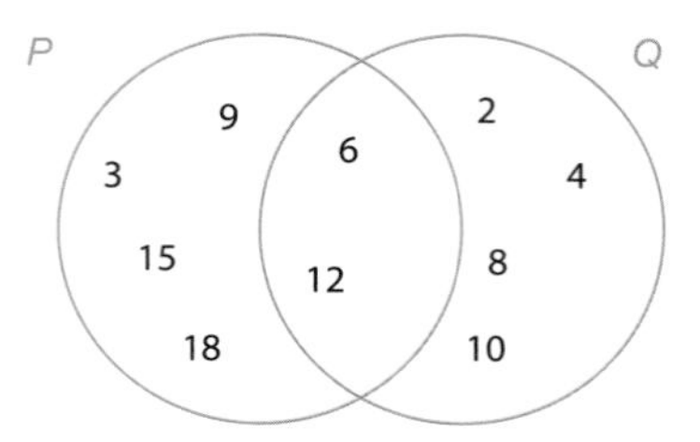


The elements which belong to both sets are placed in the region of overlap of the two circles.

As mentioned in the previous section, when two sets $P$ and $Q$ overlap as they do above, the notation $P \cap Q$ is used to denote the set of elements in the intersection, i.e. $P \cap Q = \{6, 12\}$. Note that $6 \in P \cap Q$; $8 \notin P \cap Q$.

$J = \{10, 20, 30, 40, 50, 60, 70, 80, 90, 100\}$ and $K = \{60, 70, 80\}$ can be represented as shown below; this is shown in symbols as $K \subset J$.

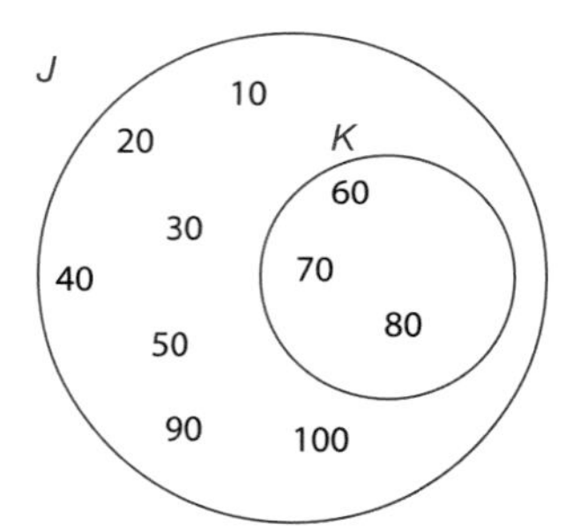


$X = \{1, 3, 6, 7, 14\}$ and $Y = \{3, 9, 13, 14, 18\}$ are represented as:

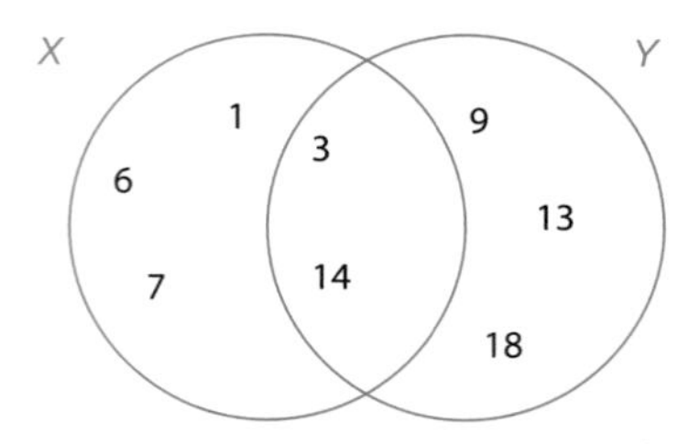


The union of two sets is everything which belongs to either or both sets and is represented by the symbol $\cup$. Therefore, in the example above, $X \cup Y = \{1, 3, 6, 7, 9, 13, 14, 18\}$.

## Exercise 3.5.3

1 Using the Venn diagram, deduce whether the following statements are true or false. ∈ means 'is an element of' and ∉ means 'is not an element of'.

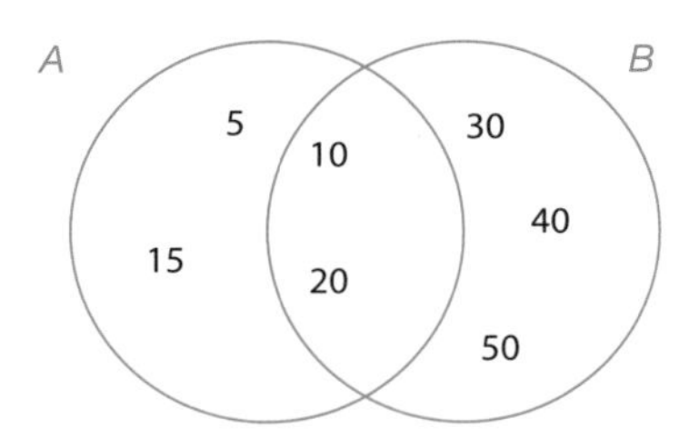


a $5 \in A$
b $20 \in B$
c $20 \notin A$
d $50 \in A$
e $50 \notin B$
f $A \cap B = \{10, 20\}$

2 Copy and complete the statement $A \cap B = \{...\}$ for each of the Venn diagrams below.

a
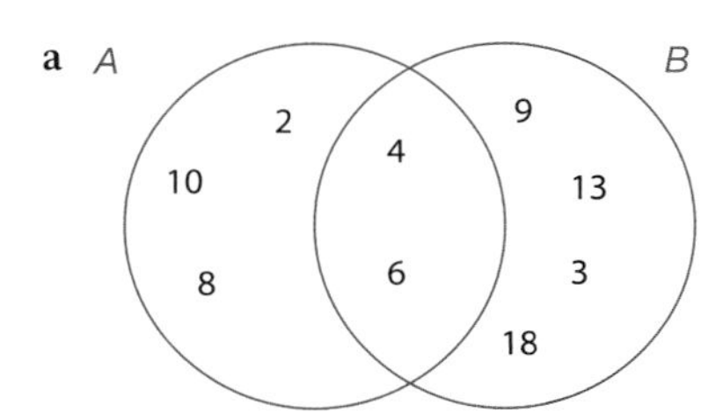


b
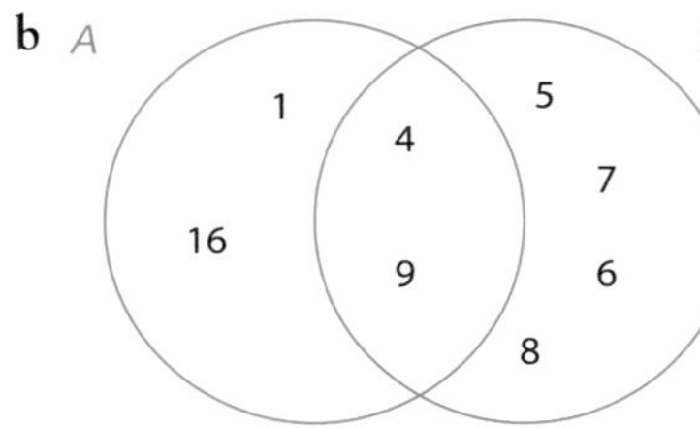


c
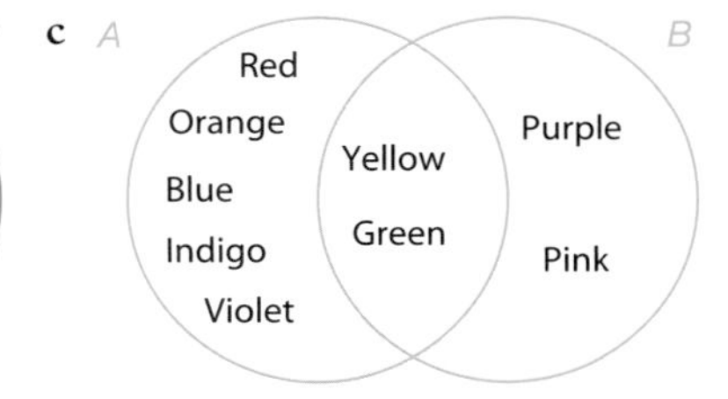


3 Copy and complete the statement $A \cup B = \{...\}$ for each of the Venn diagrams in question 2 above.

4 Using the Venn diagram, copy and complete these statements.

a $U = \{...\}$
b $A' = \{...\}$

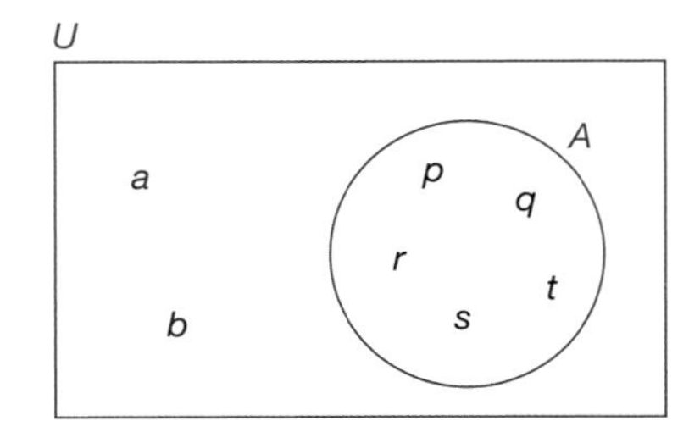


5 Using the Venn diagram, copy and complete the following statements.

a $U = \{...\}$
b $A' = \{...\}$
c $A \cap B = \{...\}$
d $A \cup B = \{...\}$
e $(A \cap B)' = \{...\}$
f $A \cap B' = \{...\}$

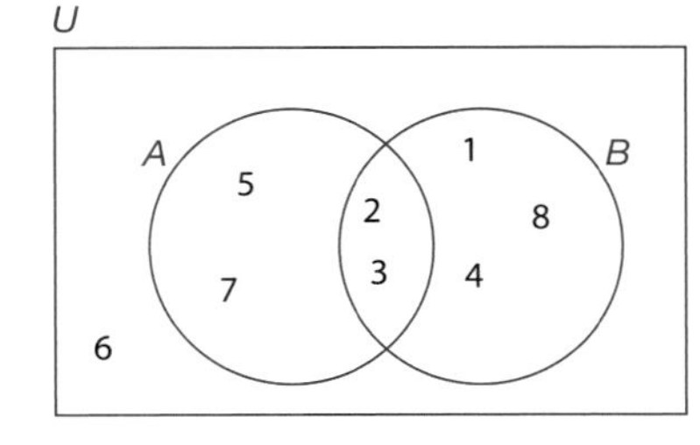

6

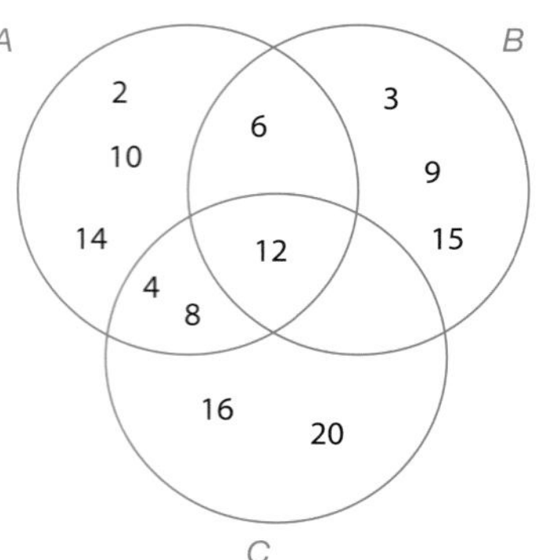


a Using the Venn diagram, describe in words the elements of:
i) set A　　ii) set B　　iii) set C.

b Copy and complete the following statements.
i) $A \cap B = \{...\}$　　ii) $A \cap C = \{...\}$　　iii) $B \cap C = \{...\}$
iv) $A \cap B \cap C = \{...\}$　　v) $A \cup B = \{...\}$　　vi) $C \cup B = \{...\}$

7

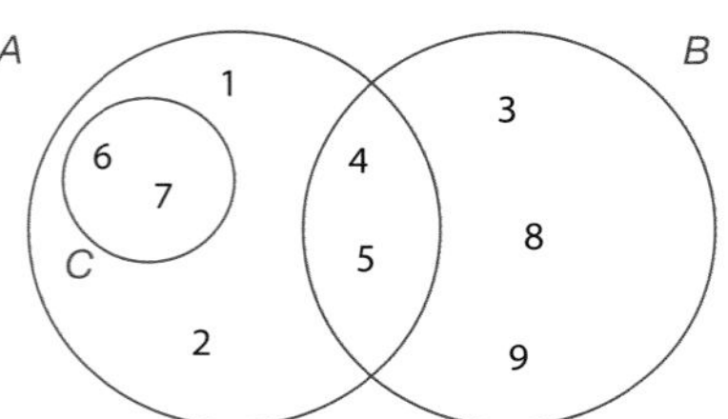


a Using the Venn diagram, copy and complete the following statements.
i) $A = \{...\}$　　ii) $B = \{...\}$　　iii) $C' = \{...\}$
iv) $A \cap B = \{...\}$　　v) $A \cup B = \{...\}$　　vi) $(A \cap B)' = \{...\}$

b State, using set notation, the relationship between C and A.

8

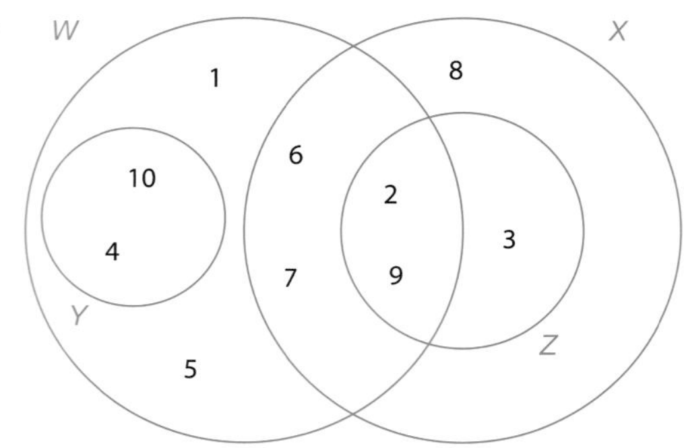


a Copy and complete the following statements.
i) $W = \{...\}$　　ii) $X = \{...\}$　　iii) $Z' = \{...\}$
iv) $W \cap Z = \{...\}$　　v) $W \cap X = \{...\}$　　vi) $Y \cap Z = \{...\}$

b Which of the named sets is a subset of $X$?

**6** In a group of 125 students who play tennis, volleyball or football, 10 play all three. Twice as many play tennis and football only. Three times as many play volleyball and football only, and 5 play tennis and volleyball only.

If $x$ play tennis only, $2x$ play volleyball only and $3x$ play football only, determine:

**a** how many play tennis
**b** how many play volleyball
**c** how many play football.

## The analogy of logic and set theory

The use of **No** or **Never** or **All ... do not** in statements (e.g. **No** French people are British people) means the sets are **disjoint**, i.e. they do not overlap.

The use of **All** or **If ... then** or **No ... not** in statements (e.g. There is **no** nurse who does **not** wear a uniform) means that one set is a subset of another.

The use of **Some** or **Most** or **Not all** in statements (e.g. **Some** televisions are very expensive) means that the sets intersect.

The validity of an argument can be tested using Venn diagrams.

If $p$, $q$ and $r$ are three statements and if $p \Rightarrow q$ and $q \Rightarrow r$, then it follows that $p \Rightarrow r$.

In terms of sets, if $A$, $B$ and $C$ are all proper subsets ($\subset$) of the universal set $U$ and if $A \subset B$ and $B \subset C$ then $A \subset C$.

Diagrammatically this can be represented as shown in the Venn diagram opposite:

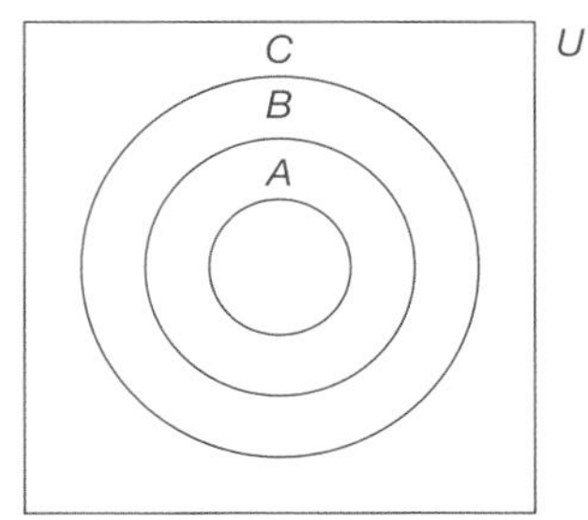


**Worked examples**

**1** $P$ is the set of French people and $Q$ is the set of British people. Draw a Venn diagram to represent the sets.

The Venn diagram is as shown,
i.e. $P \cap Q = \varnothing$
In logic this can be written $p \veebar q$,
i.e. $p$ or $q$ but not both.

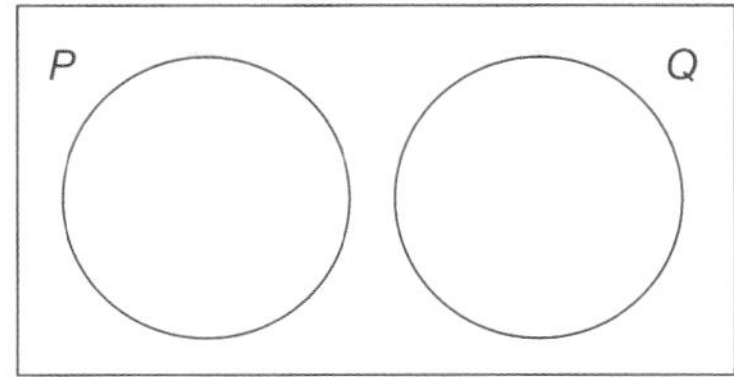


**2** $P$ is the set of nurses and $Q$ is the set of people who wear uniform. Draw a Venn diagram to represent the sets.

$P$ is a subset of $Q$ as there are other people who wear uniforms apart from nurses, i.e. $P \subset Q$
In logic this can be written $p \Rightarrow q$.

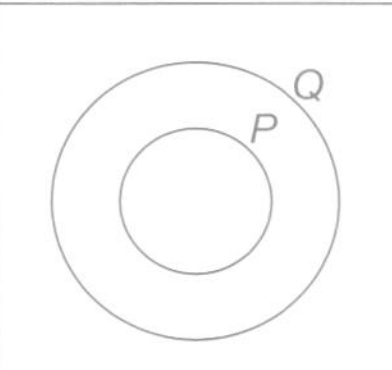

3 $P$ is the set of televisions and $Q$ is the set of expensive electrical goods. Draw a Venn diagram to represent the sets.

$P$ intersects $Q$ as there are expensive electrical goods that are not televisions and there are televisions that are not expensive.
In logic the intersection can be written $p \wedge q$.

## Exercise 3.5.5

1 Draw a Venn diagram to illustrate the following sets.
$Q$: students wearing football shirts
$P$: professional footballers wearing football shirts
Shade the region that represents the statement 'Kofi is a professional footballer and a student'. How would you write this using logic symbols?

2 Draw a Venn diagram to illustrate the following sets.
$Q$: students wearing football shirts
$P$: professional footballers wearing shirts
Shade the region that satisfies the statement 'Maanu is either a student or a professional footballer but he is not both'. How would you write this using logic symbols?

3 Draw a Venn diagram to illustrate the following sets.
$P$: maths students $U$: all students.
Shade the region that satisfies the statement 'Boamah is not a maths student'. How would you write this using logic symbols?

4 Draw a Venn diagram to illustrate the following sets.
$P$: five-sided shapes $U$: all shapes
Shade the region that satisfies the statement 'A regular pentagon is a five-sided shape'. How would you write this using logic symbols?

5 Draw a Venn diagram to illustrate the following sets.
$Q$: multiples of 5 $U$: integers
Shade the region where you would place 17. How would you write this using logic symbols?

6 Draw a Venn diagram to illustrate the following sets.
$P$: people who have studied medicine
$Q$: people who are doctors
Shade the region that satisfies the statement 'All doctors have studied medicine'. How would you write this using logic symbols?

7 Illustrate the statement 'People with too much money are never happy' using a Venn diagram with these sets.
$P$: people who have too much money
$Q$: people who are happy
Shade the region that satisfies the statement 'People with too much money are never happy'. How would you write this using logic symbols?

8 Draw a Venn diagram to illustrate the following sets.
$P$: music lessons $Q$: lessons that are expensive
Shade the region that satisfies the statement 'Some music lessons are expensive'. How would you write this using logic symbols?

## 3.6 Probability

Pierre de Fermat

Although Newton and Galileo had some thoughts about chance, it is accepted that the study of what we now call probability began when Blaise Pascal (1623–1662) and Pierre de Fermat (of Fermat's last theorem fame) corresponded about problems connected with games of chance. Later Christiaan Huygens wrote the first book on the subject, *The Value of all Chances in Games of Fortune*, in 1657. This included a chapter entitled 'Gambler's Ruin'.

In 1821 Carl Friedrich Gauss (1777–1855), one of the greatest mathematicians who ever lived, worked on the 'normal distribution', a very important contribution to the study of probability.

Probability is the study of chance, or the likelihood of an event happening. In this section we will be looking at theoretical probability. But, because probability is based on chance, what theory predicts does not necessarily happen in practice.

### Sample space

Set theory can be used to study probability.

A **sample space** is the set of all possible results of a trial or experiment. Each result or outcome is sometimes called an **event**.

#### Complementary events

A dropped drawing pin can land either pin up, $U$, or pin down, $D$. These are the only two possible outcomes and cannot both occur at the same time. The two events are therefore **mutually exclusive** (cannot happen at the same time) and **complementary** (the sum of their probabilities equal 1). The complement of an event $A$ is written $A'$.

Therefore $P(A) + P(A') = 1$. In words this is read as 'the probability of event $A$ happening added to the probability of event $A$ not happening equals 1'.

**Worked examples**

1 A fair dice is rolled once. What is its sample space?
The sample space $S$ is the set of possible outcomes or events. Therefore $S = \{1, 2, 3, 4, 5, 6\}$ and the number of outcomes or events is 6.

2 **a** What is the sample space, $S$, for two drawing pins dropped together.
**b** How many possible outcomes are there?

**a** $S = \{UU, UD, DU, DD\}$
**b** There are four possible outcomes.

3 The probability of an event $B$ happening is $P(B) = \frac{3}{5}$. Calculate $P(B')$.

$P(B)$ and $P(B')$ are complementary events, so $P(B) + P(B') = 1$.

$P(B') = 1 - \frac{3}{5} = \frac{2}{5}$.

## Exercise 3.6.1

1 What is the sample space and the number of events when three coins are tossed?

2 What is the sample space and number of events when a blue dice and a red dice are rolled? (Note: (1, 2) and (2, 1) are different events.)

3 What is the sample space and the number of events when an ordinary dice is rolled and a coin is tossed?

4 A mother gives birth to twins. What is the sample space and number of events for their sex?

5 What is the sample space if the twins in question 4 are identical?

6 Two women take a driving test.
  a What are the possible outcomes?
  b What is the sample space?

7 A tennis match is played as 'best of three sets'.
  a What are the possible outcomes?
  b What is the sample space?

8 If the tennis match in question 7 is played as 'best of five sets',
  a what are the possible outcomes?
  b what is the sample space?

### Probability of an event

A favourable outcome refers to the event in question actually happening. The total number of possible outcomes refers to all the different types of outcome one can get in a particular situation. In general:

$$\text{Probability of an event} = \frac{\text{number of favourable outcomes}}{\text{total number of equally likely outcomes}}$$

This can also be written as: $P(A) = \frac{n(A)}{n(U)}$,

where P(A) is the probability of event A, $n(A)$ is the number of ways event $A$ can occur and $n(U)$ is the total number of equally likely outcomes.

Therefore

if the probability = 0, it implies the event is impossible
if the probability = 1, it implies the event is certain to happen

**Worked example**

An ordinary, fair dice is rolled.
  a Calculate the probability of getting a 6.
  b Calculate the probability of not getting a 6.

  a Number of favourable outcomes = 1 (i.e. getting a 6)

  Total number of possible outcomes = 6 (i.e. getting a 1, 2, 3, 4, 5 or 6)
  Probability of getting a 6, $P(6) = \frac{1}{6}$

# Student assessment 2

1 If $A = \{2, 4, 6, 8\}$, write all the proper subsets of $A$ with two or more elements.

2 X = {lion, tiger, cheetah, leopard, puma, jaguar, cat}
Y = {elephant, lion, zebra, cheetah, gazelle}
Z = {anaconda, jaguar, tarantula, mosquito}
  a Draw a Venn diagram to represent the above information.
  b Copy and complete the statement $X \cap Y = \{...\}$.
  c Copy and complete the statement $Y \cap Z = \{...\}$.
  d Copy and complete the statement $X \cap Y \cap Z = \{...\}$.

3 $U$ is the set of natural numbers, M is the set of even numbers and $N$ is the set of multiples of 5.
  a Draw a Venn diagram and place the numbers 1, 2, 3, 4, 5, 6, 7, 8, 9, 10 in the appropriate places in it.
  b If $X = M \cap N$, describe set X in words.

4 A group of 40 people were asked whether they like tennis ($T$) and football ($F$). The number liking both tennis and football was three times the number liking only tennis. Adding 3 to the number liking only tennis and doubling the answer equals the number of people liking only football. Four said they did not like sport at all.
  a Draw a Venn diagram to represent this information.
  b Calculate $n(T \cap F)$.
  c Calculate $n(T \cap F')$.
  d Calculate $n(T' \cap F)$.

5 The Venn diagram below shows the number of elements in three sets $P$, $Q$ and $R$.

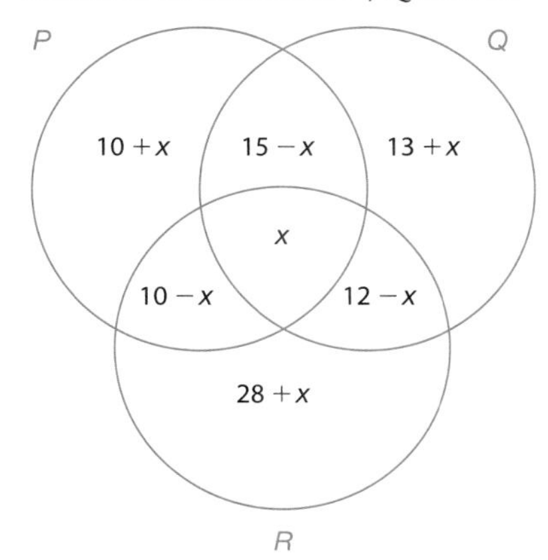


If $n(P \cup Q \cup R) = 93$ calculate:
  a $x$
  b $n(P)$
  c $n(Q)$
  d $n(R)$
  e $n(P \cap Q)$
  f $n(Q \cap R)$
  g $n(P \cap R)$
  h $n(R \cup Q)$
  i $n(P \cap Q)'$.

6 What is meant by $p \wedge q$?

7 Copy and complete the truth table below.

| $p$ | $q$ | $p \wedge q$ | $p \vee q$ |
|---|---|---|---|
| T | | | |
| T | | | |
| F | | | |
| F | | | |

8 What is a tautology? Give an example.

9 A goalkeeper expects to save one penalty out of every three. Calculate the probability that he:
  a saves one penalty out of the next three
  b fails to save any of the next three penalties
  c saves two out of the next three penalties.

## Student assessment 3

1 The probability that a student takes English is 0.8. The probability that a student takes English and Spanish is 0.25.
What is the probability that a student takes Spanish, given that he takes English?

2 A card is drawn from a standard pack of cards.
 a Draw a Venn diagram to show the following:
 A is the set of aces
 B is the set of picture cards
 C is the set of clubs
 b From your Venn diagram find the following probabilities.
 i) P(ace or picture card)
 ii) P(not an ace or picture card)
 iii) P(club or ace)
 iv) P(club and ace)
 v) P(ace and picture card)

3 Students in a school can choose to study one or more science subjects from Physics, Chemistry and Biology.
In a year group of 120 students, 60 took Physics, 60 took Biology and 72 took Chemistry; 34 took Physics and Chemistry, 32 took Chemistry and Biology and 24 took Physics and Biology; 18 took all three.
 a Draw a Venn diagram to represent this information.
 b If a student is chosen at random, what is the probability that:
 i) the student chose to study only one science subject
 ii) the student chose Physics or Chemistry, and did not choose Biology?

4 A class took an English test and a Maths test. 40% passed both tests and 75% passed the English test.
What percentage of those who passed the English test also passed the Maths test?

5 A jar contains blue and red counters. Two counters are chosen without replacement. The probability of choosing a blue then a red counter is 0.44. The probability of choosing a blue counter on the first draw is 0.5.
What is the probability of choosing a red counter on the second draw if the first counter chosen was blue?

6 In a group of children, the probability that a child has black hair is 0.7. The probability that a child has brown eyes is 0.55. The probability that a child has either black hair or brown eyes is 0.85.
What is the probability that a child chosen at random has both black hair and brown eyes?

7 A ball enters a chute at X.

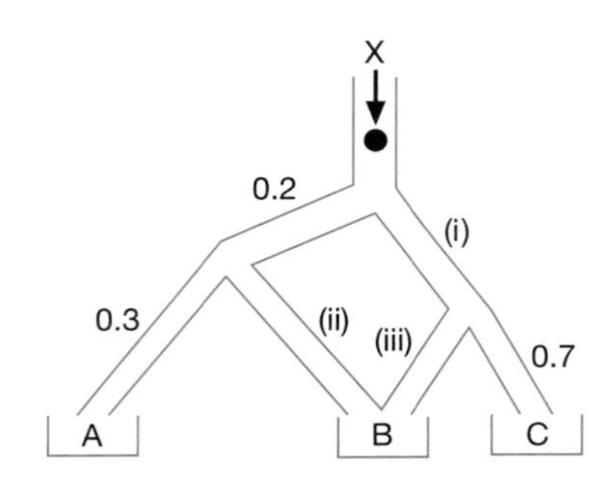


 a What are the probabilities of the ball going down each of the chutes labelled (i), (ii) and (iii)?
 b Calculate the probability of the ball landing in:
 i) tray A
 ii) tray C
 iii) tray B.

# Examination questions

1 A fitness club has 60 members. 35 of the members attend the club's aerobics course (A) and 28 members attend the club's yoga course (Y). 17 members attend both courses. A Venn diagram is used to illustrate this situation.

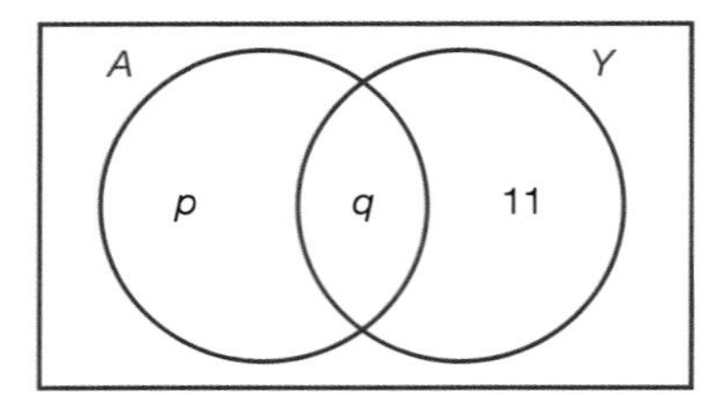

a Write down the value of $q$. [1]
b Find the value of $p$. [2]
c Calculate the number of members of the fitness club who attend neither the aerobics course ($A$) nor the yoga course ($Y$). [2]
d Shade, on a copy of the Venn diagram, $A' \cap Y$. [1]

**Paper 1, May 10, Q6**

2 Police in a town are investigating the theft of mobile phones one evening from three cafés, 'Alan's Diner', 'Sarah's Snackbar' and 'Pete's Eats'.

They interviewed two suspects, Matthew and Anna about that evening.

Matthew said: "I visited Pete's Eats and visited Alan's Diner and I did not visit Sarah's Snackbar."

Let $p$, $q$ and $r$ be the statements:

$p$: I visited Alan's Diner
$q$: I visited Sarah's Snackbar
$r$: I visited Pete's Eats

a Write down Matthew's statement in symbolic logic form. [3]

What Anna said was lost by the police, but in symbolic form it was

$$(q \vee r) \Rightarrow \neg p$$

b Write down, in words, what Anna said. [3]

**Paper 1, May 11, Q4**

3 **Part A**

The following Venn diagram represents the students studying Mathematics ($A$), Further Mathematics ($B$) and Physics ($C$) in a school.

50 students study Mathematics
38 study Physics
20 study Mathematics and Physics but not Further Mathematics
10 study Further Mathematics but not Physics
12 study Further Mathematics and Physics
6 study Physics but not Mathematics
3 study none of these three subjects.

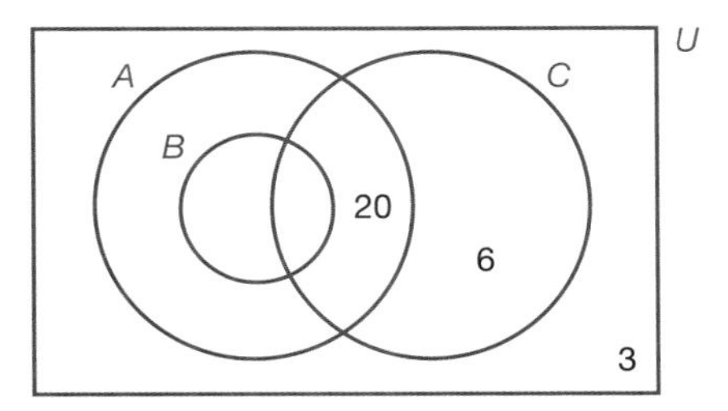


a Copy and complete the Venn diagram. [3]
b Write down the number of students who study Mathematics but not Further Mathematics. [1]
c Write down the total number of students in the school. [1]
d Write down $n(B \cup C)$. [2]

**Part B**

Three propositions are given as

$p$: It is snowing
$q$: The roads are open
$r$: We will go skiing

a Write the following compound statement in symbolic form.
'It is snowing and the roads are not open.' [2]
b Write the following compound statement in words.
$(\neg p \wedge q) \Rightarrow r$ [3]
c Copy and complete the truth table. [3]

| $p$ | $q$ | $r$ | $\neg p$ | $\neg p \wedge q$ | $(\neg p \wedge q) \Rightarrow r$ |
|---|---|---|---|---|---|
| T | T | T | | | |
| T | T | F | | | |
| T | F | T | | | |
| T | F | F | | | |
| F | T | T | | | |
| F | T | F | | | |
| F | F | T | | | |
| F | F | F | | | |

**Paper 2, Nov 09, Q2**

# Applications project ideas and theory of knowledge

**1** Research and discuss Russell's antinomy (not the element antimony).

**2** Research and discuss 'Bertrand's Box Paradox'. This could be the starting point for a project on paradoxes.

**3** Set theory is an area that could be studied as a project beyond the Mathematical Studies syllabus. Your teacher may suggest some areas of study.

**4** How do governments use probability to plan ahead?

**5** Draw a Venn diagram to represent Belief, Truth and Knowledge. Discuss the statement 'Knowledge is found where belief and truth intersect.'

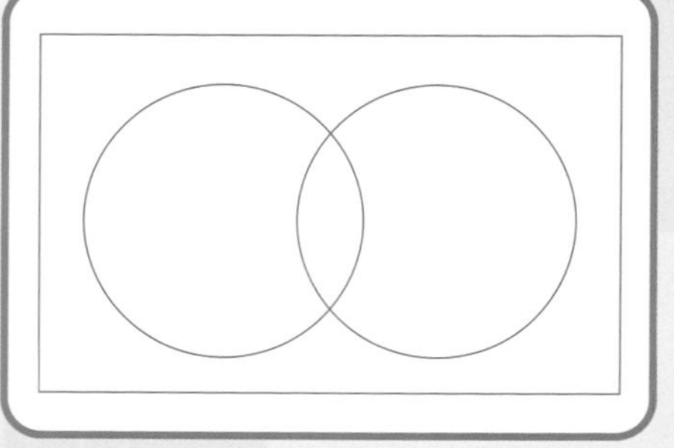

**6** The set of whole numbers and the set of square numbers have an infinite number of elements. Does this mean that there are different values of infinity?

**7** What is the difference between zero and an empty set? Is a vacuum an empty set?

3 Two different brands of batteries, X and Y, have the same mean lifespan of 45 hours when tested. However the results of brand X have a standard deviation of 8 hours, whilst those of brand Y have a standard deviation of 2 hours. Assume the lifespans of both batteries are normally distributed.
   a On the same graph, sketch the distributions of each brand, labelling them clearly.
   b A battery of brand X is picked at random. What is the probability that it lasts longer than 60 hours?
   c A battery is picked at random and tested. It lasts less than 40 hours. Which brand is it most likely to belong to? Justify your answer.
   d Is it possible for a brand Y battery to last more than 55 hours? Justify your answer.

4 The lengths of telephone calls at a call centre are assumed to be normally distributed with a mean length of 6 minutes and a standard deviation of 2.5 minutes.
   a A telephone call is picked at random. What is the probability that it lasted longer than 9 minutes?
   b What is the probability that a call lasts between 5 and 6 minutes?
   c Sketch the distribution and label the horizontal axis clearly.
   d With reference to your sketch, describe why the assumption that the lengths of calls are normally distributed is incorrect.

5 500 g cereal packets are filled by a machine. The masses (g) of cereal in the packets are normally distributed with an expected mass of 510 g and a standard deviation of 4 g.
   a What is the probability that a packet picked at random will have a smaller mass than that stated on the packet?
   b The company fills 1.8 million packets a year. How many packets would be expected to have less than 500 g of cereal in them?

6 The masses (kg) of pumpkins are recorded. The masses are normally distributed with the following parameters: $\mu = 5.6$ kg and $\sigma = 1.2$ kg.
   The pumpkins are labelled 'large', 'medium' and 'small' according to their mass. A pumpkin with a mass greater than 6.6 kg is labelled 'large'. A pumpkin with a mass less than 4 kg is labelled 'small'; the rest are labelled 'medium'.
   a What is the probability that a pumpkin selected at random is large?
   b What is the probability that a pumpkin selected at random is small?
   c If 8400 pumpkins are labelled medium, how many pumpkins are there altogether?

## Inverse normal calculations

The calculations so far for the normal distribution have involved finding the probability of an event happening (e.g. the probability that a mass is greater or less than a particular value). The probability relates to the area under the relevant part of the curve.

It is important though to be able to work backwards. These are known as inverse normal calculations.

**Worked example**

The heights of a group of boys are normally distributed with a mean of 175 cm and a standard deviation of 7 cm. The college basketball team is looking for new players. It is only considering boys whose heights fall in the top 10% of the population.

What is the minimum height that a boy must be in order to be considered for the basketball team?

It is always good practice to draw a normal distribution curve to display the information clearly.

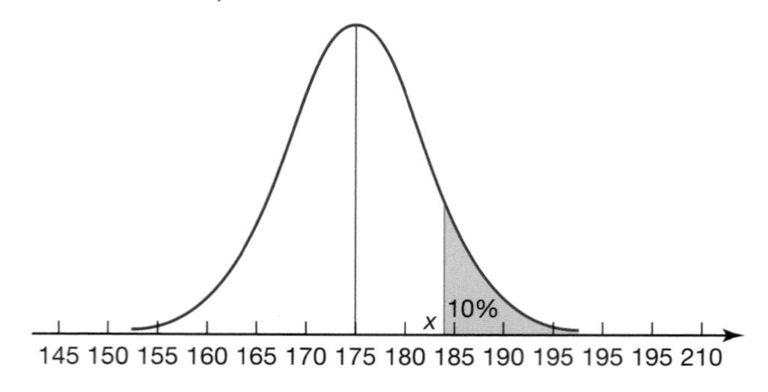


The minimum height required is indicated as '$x$' on the graph, as 10% of the data falls above this value.

Your GDC will calculate this value for you.

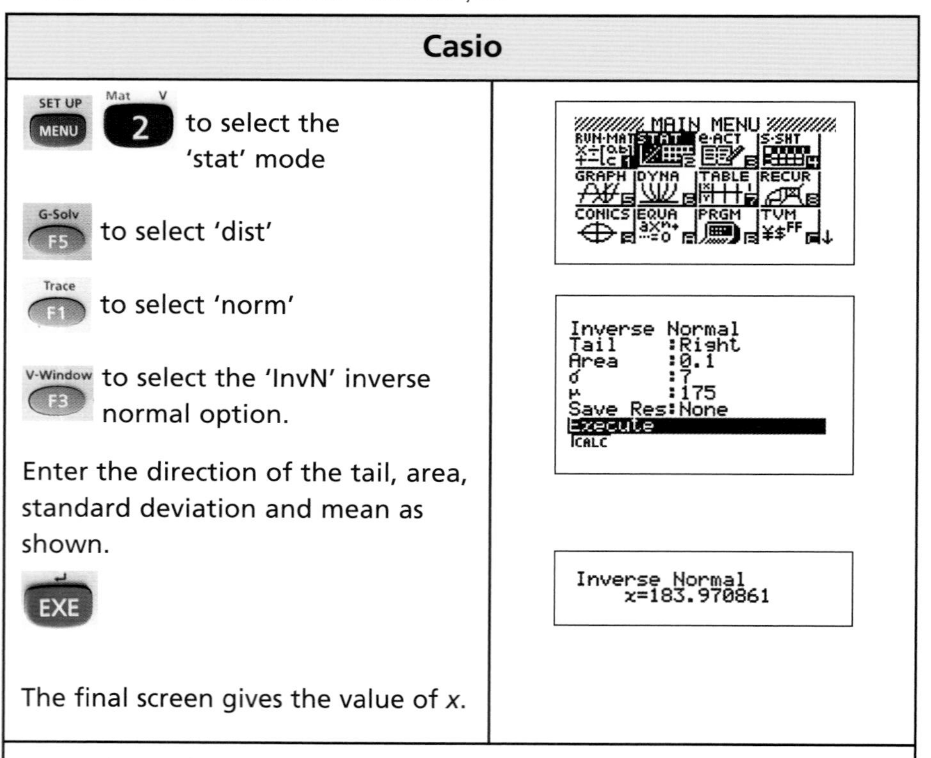


Note: the 'tail' in this instance is to the right as the top 10% is being considered. However, the same value for $x$ could be calculated using the tail in the left and an area of 90%.

3 The table shows the salary and the number of years experience of a group of firefighters.

| **Salary $(000s)** | 32 | 27 | 40 | 40 | 36 | 32 | 25 | 30 | 27 | 40 | 37 | 33 |
|---|---|---|---|---|---|---|---|---|---|---|---|---|
| **Years of experience** | 7 | 4 | 20 | 18 | 11 | 7 | 1 | 5 | 3 | 17 | 15 | 12 |

a Calculate the correlation coefficient $r$ and comment on its value.
b Calculate the equation of the regression line for $y$ on $x$.
c Estimate the salary of a firefighter with ten years' experience. Comment on the validity of your estimation.
d A firefighter has a salary of $100 000. Estimate his age using your equation of the regression line from part **b**. Comment on the validity of your answer.

4 The air temperature was taken at various heights by a meteorological balloon. The results are shown below.

| **Height (000s m)** | 4 | 8 | 12 | 16 | 20 | 24 | 28 | 32 | 36 | 40 |
|---|---|---|---|---|---|---|---|---|---|---|
| **Temperature (°C)** | 8 | 4 | −20 | −32 | −40 | −46 | −48 | −51 | −57 | −60 |

a Calculate the correlation coefficient $r$. Comment on its value.
b Calculate the equation of the regression line for $y$ on $x$.
c Estimate the height of the balloon if the outside temperature is recorded as −70 °C. Comment on the validity of your answer.

## 4.4 The $\chi^2$ test for independence

You may use this in Biology, Psychology and Geography Diploma courses.

A chi-squared ($\chi^2$) test for independence is used to assess whether or not paired observations, expressed in a contingency table (two-way table), are independent.

For example:

Volunteers are testing a new drug in a clinical trial. It is claimed that the new drug will result in a more rapid improvement rate for sick patients than would happen if they did not receive the drug.

The **observed** results of the trials are presented in the contingency table below.

| | **Improved** | **Did not improve** | **Total** |
|---|---|---|---|
| **Given drug** | 55 | 40 | 95 |
| **Not given drug** | 42 | 43 | 85 |
| **Total** | 97 | 83 | 180 |

It is difficult to tell from the results whether or not the drug had a significant positive effect on improvement rate. Although the results show that more volunteers improved than did not improve when given the drug, it is not certain whether the difference in results is significant enough to justify the claim. In order to verify the claim, a chi-squared ($\chi^2$) test for independence can be carried out.

The first step is to set up a **null hypothesis** ($H_0$). The null hypothesis is always that there is no link between the variables and it is contrasted against an **alternative hypothesis** ($H_1$) which states that there is a link between the variables. The null hypothesis is treated as valid unless the data contradicts it.

In the example above:

$H_0$: There is no link between patients being given the drug and improvement rates.
$H_1$: There is a link between patients being given the drug and improvement rates.

The observed results need to be compared with expected or theoretical population results.

Of the 180 people in the sample, 95 were given the drug. So, from this we estimate that, in the population, the probability of being given the drug is $\frac{95}{180}$.

A total of 97 patients in the trial improved. From the null hypothesis, we would expect $\frac{95}{180}$ of these to have been given the drug, i.e. the expected number of improved patients who had been given the drug is $\frac{95}{180} \times 97 = 51.19$.

The next step is to draw up a table of expected frequencies for a group of 180 patients under the null hypothesis that having the drug is independent of improvement.

| | Improved | Did not improve | Total |
|---|---|---|---|
| **Given drug** | $\frac{95 \times 97}{180} = 51.19$ | $\frac{95 \times 83}{180} = 43.81$ | 95 |
| **Not given drug** | $\frac{85 \times 97}{180} = 45.81$ | $\frac{85 \times 83}{180} = 39.19$ | 85 |
| **Total** | 97 | 83 | 180 |

Note: The full answer to each calculation should be stored in your calculator's memory.

The formula for calculating $\chi^2$ is as follows:

$$\chi^2 = \Sigma\frac{(f_o - f_e)^2}{f_e}, \text{ where } f_o \text{ are the observed frequencies, } f_e \text{ are the expected frequencies}$$

$$\text{Therefore } \chi^2 = \frac{(55 - 51.19)^2}{51.19} + \frac{(40 - 43.81)^2}{43.81} + \frac{(42 - 45.81)^2}{45.81} + \frac{(43 - 39.19)^2}{39.19}$$

$$= 1.299$$

The importance of this number depends on two other factors:

- The percentage level of significance required ($p$)
- The number of degrees of freedom of the data ($v$).

The level of significance refers to the percentage of the data you would expect to be outside the normal bounds.

The number of degrees of freedom of the data relates to the amount of data that is needed in order for the contingency table to be completed once the totals for each row and column are known. In the example above, if any one of the four pieces of data are known, the rest can be deduced.

In general, if the data in a contingency table has $c$ columns and $r$ rows, then the number of degrees of freedom, $v = (c - 1)(r - 1)$.

The result for $\chi^2$ above needs to be compared with the result in a $\chi^2$ table. (This is provided in the information booklet in your examination.)

| $p$ | 0.005 | 0.01 | 0.025 | 0.05 | 0.1 | 0.9 | 0.95 | 0.975 | 0.99 | 0.995 |
|---|---|---|---|---|---|---|---|---|---|---|
| $v = 1$ | 0.00004 | 0.0002 | 0.001 | 0.004 | 0.016 | 2.706 | **3.841** | 5.024 | 6.635 | 7.875 |
| 2 | 0.010 | 0.020 | 0.051 | 0.103 | 0.211 | 4.605 | 5.991 | 7.378 | 9.210 | 10.597 |
| 3 | 0.072 | 0.115 | 0.216 | 0.352 | 0.584 | 6.251 | 7.815 | 9.348 | 11.345 | 12.838 |
| 4 | 0.207 | 0.297 | 0.484 | 0.711 | 1.064 | 7.779 | 9.488 | 11.143 | 13.277 | 14.860 |
| 5 | 0.412 | 0.554 | 0.831 | 1.145 | 1.610 | 9.236 | 11.070 | 12.883 | 15.086 | 16.750 |
| 6 | 0.676 | 0.872 | 1.237 | 1.635 | 2.204 | 10.645 | 12.592 | 14.449 | 16.812 | 18.548 |
| 7 | 0.989 | 1.239 | 1.690 | 2.167 | 2.833 | 12.017 | 14.67 | 16.013 | 18.475 | 20.278 |
| 8 | 1.344 | 1.646 | 2.180 | 2.733 | 3.490 | 13.362 | 15.507 | 17.535 | 20.090 | 21.955 |
| 9 | 1.735 | 2.088 | 2.700 | 3.325 | 4.168 | 14.864 | 16.919 | 19.203 | 21.666 | 23.589 |
| 10 | 2.156 | 2.558 | 3.247 | 3.940 | 4.865 | 15.987 | 18.307 | 20.483 | 23.209 | 25.188 |
| 11 | 2.603 | 3.053 | 3.816 | 4.575 | 5.578 | 17.275 | 19.675 | 21.920 | 24.725 | 26.757 |
| 12 | 3.074 | 3.571 | 4.404 | 5.266 | 6.304 | 18.549 | 21.026 | 23.337 | 26.217 | 28.300 |

$p = P(X \leq c)$

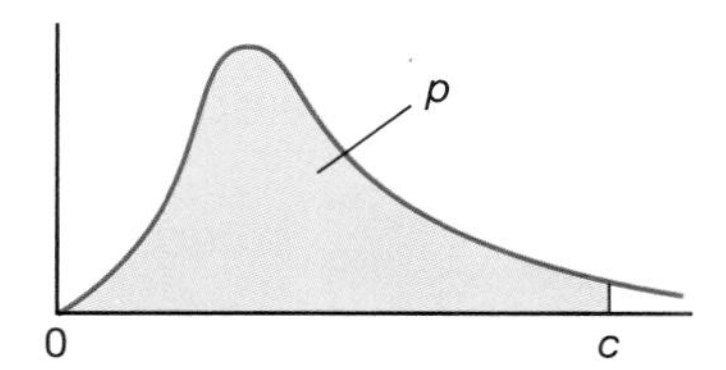


All the values in this table are known as critical values. The highlighted value of 3.841 is the critical value for the 5% level of significance for data with 1 degree of freedom.

Our calculation of $\chi^2$ gave a value of 1.299. As $1.299 < 3.841$, the null hypothesis is supported, i.e. there is no evidence of a link between being given the drug and improvement rates in patients.

If the calculated value of $\chi^2$ is greater than the critical value then the null hypothesis is rejected.

The table above has values for 1 degree of freedom ($v = 1$). Sometimes it is felt that these estimates are not sufficiently accurate and a different method known as Yates' continuity correction can be used. However, this method is beyond the scope of this book.

The value of $\chi^2$ can be calculated using your GDC for the example above.

## Casio

The observed data needs to be entered as a matrix.

   to enter the matrix menu.  to select matrix A. The dimensions refer to the number of rows × columns of the matrix, i.e. 2 × 2 EXE .

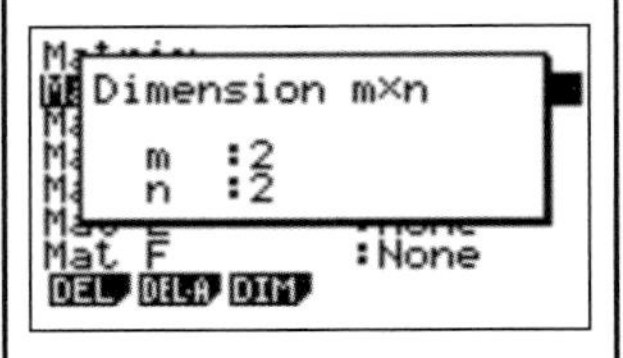


Enter the observed data in matrix A.

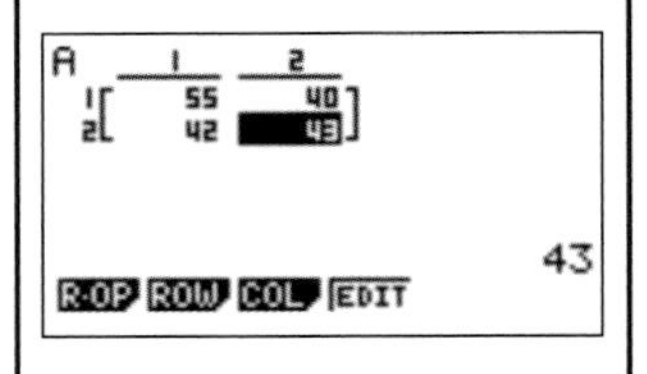


Now the $\chi^2$ test can be applied:

  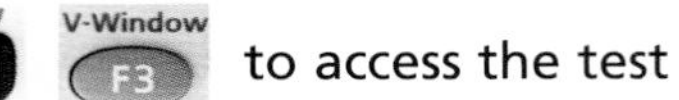 to access the test menu within the statistics mode.

 to select 'CHI'.

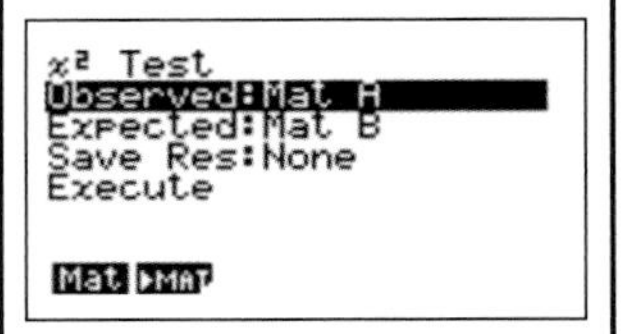


Select the observed matrix as A. Select the destination of the expected matrix as B. EXE

The value of $\chi^2$ is displayed as 1.29915596.

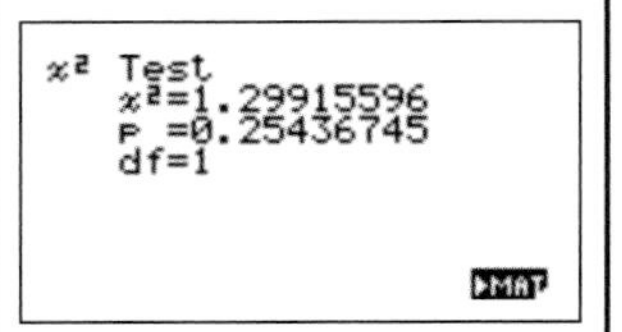


To check the expected values matrix B from this screen:

  to select matrix B

 to display the expected matrix B.

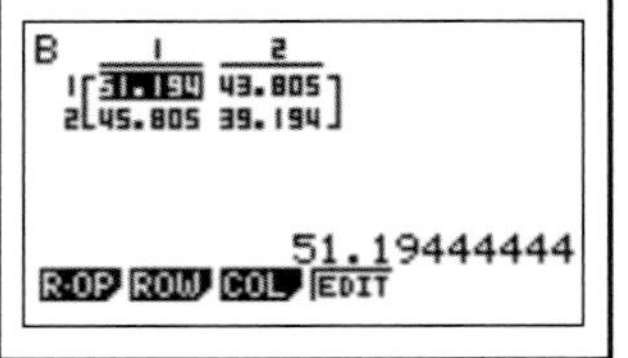

# Student assessment 2

1 State what type of correlation you might expect if the following data was collected and plotted on a scatter diagram. Justify your answers.
  a The amount someone earns and the amount they spend
  b A person's age and the number of brother and sisters they have

2 The durations of songs on a girl's mp3 player are normally distributed with an expected value of 3.8 minutes and a standard deviation of 0.4 minutes.
  a What percentage of songs last between 3.6 and 4.1 minutes?
  b Sketch a fully labelled normal distribution curve to show the results obtained in part **a** above.

3 In a school long jump competition, the lengths jumped are considered normally distributed with a mean of 2.8 m and a standard deviation of 1.1 m.
  a What percentage of students manage to jump further than 4.0 m?
  b Sketch the normal distribution curve, labelling the horizontal axis clearly and shading the region of the curve represented in part **a** above.
  c Explain, with reference to your sketch and the data given, why the data cannot be truly normally distributed.

4 The popularity of a group of professional football players and their yearly salary is given in the table below.

| **Popularity** | 1 | 2 | 3 | 4 | 5 | 6 | 7 | 8 | 9 | 10 |
|---|---|---|---|---|---|---|---|---|---|---|
| **Salary ($ million)** | 4.8 | 3.6 | 4.5 | 3.1 | 7.7 | 6.3 | 2.9 | 3.1 | 4.1 | 1..8 |
| **Popularity** | 11 | 12 | 13 | 14 | 15 | 16 | 17 | 18 | 19 | 20 |
| **Salary ($ million)** | 4.5 | 3.1 | 2.7 | 3.9 | 6.2 | 5.8 | 4.1 | 5.3 | 7.2 | 6.5 |

  a Calculate the equation of the regression line for $y$ on $x$.
  b Calculate the value of the correlation coefficient $r$.
  c The statement is made in a newspaper 'Big money footballers are not popular with fans'. Comment on this statement in the light of your results above.

5 A vote was taken for or against a ban on foxhunting. The results for town and country dwellers are recorded in the contingency table below.

| | **Town dwellers** | **Country dwellers** | **Total** |
|---|---|---|---|
| **Ban foxhunting** | 4240 | 1263 | 5503 |
| **Allow foxhunting** | 1360 | 2537 | 3897 |
| **Total** | 5600 | 3800 | 9400 |

  a Set up a null and alternative hypothesis.
  b Calculate the value of $\chi^2$.
  c State whether the null hypothesis is supported or rejected at a 1% level of significance. Justify your answers.

# Examination questions

1 Tony wants to carry out a $\chi^2$ test to determine whether or not a person's choice of one of the three professions; engineering, medicine or law is influenced by the person's sex (gender).

a State the null hypothesis, $H_0$, for this test. [1]

b Write down the number of degrees of freedom. [1]

Of the 400 people Tony interviewed, 220 were male and 180 were female. 80 of the people had chosen engineering as a profession.

c Calculate the expected number of female engineers. [2]

Tony used a 5% level of significance for his test and obtained a $p$-value of 0.0634 correct to 3 significant figures.

d State Tony's conclusion to the test. Give a reason for this conclusion. [2]

**Paper 1, May 10, Q10**

2 The heat output in thermal units from burning 1 kg of wood changes according to the wood's percentage moisture content. The moisture content and heat output of 10 blocks of the same type of wood each weighing 1 kg were measured. These are shown in the table.

| **Moisture content % (*x*)** | 8 | 15 | 22 | 30 | 34 | 45 | 50 | 60 | 74 | 82 |
|---|---|---|---|---|---|---|---|---|---|---|
| **Heat output (*y*)** | 80 | 77 | 74 | 69 | 68 | 61 | 61 | 55 | 50 | 45 |

a Draw a scatter diagram to show the above data. Use a scale of 2 cm to represent 10% on the $x$-axis and a scale of 2 cm to represent 10 thermal units on the $y$-axis. [4]

b Write down

i) the mean percentage moisture content, $\bar{x}$;

ii) the mean heat output, $\bar{y}$. [2]

c Plot the point $(\bar{x}, \bar{y})$ on your scatter diagram and label this point M. [2]

d Write down the product–moment correlation coefficient, $r$. [2]

The equation of the regression line $y$ on $x$ is $y = -0.470x + 83.7$.

e Draw the regression line $y$ on $x$ on your scatter diagram. [2]

f Estimate the heat output in thermal units of a 1 kg block of wood that has 25% moisture content. [2]

g State, with a reason, whether it is appropriate to use the regression line $y$ on $x$ to estimate the heat output in part **f**. [2]

**Paper 2, May 11, Q1**

3 **Part A**

In a mountain region there appears to be a relationship between the number of trees growing in the region and the depth of snow in winter. A set of 10 areas was chosen, and in each area the number of trees was counted and the depth of snow measured.

The results are given in the table opposite.

| **Number of trees (*x*)** | **Depth of snow in cm (*y*)** |
|---|---|
| 45 | 30 |
| 75 | 50 |
| 66 | 40 |
| 27 | 25 |
| 44 | 30 |
| 28 | 5 |
| 60 | 35 |
| 35 | 20 |
| 73 | 45 |
| 47 | 25 |

a Use your graphic display calculator to find
 i) the mean number of trees;
 ii) the standard deviation of the number of trees;
 iii) the mean depth of snow;
 iv) the standard deviation of the depth of snow. [4]

The covariance, $S_{xy} = 188.5$.

b Write down the product–moment correlation coefficient, $r$. [2]

c Write down the equation of the regression line of $y$ on $x$. [2]

d If the number of trees in an area is 55, estimate the depth of the snow. [2]

e i) Use the equation of the regression line to estimate the depth of snow in an area with 100 trees.
 ii) Decide whether the answer in **e i)** is a valid estimate of the depth of snow in the area. Give a reason for your answer. [2]

**Part B**

In a study on 100 students there seemed to be a difference between males and females in their choice of favourite car colour. The results are given in the table below. A $\chi^2$ test was conducted.

| | **Blue** | **Red** | **Green** |
|---|---|---|---|
| **Males** | 14 | 6 | 8 |
| **Females** | 31 | 24 | 17 |

a Write down the total number of male students. [1]

b Show that the expected frequency for males, whose favourite car colour is blue, is 12.6. [2]

The calculated value of $\chi^2$ is 1.367.

c i) Write down the null hypothesis for this test.
 ii) Write down the number of degrees of freedom.
 iii) Write down the critical value of $\chi^2$ at the 5% significance level.
 iv) Determine whether the null hypothesis should be accepted. Give a reason for your answer. [5]

**Paper 2, Nov 09, Q4**

Topic 4

# Applications, project ideas and theory of knowledge

1 What is a 'Black Swan Event'? What lessons does it give students of probability?

2 Why is statistics not part of many universities' Mathematics departments?

3 Can we reliably use the equation of the regression line to make predictions?

4 It may help you to understand Pearson's product–moment correlation coefficient, r, more clearly if you calculate it without using your GDC.

e

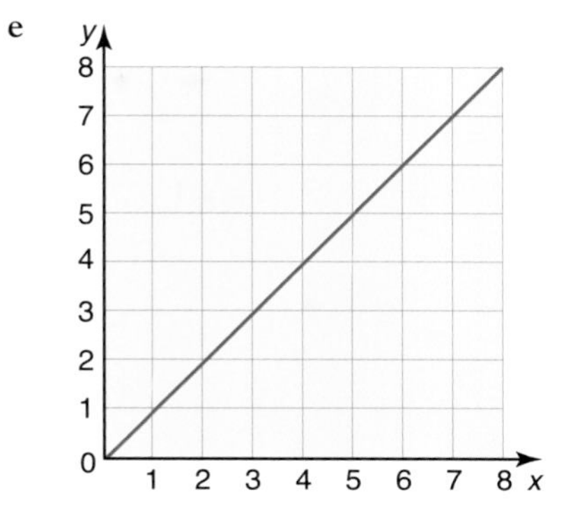


f

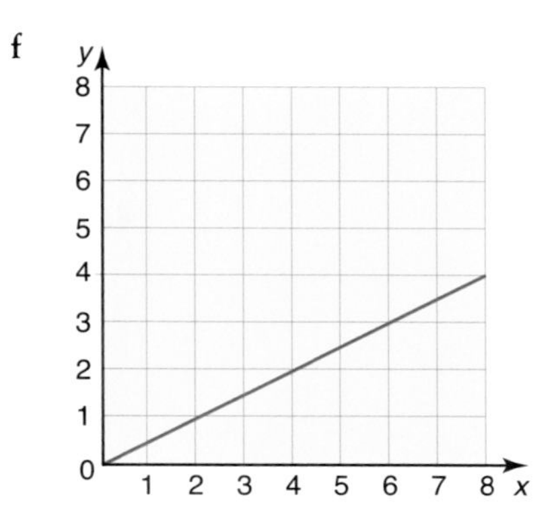


g

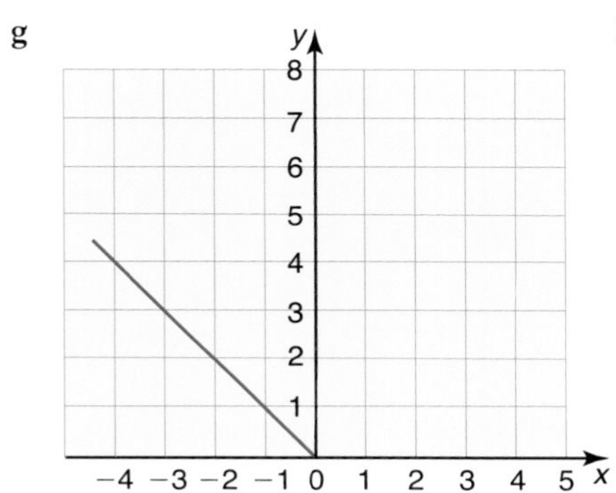


h

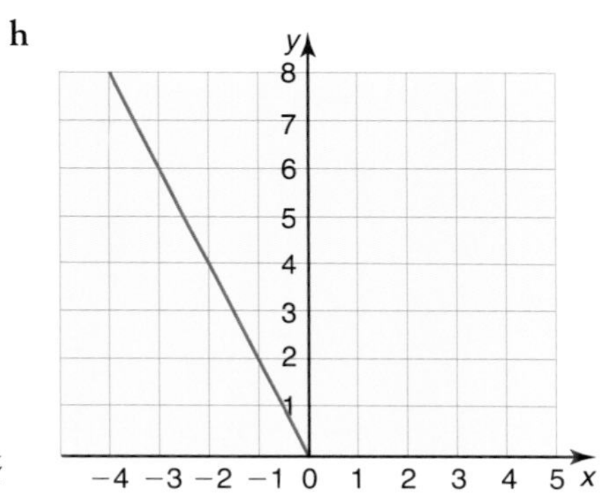


2 For each of the following, find the coordinates of some of the points on the line and use these to find the equation of the straight line.

a

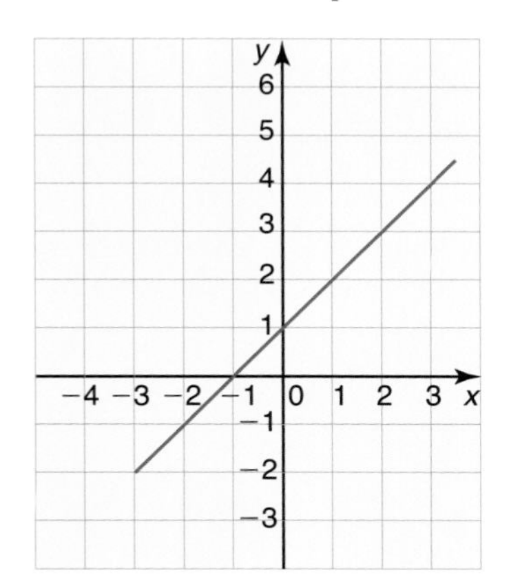


b

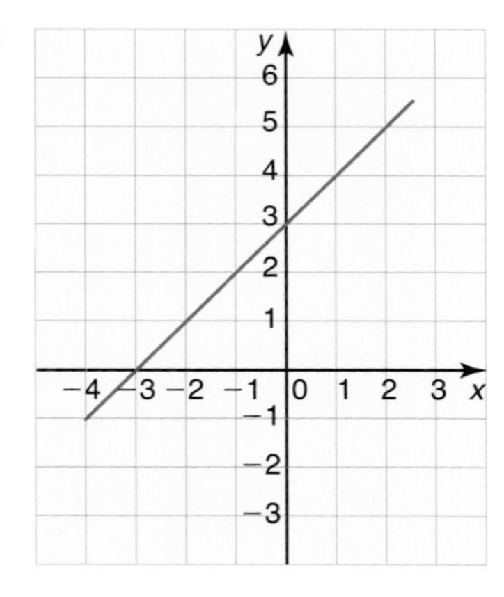

c

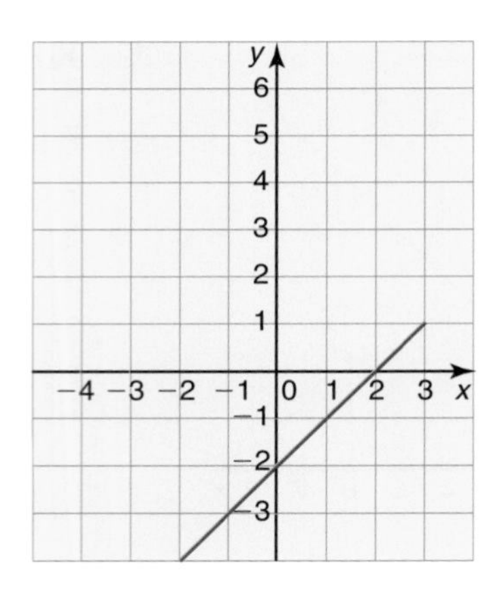


d

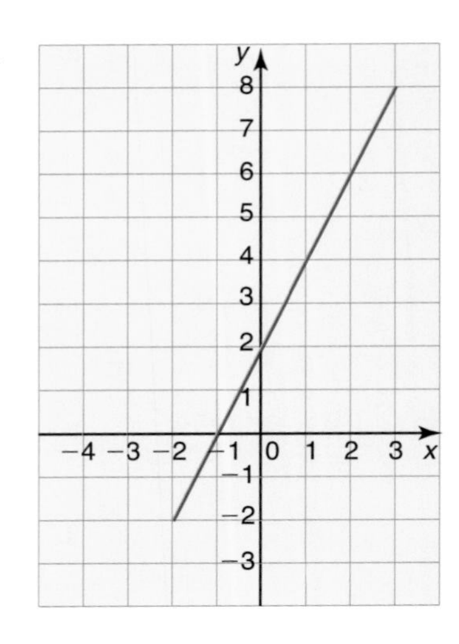


e

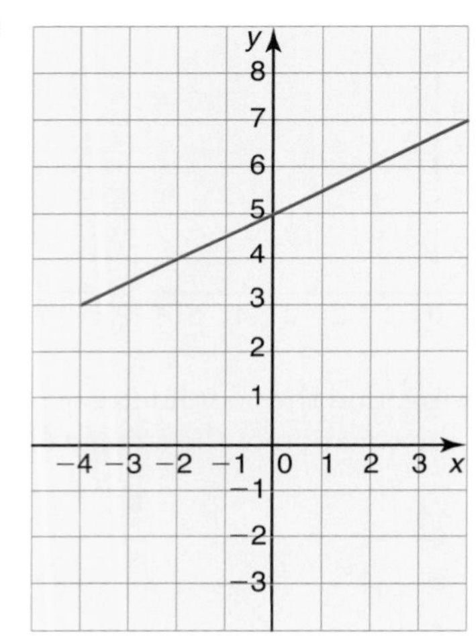


f

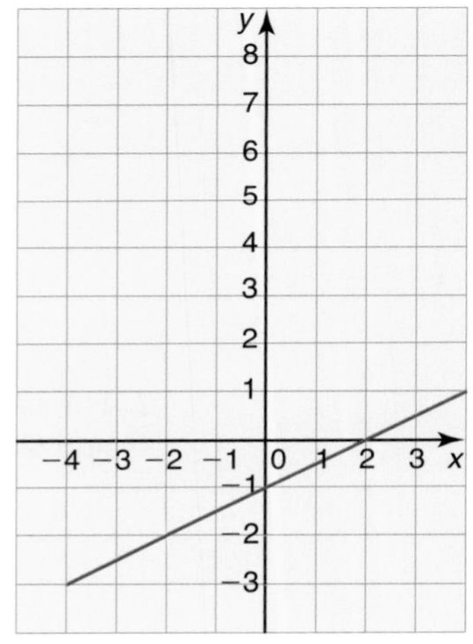


3 For each of the following, find the coordinates of some of the points on the line and use these to find the equation of the straight line.

a

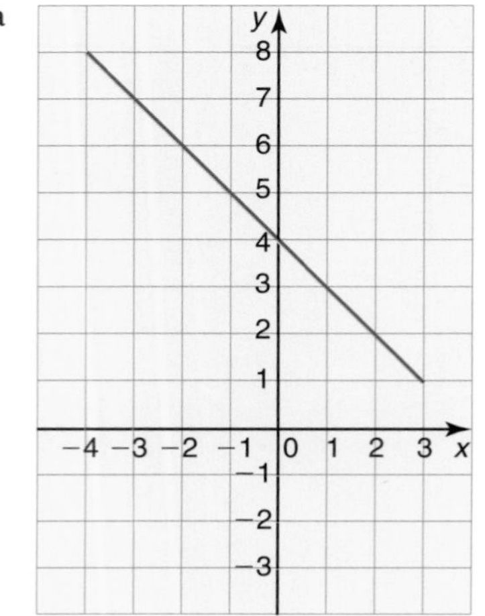


b

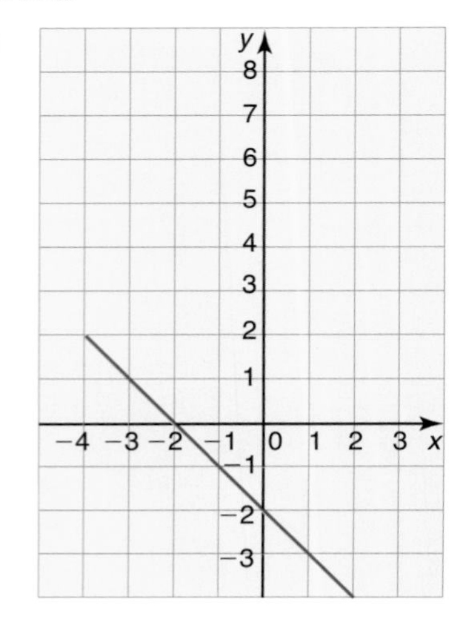

## Autograph

Select and enter the equations $3x + y = 9$ and $5x - y = 7$.

Select both graphs and click on 'Object' followed by 'Solve $f(x) = g(x)$'.

The results are displayed in the results box by selecting .

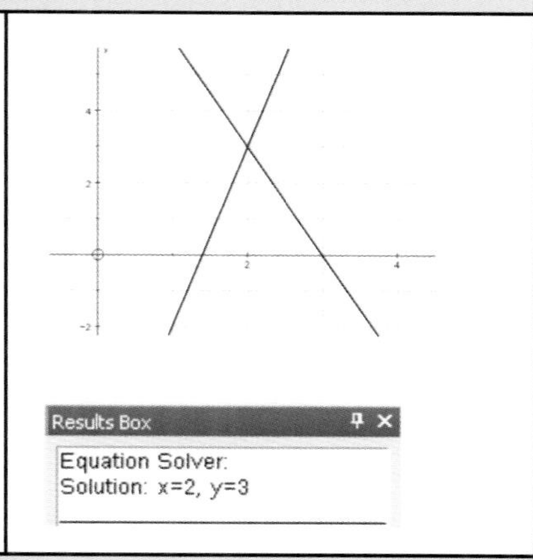


## GeoGebra

Enter the equations $3x + y = 9$ and $5x - y = 7$ in turn into the input box.

Type 'Intersect [a, b]' in the input box.

The point of intersection is marked on the graph and its coordinates appear in the algebra window.

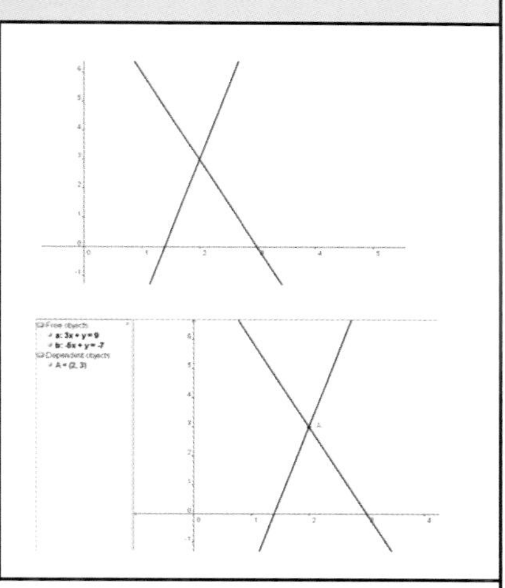

Note: The letters 'a' and 'b' are in the command 'Intersect [a, b]' as GeoGebra automatically assigns the lines the labels 'a' and 'b'.

The point of intersection of the lines with equations $3x + y = 9$ and $5x - y = 7$ is $(2, 3)$.

## Exercise 5.1.6

Solve these simultaneous equations either algebraically or graphically.

1 a $x + y = 6$
$x - y = 2$

b $x + y = 11$
$x - y - 1 = 0$

c $x + y = 5$
$x - y = 7$

d $2x + y = 12$
$2x - y = 8$

e $3x + y = 17$
$3x - y = 13$

f $5x + y = 29$
$5x - y = 11$

2 a $3x + 2y = 13$
$4x = 2y + 8$

b $6x + 5y = 62$
$4x - 5y = 8$

c $x + 2y = 3$
$8x - 2y = 6$

d $9x + 3y = 24$
$x - 3y = -14$

e $7x - y = -3$
$4x + y = 14$

f $3x = 5y + 14$
$6x + 5y = 58$

3 a $2x + y = 14$
$x + y = 9$

b $5x + 3y = 29$
$x + 3y = 13$

c $4x + 2y = 50$
$x + 2y = 20$

d $x + y = 10$
$3x = -y + 22$

e $2x + 5y = 28$
$4x + 5y = 36$

f $x + 6y = -2$
$3x + 6y = 18$

4 a $x - y = 1$
$2x - y = 6$

b $3x - 2y = 8$
$2x - 2y = 4$

c $7x - 3y = 26$
$2x - 3y = 1$

d $x = y + 7$
$3x - y = 17$

e $8x - 2y = -2$
$3x - 2y = -7$

f $4x - y = -9$
$7x - y = -18$

5 a $x + y = -7$
$x - y = -3$

b $2x + 3y = -18$
$2x = 3y + 6$

c $5x - 3y = 9$
$2x + 3y = 19$

d $7x + 4y = 42$
$9x - 4y = -10$

e $4x - 4y = 0$
$8x + 4y = 12$

f $x - 3y = -25$
$5x - 3y = -17$

6 a $2x + 3y = 13$
$2x - 4y + 8 = 0$

b $2x + 4y = 50$
$2x + y = 20$

c $x + y = 10$
$3y = 22 - x$

d $5x + 2y = 28$
$5x + 4y = 36$

7 a $-4x = 4y$
$4x - 8y = 12$

b $3x = 19 + 2y$
$-3x + 5y = 5$

c $3x + 2y = 12$
$-3x + 9y = -12$

d $3x + 5y = 29$
$3x + y = 13$

e $-5x + 3y = 14$
$5x + 6y = 58$

f $-2x + 8y = 6$
$2x = 3 - y$

## Further simultaneous equations

If neither $x$ nor $y$ can be eliminated by simply adding or subtracting the two equations then it is necessary to multiply one or both of the equations. The equations are multiplied by a number in order to make the coefficients of $x$ (or $y$) numerically equal.

**Worked example**

Solve these simultaneous equations

**a** $3x + 2y = 22$
$x + y = 9$

**b** $5x - 3y = 1$
$3x + 4y = 18$

**a** $3x + 2y = 22$ (1)
$x + y = 9$ (2)

To eliminate $y$, equation (2) is multiplied by 2:

$3x + 2y = 22$ (1)
$2x + 2y = 18$ (3)

By subtracting (3) from (1), the variable $y$ is eliminated:

$$x = 4$$

Substituting $x = 4$ into equation (2), we have:

$x + y = 9$
$4 + y = 9$
$y = 5$

Check by substituting both values into equation (1):

$3x + 2y = 22$
$12 + 10 = 22$
$22 = 22$

**b** $5x - 3y = 1$ (1)
$3x + 4y = 18$ (2)

To eliminate the variable $y$, equation (1) is multiplied by 4 and equation (2) is multiplied by 3.

$20x - 12y = 4$ (3)
$9x + 12y = 54$ (4)

By adding equations (3) and (4) the variable $y$ is eliminated:

$29x = 58$
$x = 2$

Substituting $x = 2$ into equation (2) gives:

$3x + 4y = 18$
$6 + 4y = 18$
$4y = 12$
$y = 3$

Check by substituting both values into equation (1):

$5x - 3y = 1$
$10 - 9 = 1$
$1 = 1$

## Exercise 5.1.7

Solve these equations either algebraically or graphically.

1 a $2x + y = 7$
$3x + 2y = 12$

b $5x + 4y = 21$
$x + 2y = 9$

c $x + y = 7$
$3x + 4y = 23$

d $2x - 3y = -3$
$3x + 2y = 15$

e $4x = 4y + 8$
$x + 3y = 10$

f $x + 5y = 11$
$2x - 2y = 10$

2 a $x + y = 5$
$3x - 2y + 5 = 0$

b $2x - 2y = 6$
$x - 5y = -5$

c $2x + 3y = 15$
$2y = 15 - 3x$

d $x - 6y = 0$
$3x - 3y = 15$

e $2x - 5y = -11$
$3x + 4y = 18$

f $x + y = 5$
$2x - 2y = -2$

3 a $3y = 9 + 2x$
$3x + 2y = 6$

b $x + 4y = 13$
$3x - 3y = 9$

c $2x = 3y - 19$
$3x + 2y = 17$

d $2x - 5y = -8$
$-3x - 2y = -26$

e $5x - 2y = 0$
$2x + 5y = 29$

f $8y = 3 - x$
$3x - 2y = 9$

4 a $4x + 2y = 5$
$3x + 6y = 6$

b $4x + y = 14$
$6x - 3y = 3$

c $10x - y = -2$
$-15x + 3y = 9$

d $-2y = 0.5 - 2x$
$6x + 3y = 6$

e $x + 3y = 6$
$2x - 9y = 7$

f $5x - 3y = -0.5$
$3x + 2y = 3.5$

# 5.2 Right-angled trigonometry

Lord Nelson would have used trigonometry in navigation

Trigonometry and the trigonometric ratios developed from the ancient study of the stars. The study of right-angled triangles probably originated with the Egyptians and the Babylonians, who used them extensively in construction and engineering.

The ratios, which are introduced in this chapter, were set out by Hipparchus of Rhodes about 150 BC.

Trigonometry was used extensively in navigation at sea, particularly in the sailing ships of the eighteenth and nineteenth centuries, when it formed a major part of the examination to become a lieutenant in the Royal Navy.

## The trigonometric ratios

There are three basic trigonometric ratios: sine, cosine and tangent and you should already be familiar with these.

Each of the trigonometric ratios relates an angle of a right-angled triangle to a ratio of the lengths of two of its sides.

The sides of the triangle have names, two of which are dependent on their position in relation to a specific angle.

The longest side (always opposite the right angle) is called the **hypotenuse**. The side opposite the angle is called the **opposite** side and the side next to the angle is called the **adjacent** side.

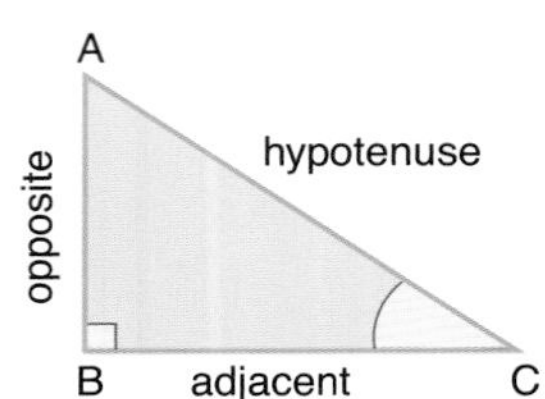

Note that, when the chosen angle is at A, the sides labelled opposite and adjacent change.

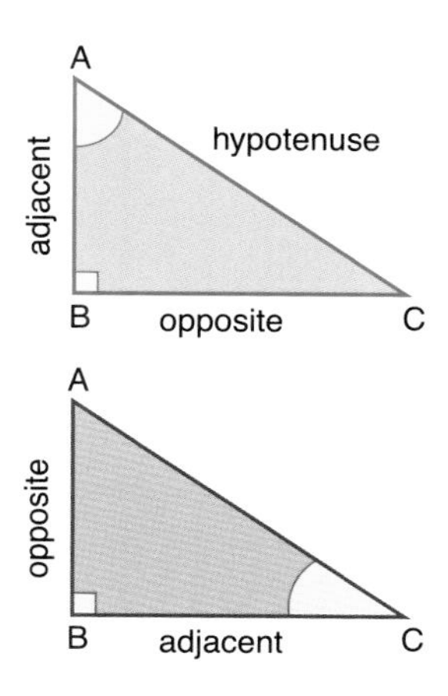


## Tangent

The tangent ratio is:

$$\tan C = \frac{\text{length of opposite side}}{\text{length of adjacent side}}$$

**Worked examples**

1 Calculate the size of angle BAC.

$$\tan x° = \frac{\text{opposite}}{\text{adjacent}} = \frac{4}{5}$$

$$x = \tan^{-1}\left(\frac{4}{5}\right)$$

$$x = 38.7 \text{ (3 s.f.)}$$

$$\angle BAC = 38.7° \text{ (3 s.f.)}$$

2 Calculate the length of QR.

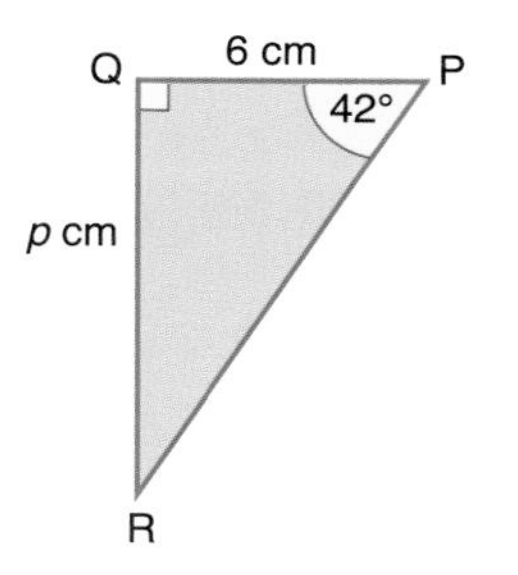


$$\tan 42° = \frac{p}{6}$$

$$6 \times \tan 42° = p$$

$$p = 5.40 \text{ (3 s.f.)}$$

$$QR = 5.40 \text{ cm (3 s.f.)}$$

## Sine

The sine ratio is:

$$\sin N = \frac{\text{length of opposite side}}{\text{length of hypotenuse}}$$

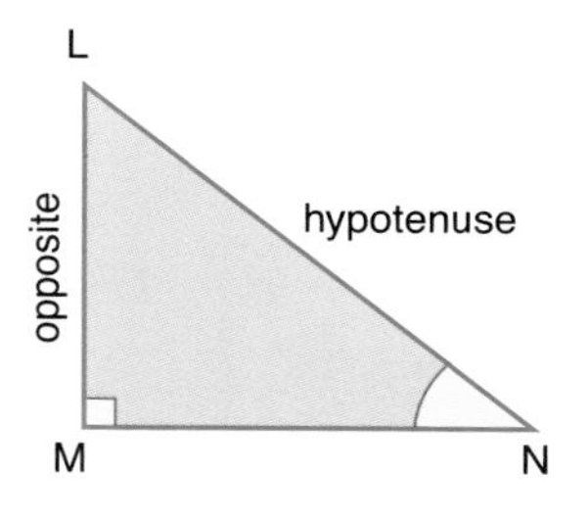

**Worked examples**

1 Calculate the size of angle BAC.

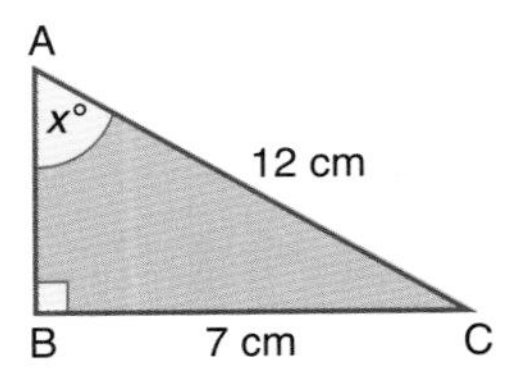


$$\sin x = \frac{\text{opposite}}{\text{hypotenuse}} = \frac{7}{12}$$

$$x = \sin^{-1}\left(\frac{7}{12}\right)$$

$$x = 35.7 \text{ (3 s.f)}$$

$$\angle\text{BAC} = 35.7° \text{ (3 s.f)}$$

2 Calculate the length of PR.

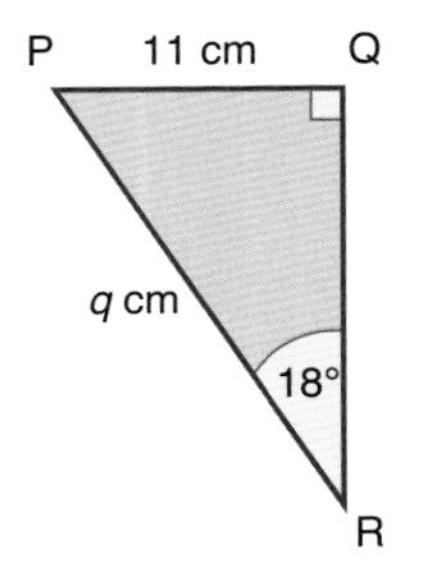


$$\sin 18° = \frac{11}{q}$$

$$q \times \sin 18° = 11$$

$$q = \frac{11}{\sin 18°}$$

$$q = 35.6 \text{ (3 s.f.)}$$

$$\text{PR} = 35.6\,\text{cm (3 s.f.)}$$

## Cosine

The cosine ratio is:

$$\cos Z = \frac{\text{length of adjacent side}}{\text{length of hypotenuse}}$$

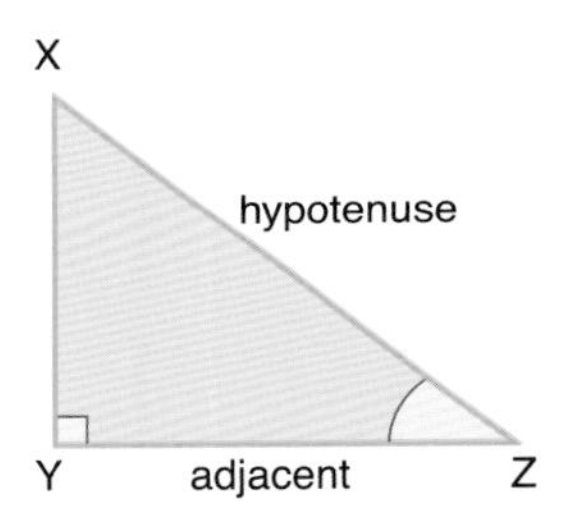


**Worked examples**

1 Calculate the length XY.

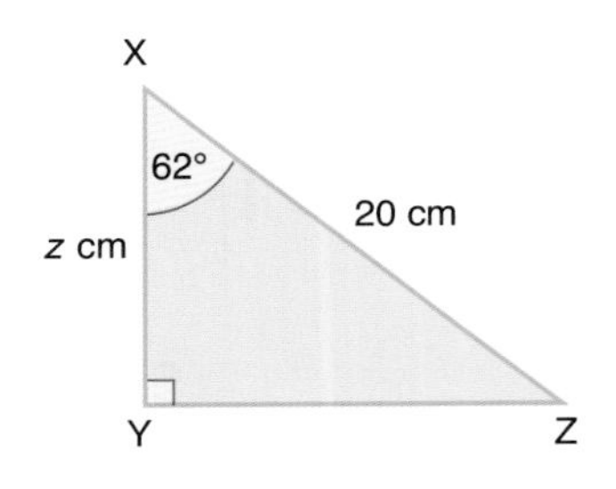


$$\cos 62° = \frac{\text{adjacent}}{\text{hypotenuse}} = \frac{z}{20}$$

$$z = 20 \times \cos 62°$$

$$z = 9.39 \text{ (3 s.f.)}$$

$$\text{XY} = 9.39\,\text{cm (3 s.f.)}$$

2 Calculate the size of angle ABC.

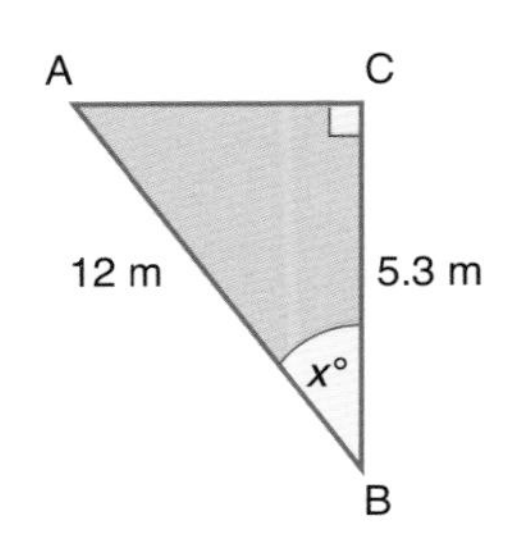


$$\cos x = \frac{\text{adjacent}}{\text{hypotenuse}}$$

$$\cos x = \frac{5.3}{12}$$

$$x = \cos^{-1}\left(\frac{5.3}{12}\right)$$

$$x = 63.8 \text{ (3 s.f.)}$$

$$\angle\text{ABC} = 63.8° \text{ (3 s.f.)}$$

2 The cosine of which angle between 0° and 180° is equal to the negative of cos 50°?

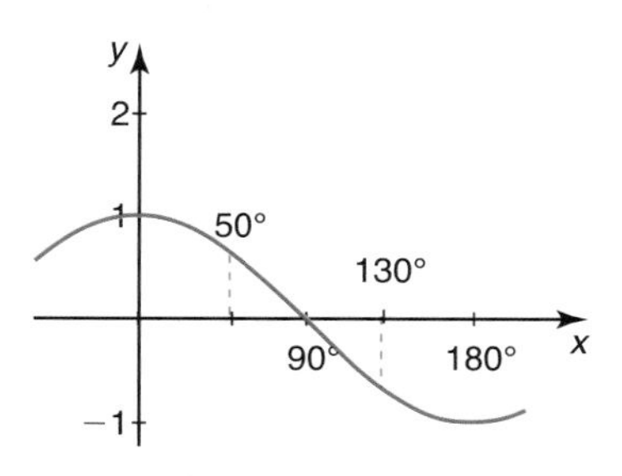


cos 130° = −cos 50°

## Exercise 5.3.2

1 Write down each of the following in terms of the cosine of another angle between 0° and 180°.

**a** cos 20° **b** cos 85° **c** cos 32°
**d** cos 95° **e** cos 147° **f** cos 106°

2 Write down each of the following in terms of the cosine of another angle between 0° and 180°.

**a** cos 98° **b** cos 144° **c** cos 160°
**d** cos 143° **e** cos 171° **f** cos 123°

3 Write down each of the following in terms of the cosine of another angle between 0° and 180°.

**a** −cos 100° **b** cos 90° **c** −cos 110°
**d** −cos 45° **e** −cos 122° **f** −cos 25°

4 The cosine of which acute angle has the same value as:

**a** cos 125° **b** cos 107° **c** −cos 120°
**d** −cos 98° **e** −cos 92° **f** −cos 110°

## The sine rule

With right-angled triangles we can use the basic trigonometric ratios of sine, cosine and tangent. The **sine rule** is a relationship which can be used with non-right-angled triangles.

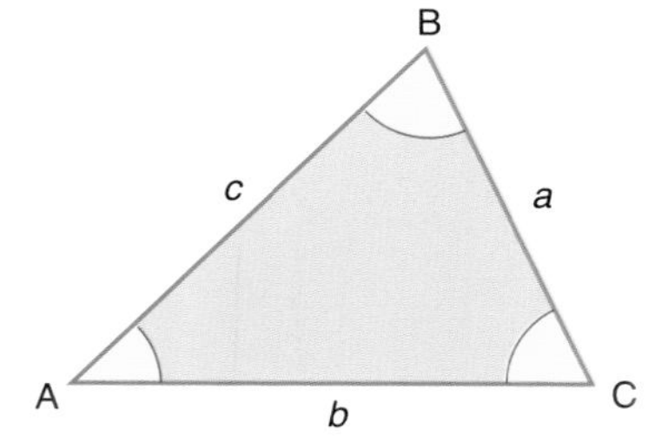


The sine rule states that:

$$\frac{a}{\sin A} = \frac{b}{\sin B} = \frac{c}{\sin C}$$

It can also be written as:

$$\frac{\sin A}{a} = \frac{\sin B}{b} = \frac{\sin C}{c}$$

**Worked examples**

1 Calculate the length of side BC.

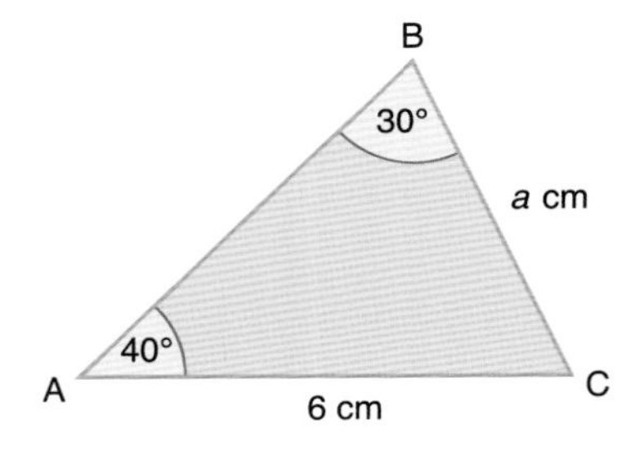


Using the sine rule:

$$\frac{a}{\sin A} = \frac{b}{\sin B}$$

$$\frac{a}{\sin 40°} = \frac{6}{\sin 30°}$$

$$a = \frac{6}{\sin 30°} \times \sin 40°$$

$$a = 7.71 \text{ (3 s.f.)}$$

BC = 7.71 cm (3 s.f.)

2 Calculate the size of angle C.

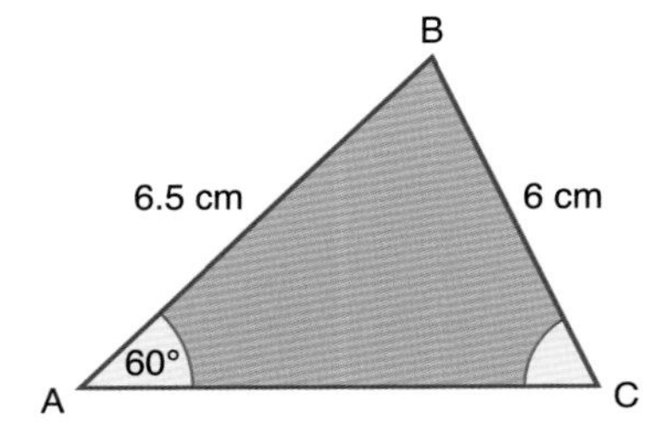


Using the sine rule:

$$\frac{\sin A}{a} = \frac{\sin C}{c}$$

$$\frac{\sin 60°}{6} = \frac{\sin C}{6.5}$$

$$\sin C = \frac{6.5 \times \sin 60°}{6}$$

$$C = \sin^{-1}(0.94)$$

$$C = 69.8° \text{ (3 s.f.)}$$

## Exercise 5.3.3

1 Calculate the length of the side marked $x$ in each of the following. Give your answers to one decimal place.

a

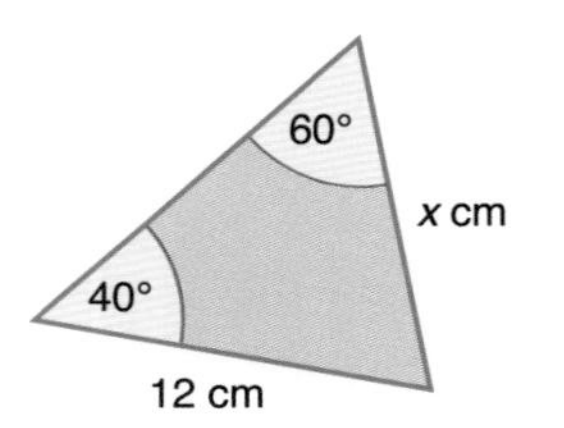


b

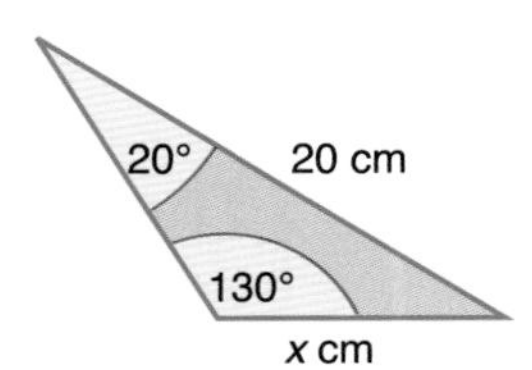


c

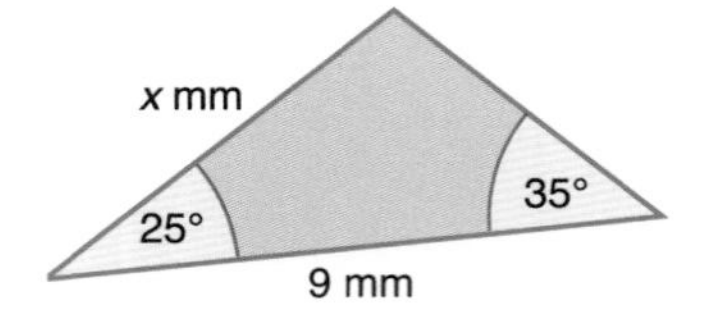


d

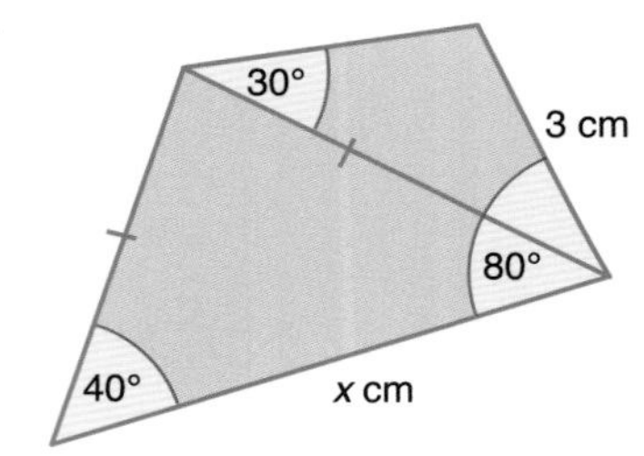

2 Calculate the size of the angle marked $\theta°$ in each of the following. Give your answers to one decimal place.

a

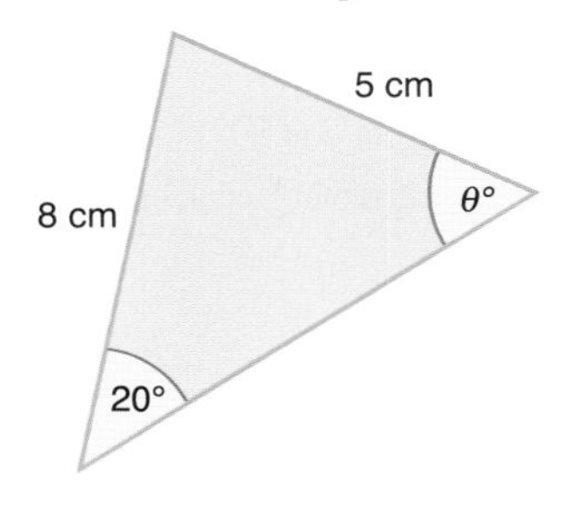


b

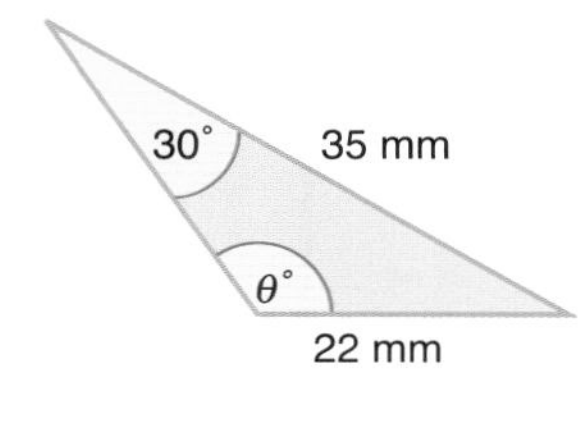


c

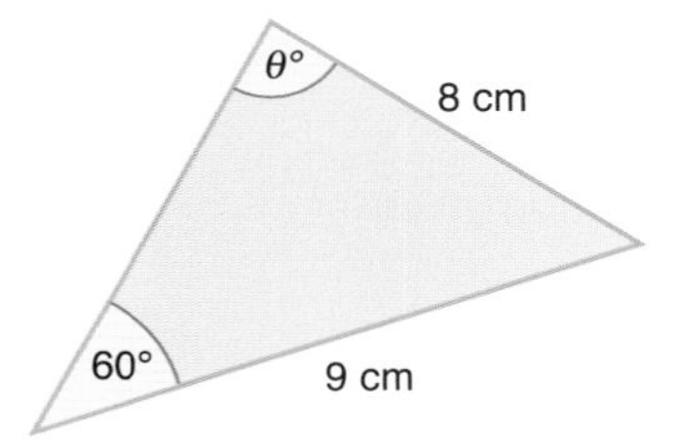


d

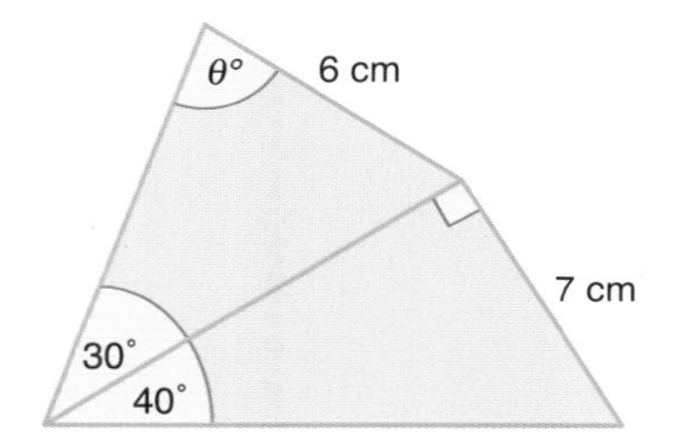


3 Triangle ABC has the following dimensions:

AC = 10 cm, AB = 8 cm and angle ACB = 20°.

a Calculate the two possible values for angle ABC.
b Sketch and label the two possible shapes for triangle ABC.

4 Triangle PQR has the following dimensions:

PQ = 6 cm, PR = 4 cm and angle PQR = 40°.

a Calculate the two possible values for angle QRP.
b Sketch and label the two possible shapes for triangle PQR.

## The cosine rule

The **cosine rule** is another relationship which can be used with non-right-angled triangles.

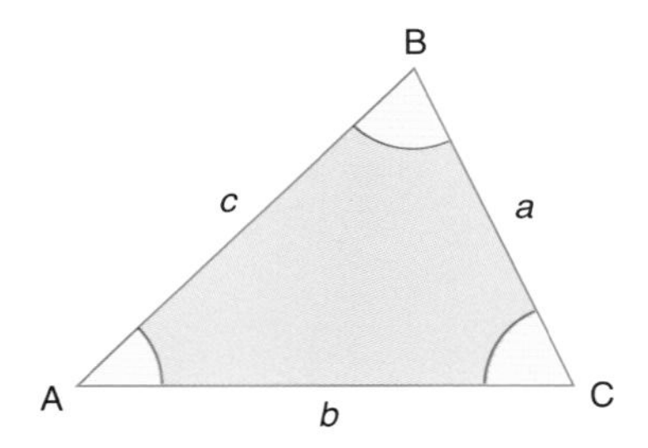


The cosine rule states that:

$a^2 = b^2 + c^2 - 2bc \cos A$

**Worked examples**

1 Calculate the length of the side BC.

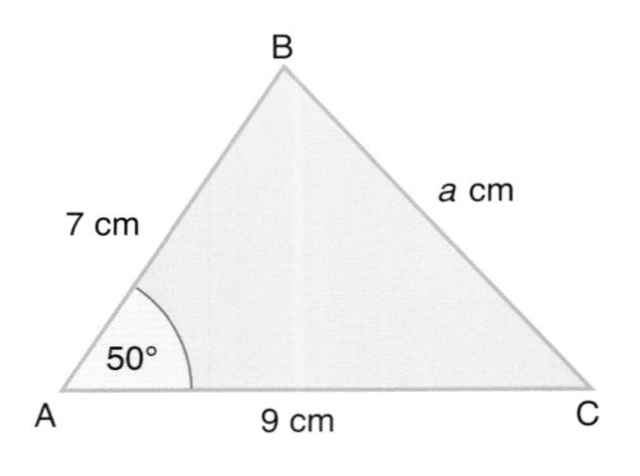


Using the cosine rule:

$$a^2 = b^2 + c^2 - 2bc\cos A$$
$$a^2 = 9^2 + 7^2 - (2 \times 9 \times 7 \times \cos 50°)$$
$$= 81 + 49 - (126 \times \cos 50°) = 49.0$$
$$a = \quad 49.0$$
$$a = 7.00 \text{ (3 s.f.)}$$

BC = 7.00 cm (3 s.f.)

2 Calculate the size of angle A.

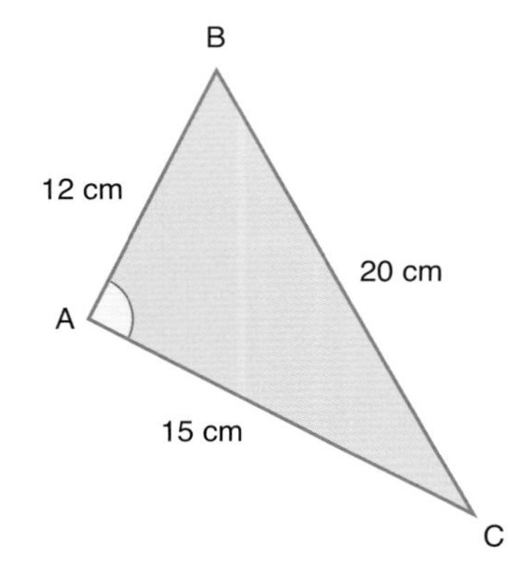


Using the cosine rule:

$$a^2 = b^2 + c^2 - 2bc\cos A$$

Rearranging the equation gives:

$$\cos A = \frac{b^2 + c^2 - a^2}{2bc}$$

$$\cos A = \frac{15^2 + 12^2 - 20^2}{2 \times 15 \times 12} = 0.086$$

$$A = \cos^{-1}(-0.086)$$
$$A = 94.9° \text{ (3 s.f.)}$$

## Exercise 5.3.4

1 Calculate the length of the side marked $x$ in each of the following. Give your answers to one decimal place.

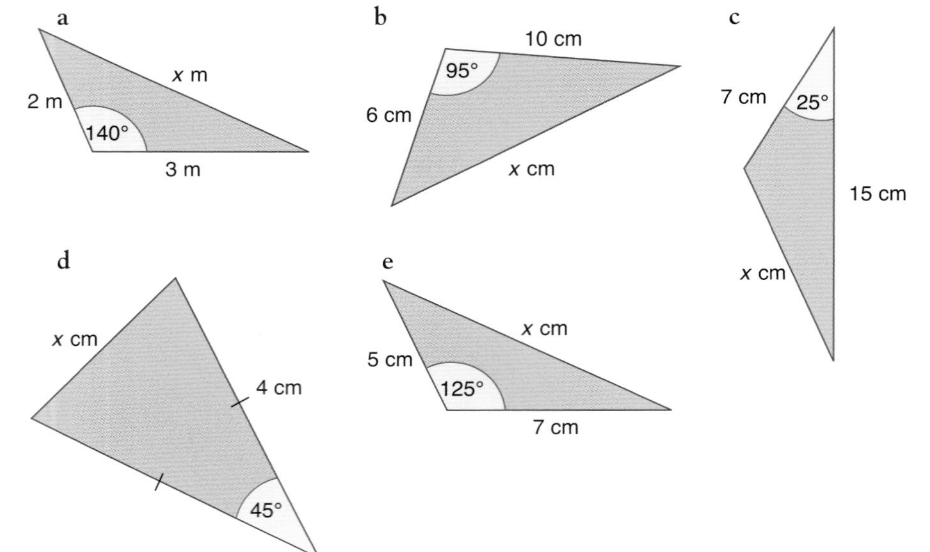

2 Calculate the angle marked $\theta°$ in each of the following. Give your answers to one decimal place.

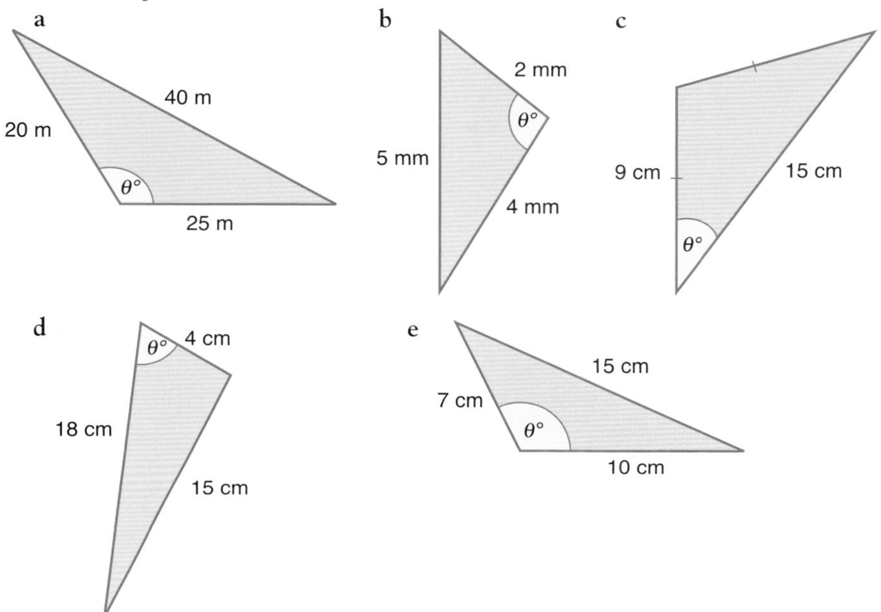


3 Four players W, X, Y and Z are on a rugby pitch. The diagram shows a plan view of their relative positions.

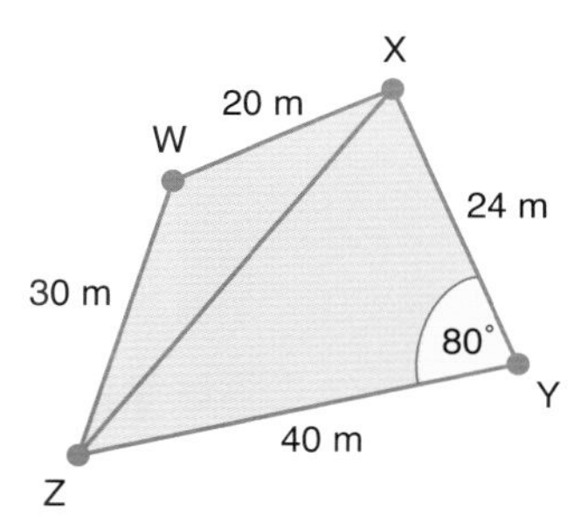


Calculate:

a the distance between players X and Z
b angle ZWX
c angle WZX
d angle YZX
e the distance between players W and Y.

4 Three yachts A, B and C are racing off the 'Cape'. Their relative positions are shown in the diagram.

Calculate the distance between B and C to the nearest 10 m.

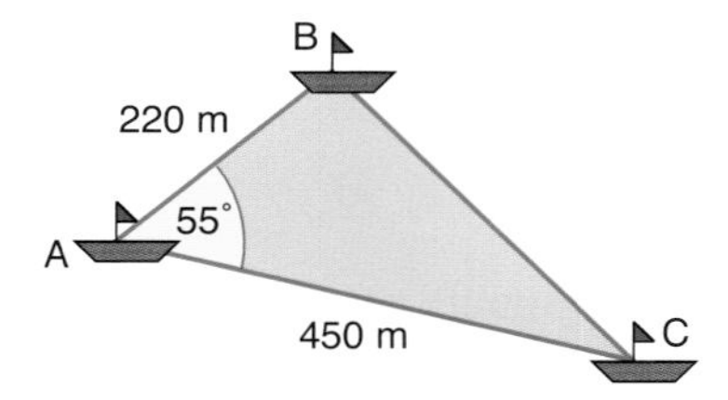

5 A girl standing on a cliff top at A can see two buoys X and Y, 200 m apart, floating on the sea. The angle of depression of Y from A is 45°, and the angle of depression of X from A is 60° (see diagram).

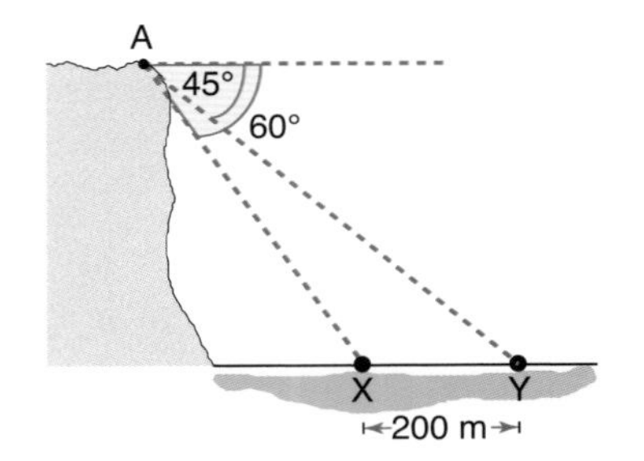


If A, X, Y are in the same vertical plane, calculate:

**a** the distance AY
**b** the distance AX
**c** the vertical height of the cliff.

6 There are two trees standing on one side of a river bank. On the opposite side is a boy standing at X.

Using the information given, calculate the distance between the two trees.

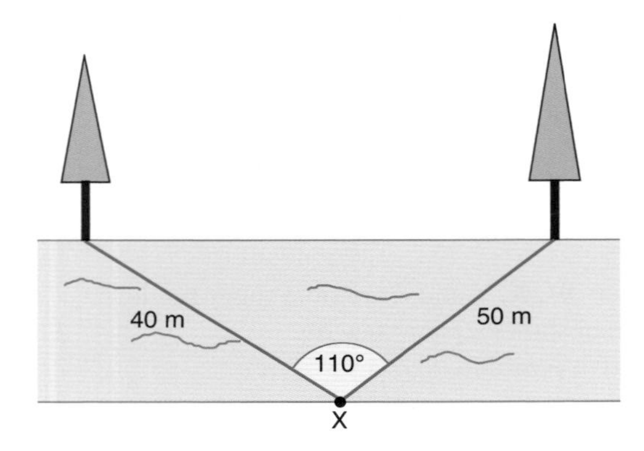


## The area of a triangle

The area of a triangle is given by the formula:

$\text{Area} = \frac{1}{2}bh$

where $b$ is the base and $h$ is the vertical height of the triangle.

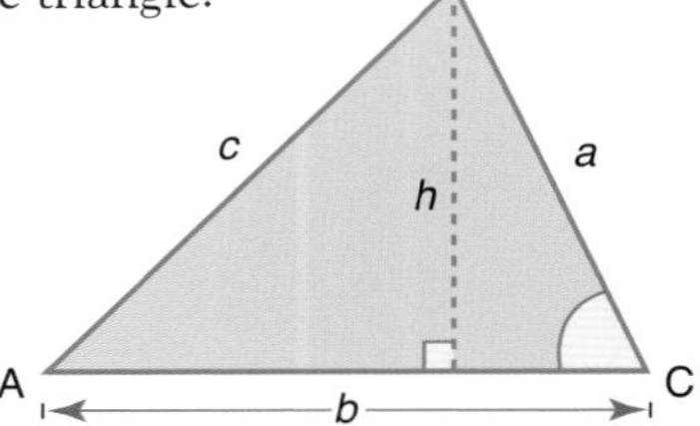


From trigonometric ratios we also know that:

$\sin C = \dfrac{h}{a}$

Rearranging, we have:

$h = a \sin C$

Substituting for $h$ in the original formula gives another formula for the area of a triangle:

$\text{Area} = \frac{1}{2}ab \sin C$

## Exercise 5.3.5

1 Calculate the area of the following triangles. Give your answers to one decimal place.

a

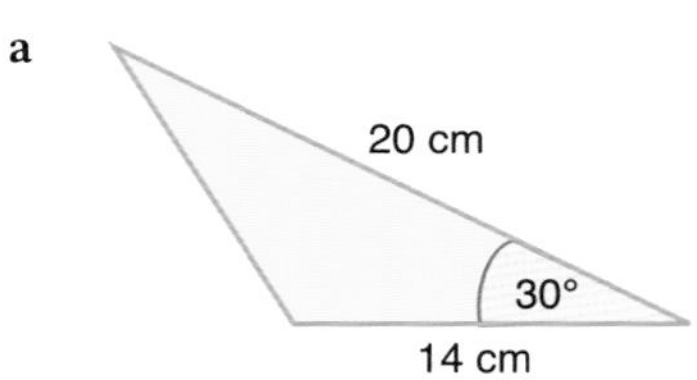


b

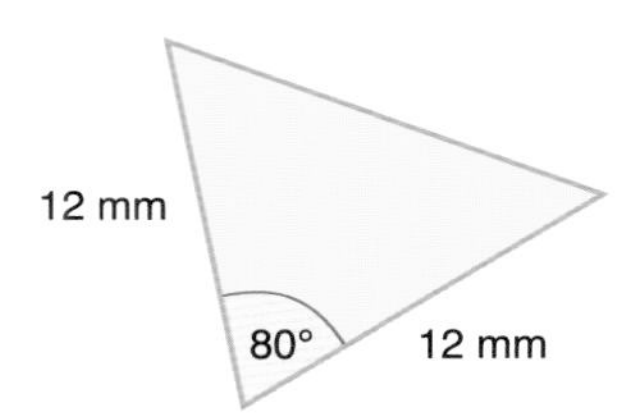


c

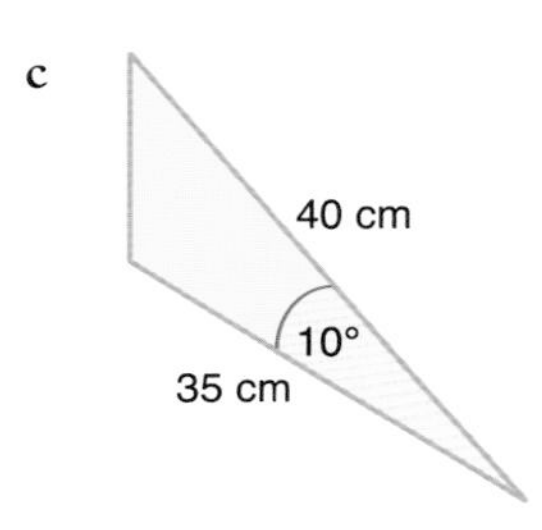


d

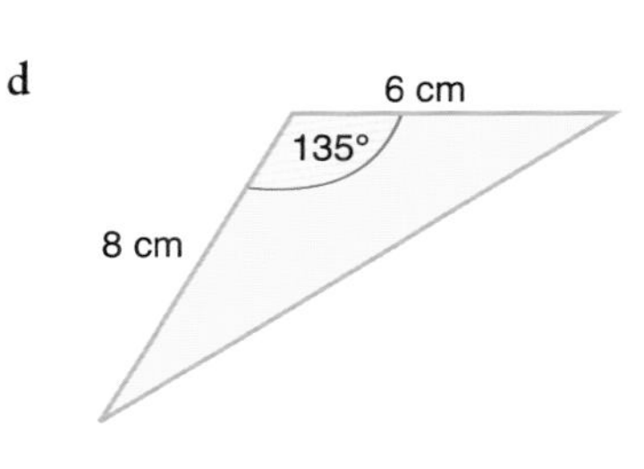


2 Calculate the value of $x$ in each of the following. Give your answers correct to one decimal place.

a

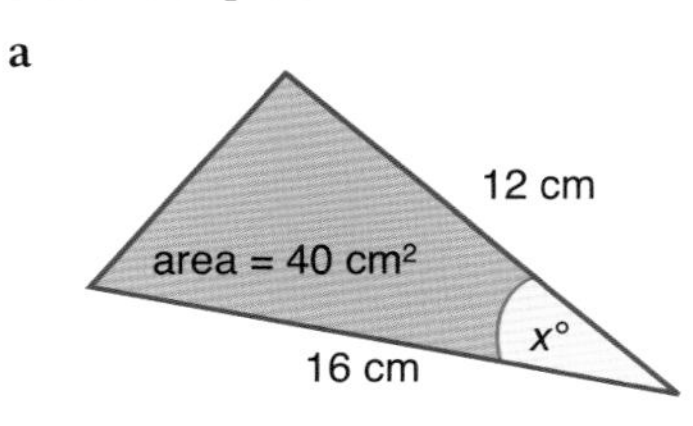


b

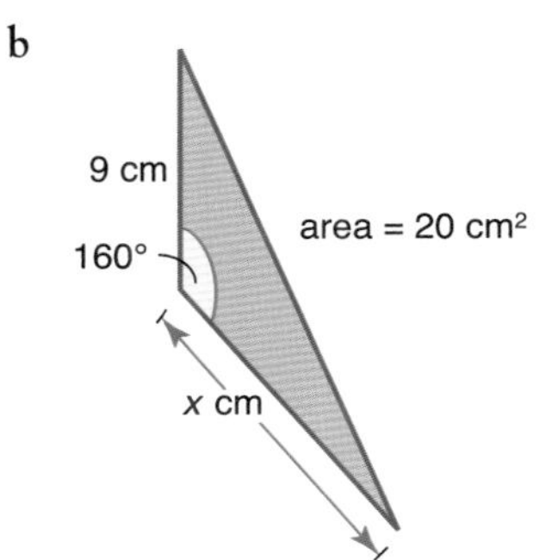


c

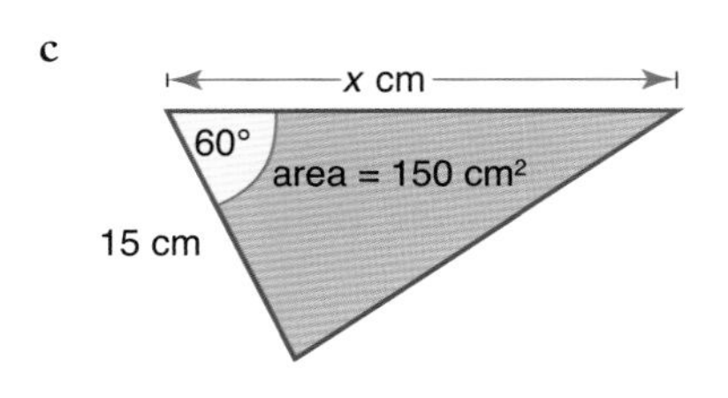


d

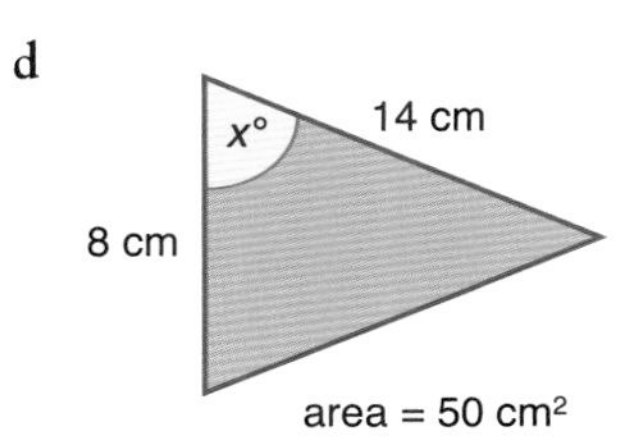

3 ABCD is a school playing field. The following lengths are known:

OA = 83 m, OB = 122 m, OC = 106 m, OD = 78 m

Calculate the area of the school playing field to the nearest 100 $m^2$.

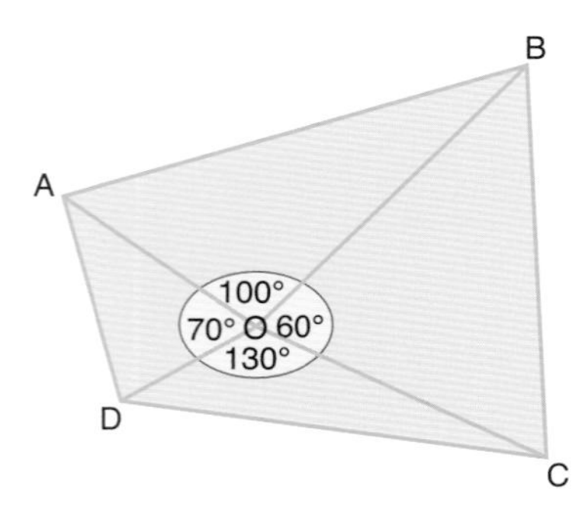


4 The roof of a garage has a slanting length of 3 m and makes an angle of 120° at its vertex. The height of the garage is 4 m and its depth is 9 m.

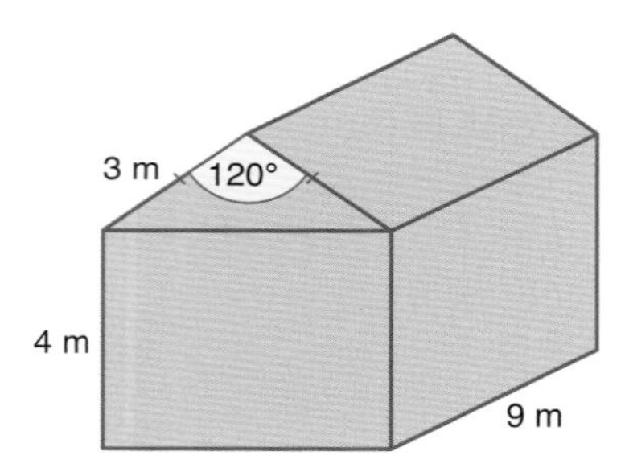


Calculate:

a the cross-sectional area of the roof

b the volume occupied by the whole garage.

# 5.4 Geometry of three-dimensional solids

Egyptian society around 2000 BC was very advanced, particularly in its understanding and development of new mathematical ideas and concepts. One of the most important pieces of Egyptian work is called the 'Moscow Papyrus' – so called because it was taken to Moscow in the middle of the nineteenth century.

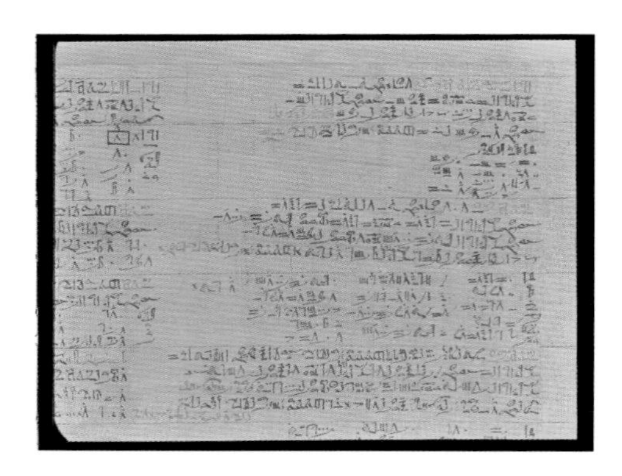

The Moscow Papyrus was written in about 1850 BC and is important because it contains 25 mathematical problems. One of the key problems is the solution to finding the volume of a truncated pyramid. Although the solution was not written in the way we write it today, it was mathematically correct and translates into the formula:

$$V = \frac{(a^2 + ab + b^2)h}{3}$$

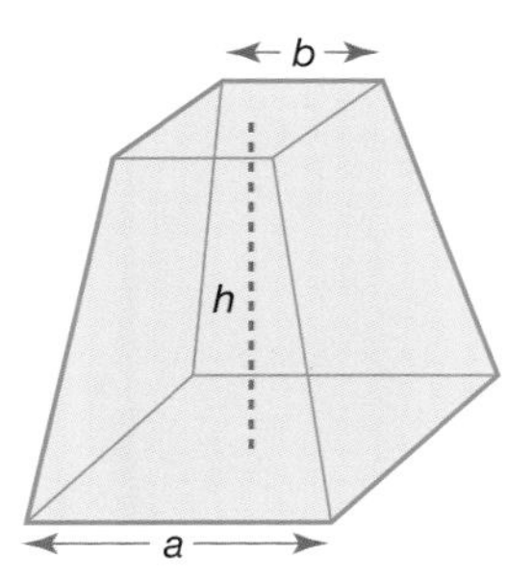


The application of trigonometry first to triangles and then to other two-dimensional shapes led to the investigation of angles between a line and a plane, and then further to the application of trigonometry to three-dimensional solids such as the cuboid, pyramid, cylinder, cone and sphere.

## Trigonometry in three dimensions

**Worked examples**

1 The diagram shows a cube of edge length 3 cm.

**a** Calculate the length EG.

**b** Calculate the length AG.

**c** Calculate the angle EGA.

**d** Calculate the distance from the midpoint X of AB to the midpoint Y of BC.

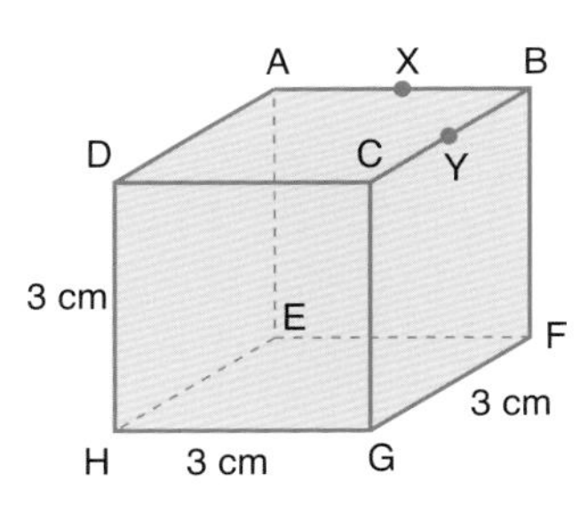


**a** Triangle EHG (below) is right angled. Use Pythagoras' theorem to calculate the length EG.

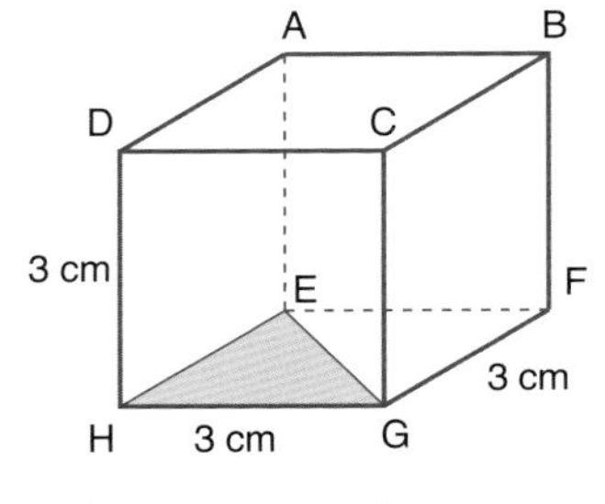


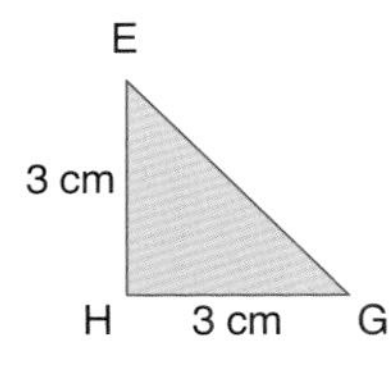


$EG^2 = EH^2 + HG^2$

$EG^2 = 3^2 + 3^2 = 18$

EG = 18 cm

**b** Triangle AEG (below) is right angled. Use Pythagoras' theorem to calculate the length AG.

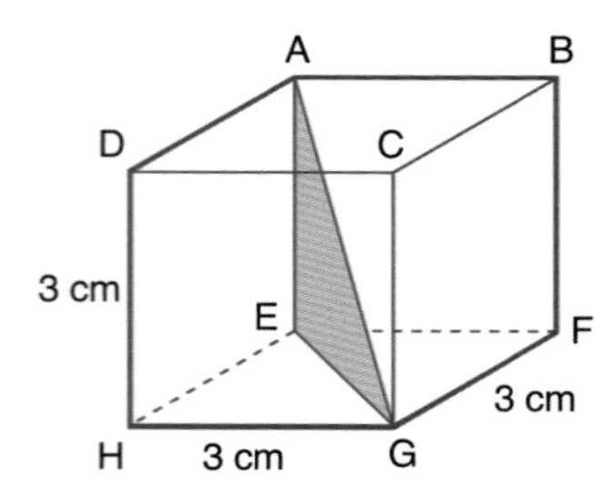


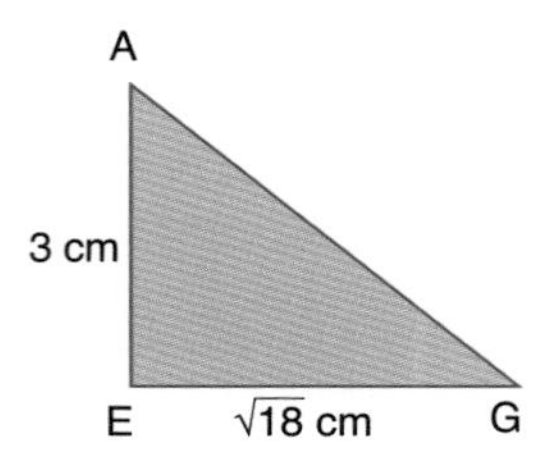


$AG^2 = AE^2 + EG^2$

$AG^2 = 3^2 + (\sqrt{18})^2$

$AG^2 = 9 + 18$

$AG = \sqrt{27}\,\text{cm}$

**c** To calculate angle EGA, use the triangle EGA:

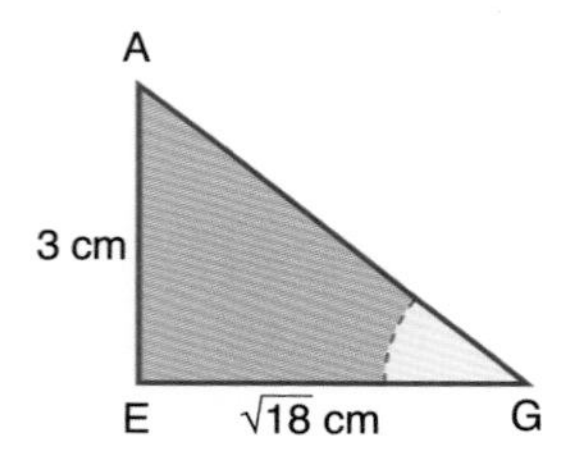


$\tan G = \dfrac{3}{\sqrt{18}}$

Angle G (EGA) = 35.3° (3 s.f.)

**d** To calculate the distance from X to Y, use Pythagoras' theorem:

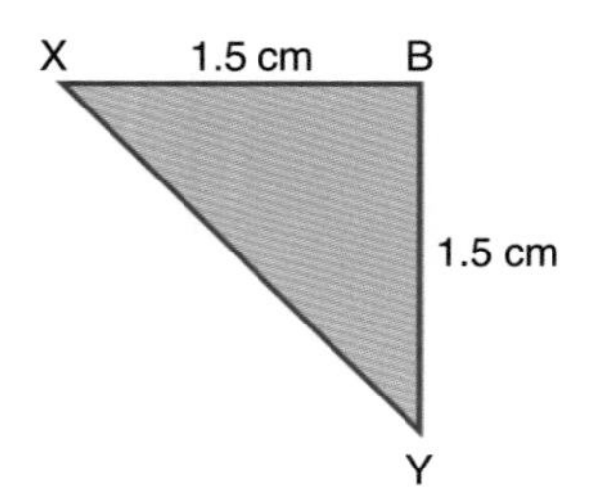


$XY^2 = XB^2 + BY^2$
$XY^2 = 1.5^2 + 1.5^2$
$XY^2 = 4.5$
$XY = 2.12\,\text{cm}$

**2** In the cuboid ABCDEFGH, AB is 12 cm, AD = 9 cm and BF = 5 cm.

**a** **i)** Calculate the length AC.

**ii)** Calculate the length BE.

**iii)** Calculate the length HB.

**b** X is the midpoint of CG and Y is the midpoint of GH.

**i)** Calculate the length XY.

**ii)** Calculate the length XA.

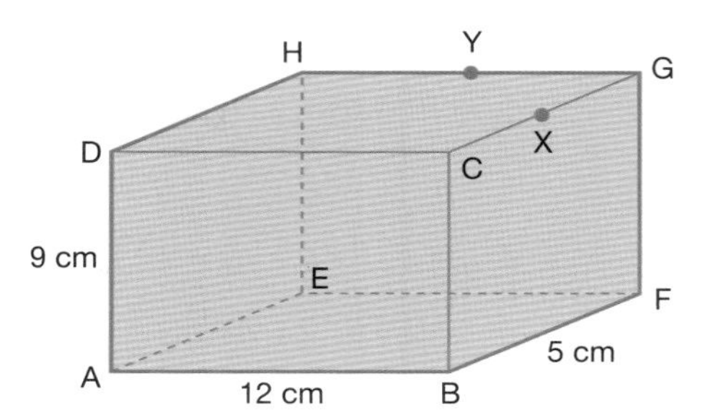


**a** **i)** To calculate AC, consider triangle ABC.
Triangle ABC is right angled, so use Pythagoras'theorem:

$AC^2 = AB^2 + AC^2$
$AC^2 = 12^2 + 9^2$
$AC^2 = 144 + 81$
$AC^2 = 225$
$AC = 15\,\text{cm}$

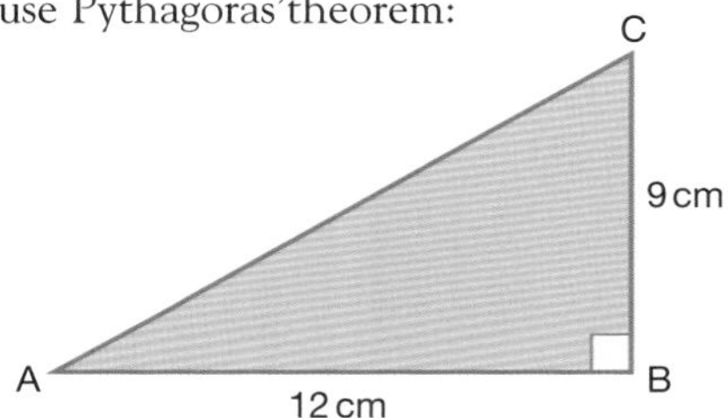


**ii)** To calculate BE, consider triangle ABE.
Triangle ABE is right angled, so use Pythagoras'theorem:

$BE^2 = AB^2 + BE^2$
$BE^2 = 12^2 + 5^2$
$BE^2 = 169$
$BE = 13\,\text{cm}$

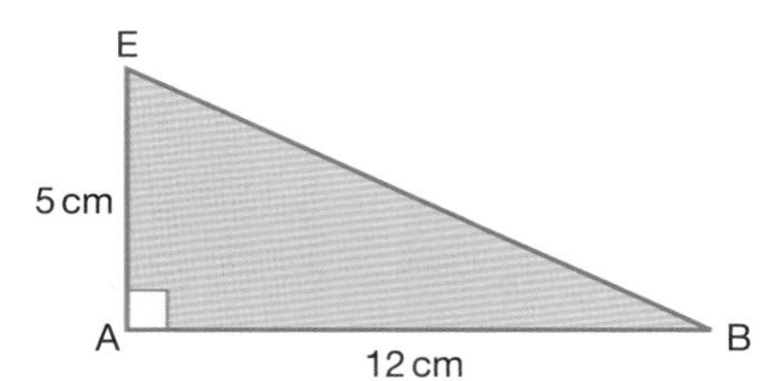


**iii)** To calculate HB, consider triangle EBH.
Triangle EBH is right angled, so use Pythagoras'theorem:

$HB^2 = EH^2 + BE^2$
$HB^2 = 9^2 + 13^2$
$HB^2 = 81 + 169$
$HB = 15.8\,\text{cm}$

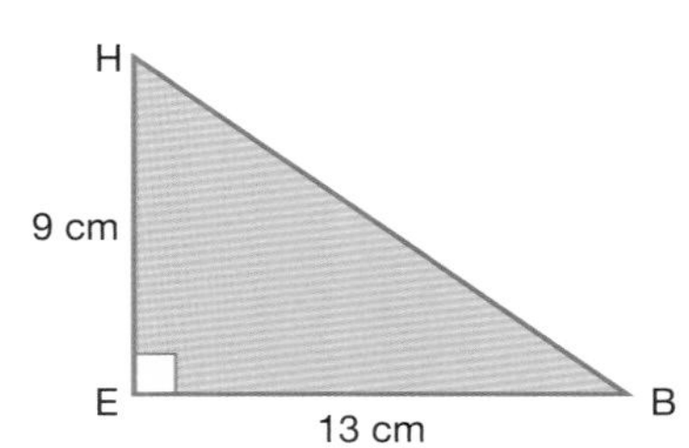

**b i)** To calculate XY, consider the triangle XYG with a right angle at G.

$XY^2 = XG^2 + YG^2$
$XY^2 = 2.5^2 + 6^2$
$XY^2 = 6.25 + 36$
$XY = \sqrt{42.25} = 6.5\,\text{cm}$

**ii)** To calculate XA consider the triangle AXZ.

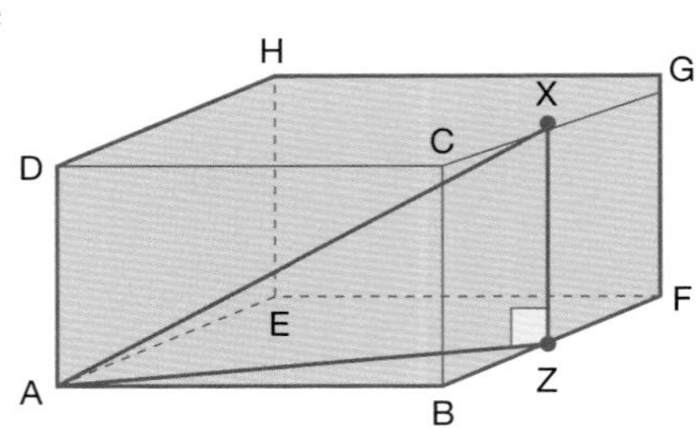


$XZ = 9\,\text{cm}$

Calculate AZ using triangle ABZ.

$AZ^2 = AB^2 + BZ^2$
$AZ^2 = 12^2 + 2.5^2$
$AZ^2 = 144 + 6.25$
$AZ = \sqrt{150.25} = 12.3\,\text{cm}$

Then find XA, using triangle AXZ.

$AX^2 = AZ^2 + XZ^2$
$AX^2 = 150.25 + 81$
$AX^2 = 231.25$
$AX = 15.2\,\text{cm}$

## Exercise 5.4.1

**Give all your answers to one decimal place.**

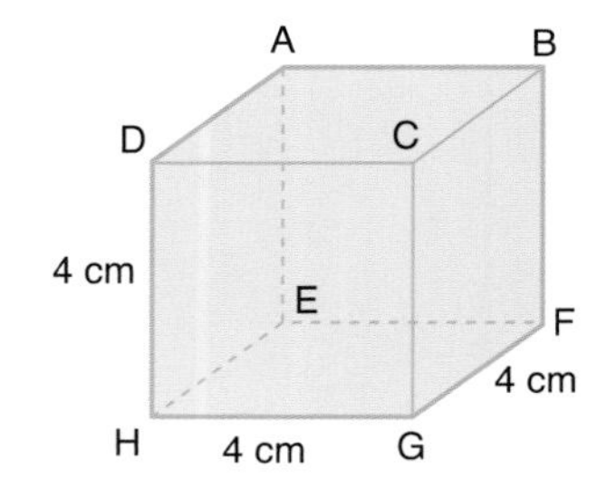


1 a Calculate the length HF.
  b Calculate the length of HB.
  c Calculate the angle BHG.
  d Calculate XY, where X and Y are the midpoints of HG and FG respectively.

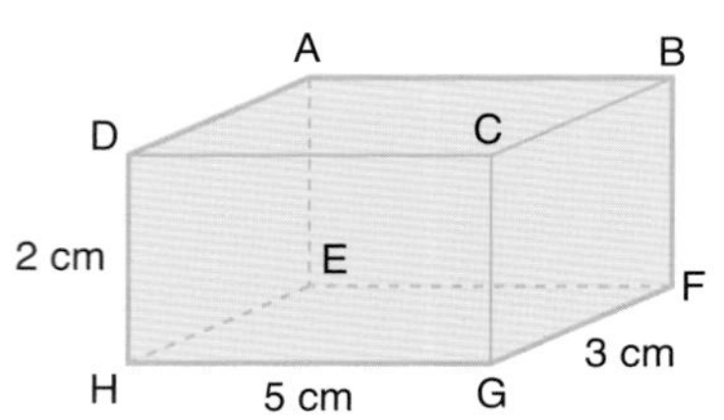


2 a Calculate the length CA.
  b Calculate the length CE.
  c Calculate the angle ACE.

3 a Calculate the length EG.
  b Calculate the length AG.
  c Calculate the angle AGE.

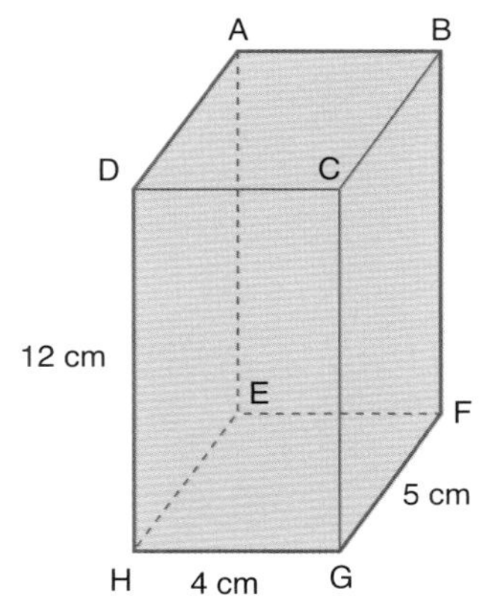


4 a Calculate the angle BCE.
  b Calculate the angle GFH.

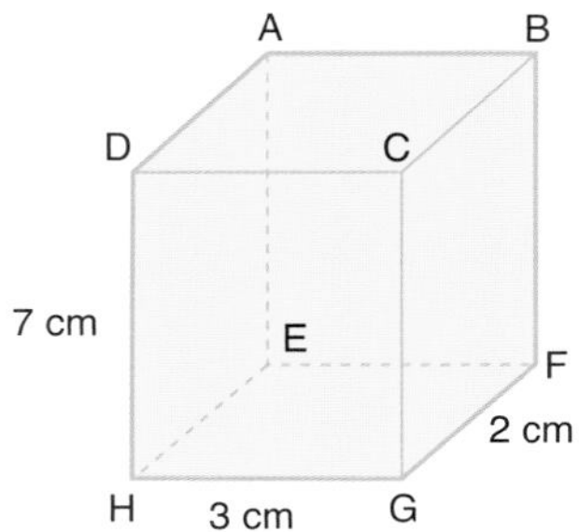


5 The diagram shows a right pyramid where A is vertically above X.
  a i) Calculate the length DB.
    ii) Calculate the angle DAX.
  b i) Calculate the angle CED.
    ii) Calculate the angle DBA.

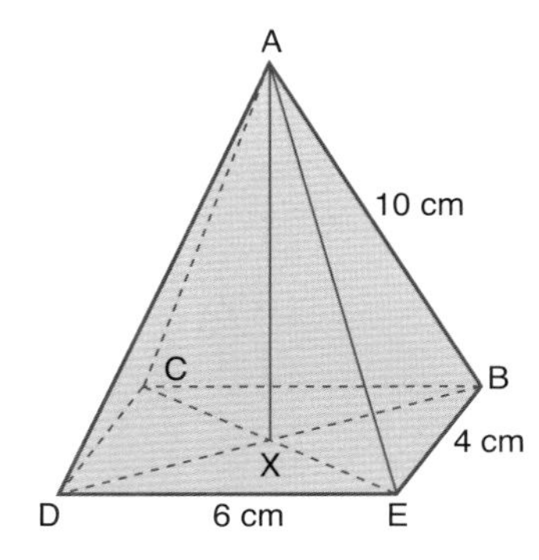


6 The diagram shows a right pyramid where A is vertically above X.
  a i) Calculate the length CE.
    ii) Calculate the angle CAX.
  b i) Calculate the angle BDE.
    ii) Calculate the angle ADB.

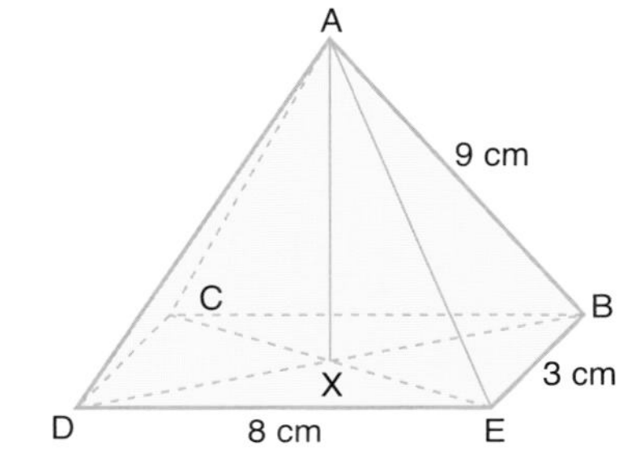

**7** In this cone the angle YXZ = 60°.

  **a** Calculate the length XY.
  **b** Calculate the length YZ.
  **c** Calculate the circumference of the base.

**8** In this cone, the angle XZY = 40°.

  **a** Calculate the length XZ
  **b** Calculate the length XY.

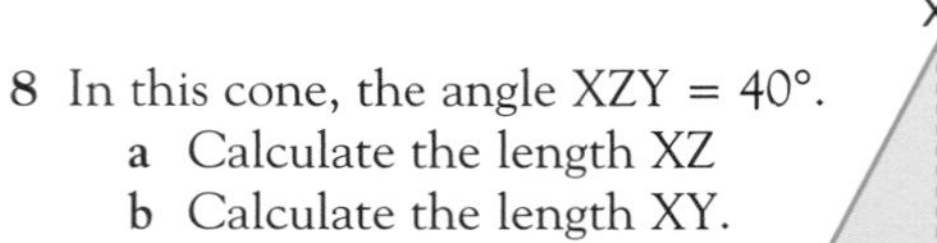

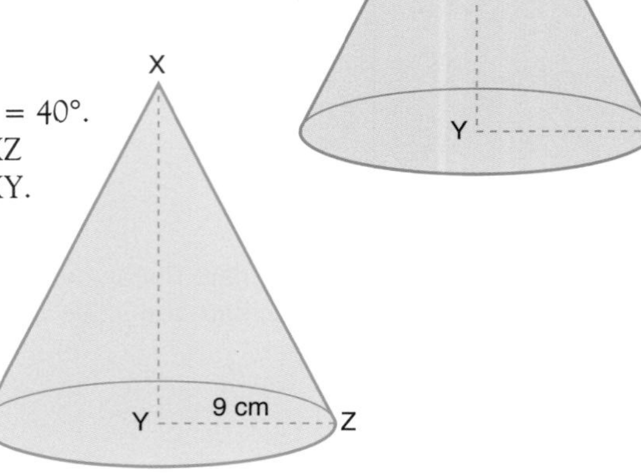


**9** In the diagram, RS = 19.2 cm, SP = 16 cm and TQ = 7.2 cm. X is the midpoint of VW. Y is the midpoint of TW.

Calculate the following, drawing diagrams of right-angled triangles to help you.

  **a** PR  **b** RV
  **c** PW  **d** XY
  **e** SY

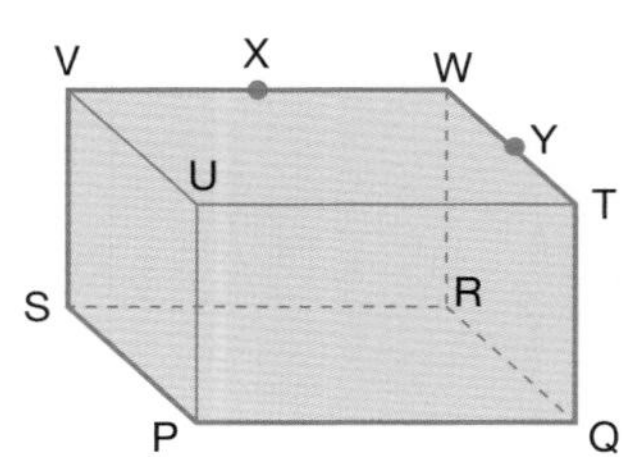


**10** One corner of this cuboid has been sliced off along the plane QTU.

WU = 4 cm.

  **a** Calculate the length of the three sides of the triangle QTU.
  **b** Calculate the three angles Q, T and U in triangle QTU.
  **c** Calculate the area of triangle QTU.

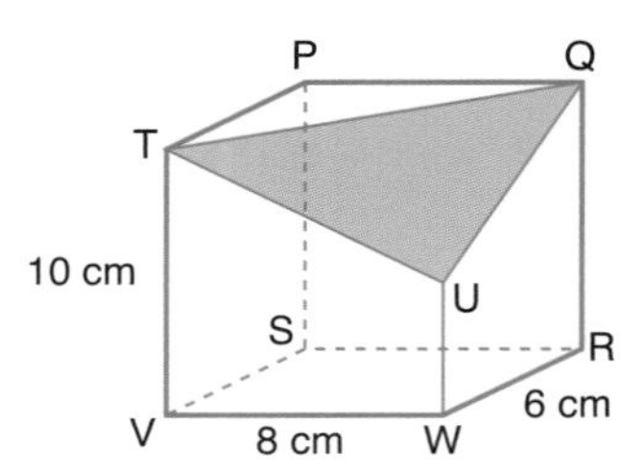


## The angle between a line and a plane

To calculate the size of the angle between the line AB and the shaded plane, drop a perpendicular from B. It meets the shaded plane at C. Then join AC.

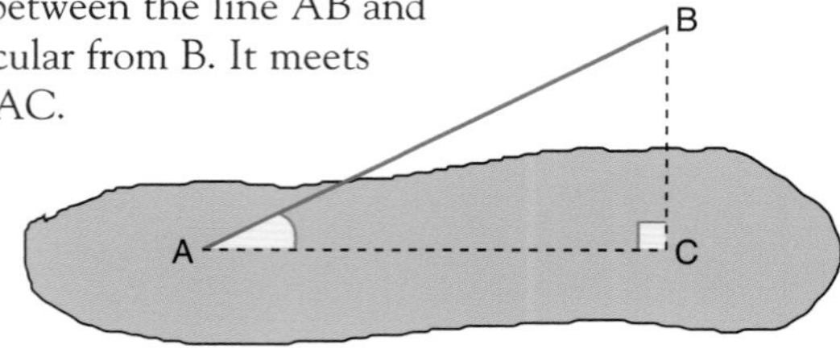

The angle between the lines AB and AC represents the angle between the line AB and the shaded plane.

The line AC is the **projection** of the line AB on the shaded plane.

**Worked example**

ABCDEFGH is a cuboid.

**a** Calculate the length CE.

**b** Calculate the angle between the line CE and the plane ADHE.

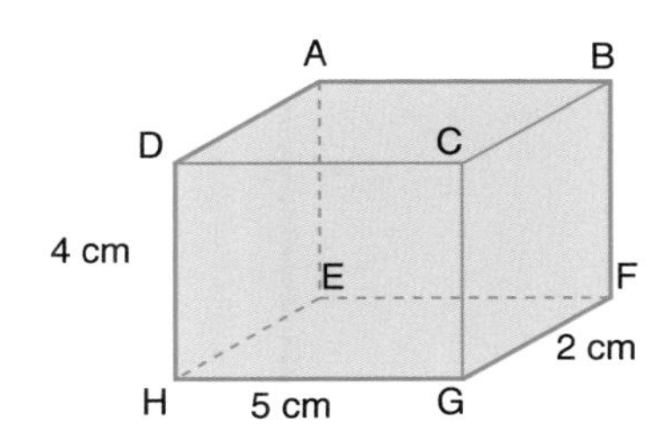


**a** First use Pythagoras' theorem to calculate the length EG:

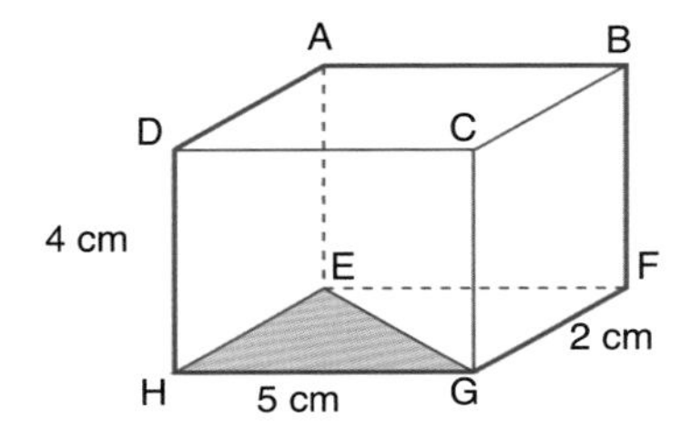


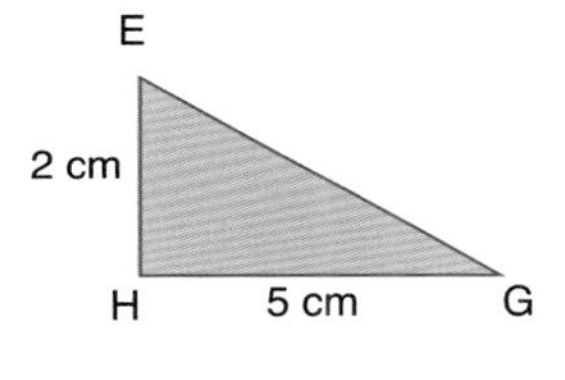


$EG^2 = EH^2 + HG^2$

$EG^2 = 2^2 + 5^2$

$EG^2 = 29$

$EG = \sqrt{29}$ cm

Now use Pythagoras' theorem to calculate CE:

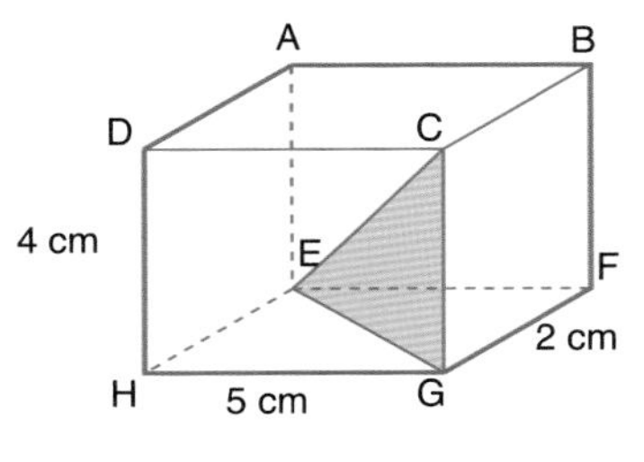


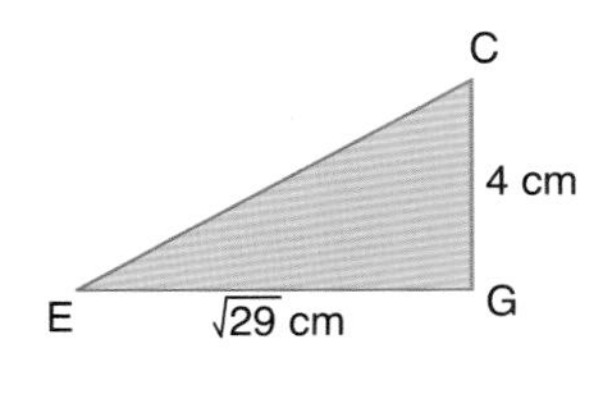


$CE^2 = EG^2 + CG^2$

$CE^2 = \left(\sqrt{29}\right)^2 + 4^2$

$CE^2 = 29 + 16$

$CE = \sqrt{45}$ cm

$CE = 6.71$ cm (3 s.f.)

b

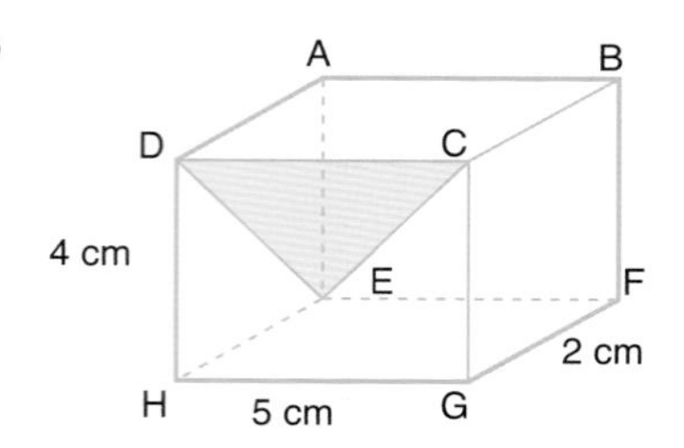


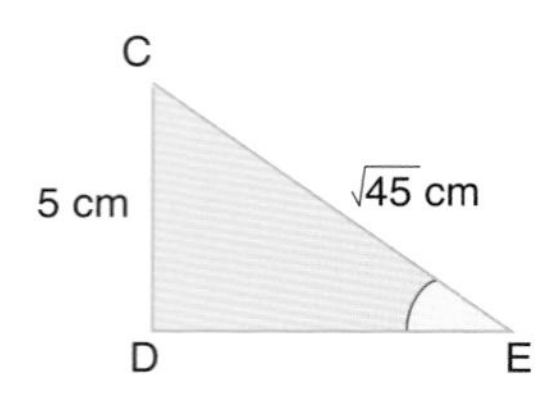


To calculate the angle between the line CE and the plane ADHE use the right-angled triangle CED and calculate the angle CED.

$$\sin E = \frac{CD}{CE}$$

$$\sin E = \frac{5}{\sqrt{45}}$$

$$E = \sin^{-1}\left(\frac{5}{\sqrt{45}}\right)$$

$$E = 48.2° \text{ (3 s.f.)}$$

## Exercise 5.4.2

1 Write down the projection of each line onto the given plane.

a TR onto RSWV
b TR onto PQUT
c SU onto PQRS
d SU onto TUVW
e PV onto QRVU
f PV onto RSWV

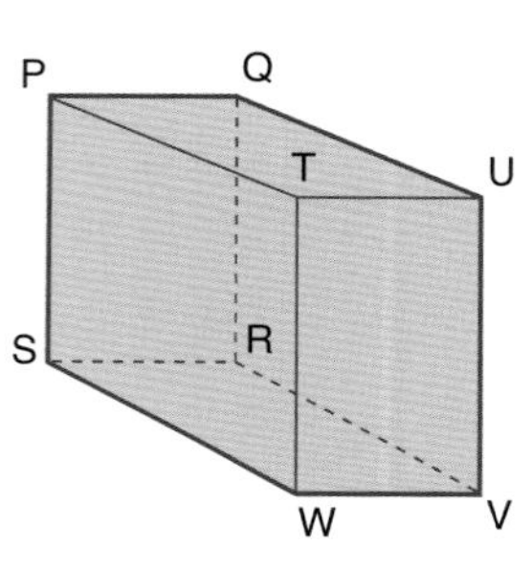


2 Write down the projection of each line onto the given plane.

a KM onto IJNM
b KM onto JKON
c KM onto HIML
d IO onto HLOK
e IO onto JKON
f IO onto LMNO

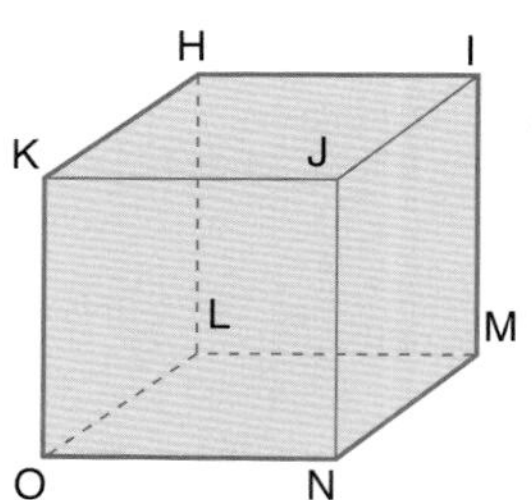


3 Write down the angle between the given line and plane.

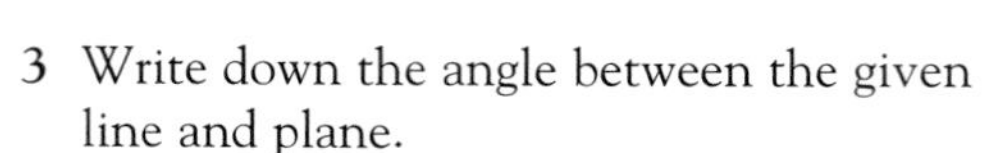

a PT and PQRS
b PU and PQRS
c SV and PSWT
d RT and TUVW
e SU and QRVU
f PV and PSWT

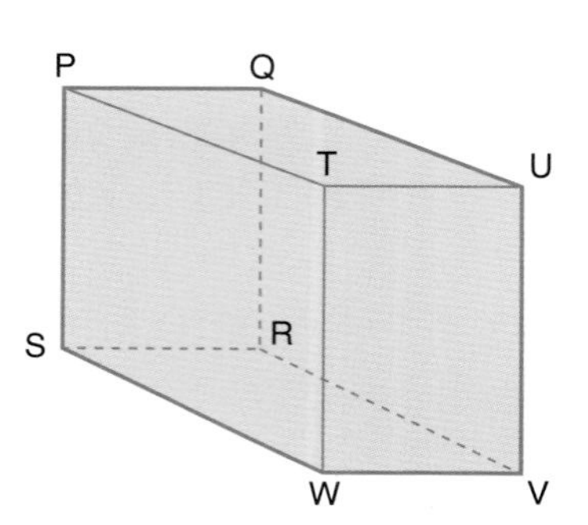

6 Two cubes are placed next to each other. The length of each of the edges of the larger cube is 4 cm. If the ratio of their surface areas is 1 : 4, calculate:
  a the surface area of the small cube
  b the length of an edge of the small cube.

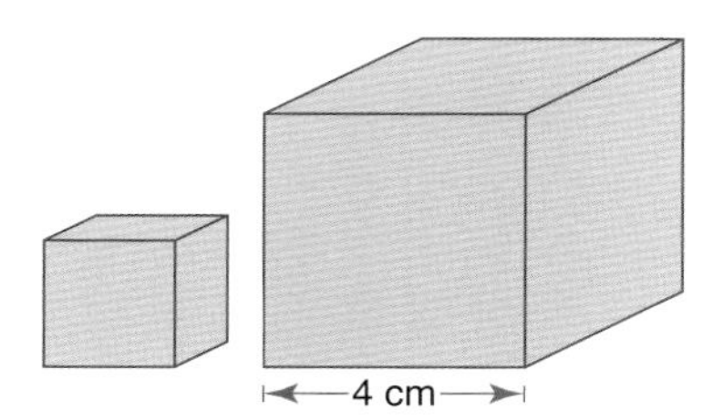


7 A cube and a cylinder have the same surface area. If the cube has an edge length of 6 cm and the cylinder a radius of 2 cm calculate:
  a the surface area of the cube
  b the height of the cylinder
  c the difference between the volumes of the cube and the cylinder.

8 Two cylinders have the same surface area. The shorter of the two has a radius of 3 cm and a height of 2 cm, and the taller cylinder has a radius of 1 cm. Calculate:
  a the surface area of (one of) the cylinders
  b the height of the taller cylinder
  c the difference between the volumes of the two cylinders.

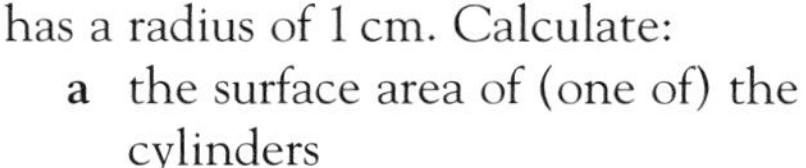


9 Two cuboids have the same surface area. The dimensions of one of the cuboids are

length = 3 cm, width = 4 cm and height = 2 cm.

Calculate the height of the other cuboid if its length is 1 cm and its width is 4 cm.

## Volume and surface area of a sphere

The volume of a sphere is given by the following formula:

Volume of sphere $= \frac{4}{3}\pi r^3$

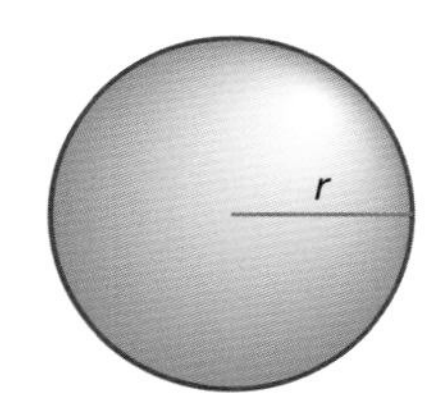


The surface area of a sphere is given by the following formula:

Surface area of sphere $= 4\pi r^2$

**Worked example**

1 Calculate the volume and surface area of the sphere shown, giving your answers to one decimal place.

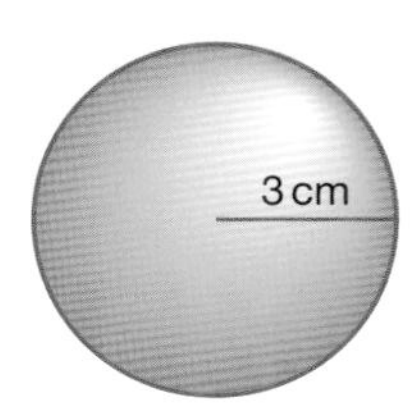

Volume of sphere $= \frac{4}{3}\pi r^3$

$= \frac{4}{3} \times \pi \times 3^3$

$= 113.1$

The volume is 113.1 cm$^3$.

Surface area of sphere $= 4\pi r^2$

$= 4 \times \pi \times 3^2$

$= 113.1$

The surface area is 113.1 cm$^2$.

2 Given that the volume of a sphere is 150 cm$^3$, calculate its radius to one decimal place.

$V = \frac{4}{3}\pi r^3$

$r^3 = \frac{3V}{4\pi}$

$r^3 = \frac{3 \times 150}{4 \times \pi} = 35.8$

$r = \sqrt[3]{35.8} = 3.3$

The radius is 3.3 cm.

## Exercise 5.5.2

1 Calculate the volume and surface area of each of the following spheres. The radius $r$ is given in each case.

**a** $r = 6$ cm  **b** $r = 9.5$ cm  **c** $r = 8.2$ cm  **d** $r = 0.7$ cm

2 Calculate the radius of each of the following spheres. Give your answers in centimetres and to one decimal place. The volume $V$ is given in each case.

**a** $V = 720$ cm$^3$  **b** $V = 0.2$ m$^3$

3 Calculate the radius of each of the following spheres, given the surface area in each case.

**a** $A = 16.5$ cm$^2$  **b** $A = 120$ mm$^2$

4 Given that sphere B has twice the volume of sphere A, calculate the radius of sphere B.

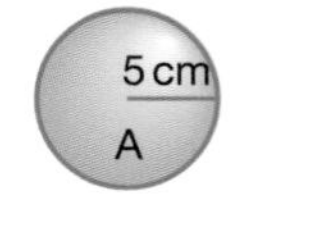


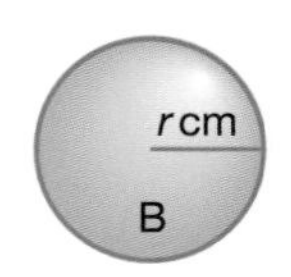


5 Calculate the volume of material used to make the hemispherical bowl shown, if the inner radius of the bowl is 5 cm and its outer radius 5.5 cm.

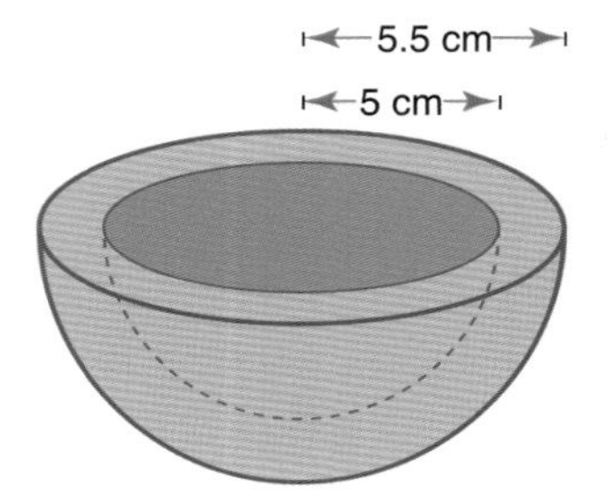

**6** The volumes of the material used to make the sphere and hemispherical bowl shown are the same. Given that the radius of the sphere is 7 cm and the inner radius of the bowl is 10 cm, calculate the outer radius $r$ cm of the bowl.

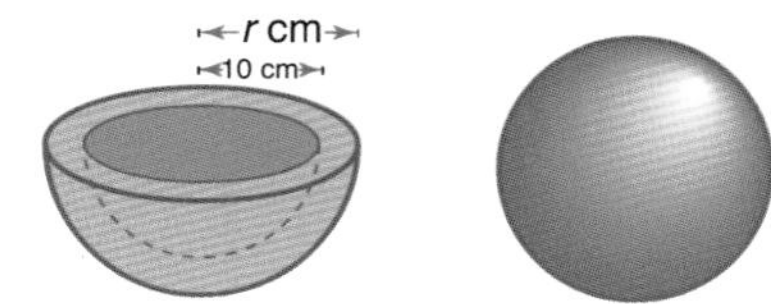


**7** A ball is placed inside a box into which it will fit tightly. If the radius of the ball is 10 cm, calculate:

- **a** the volume of the ball
- **b** the volume of the box
- **c** the percentage volume of the box not occupied by the ball.

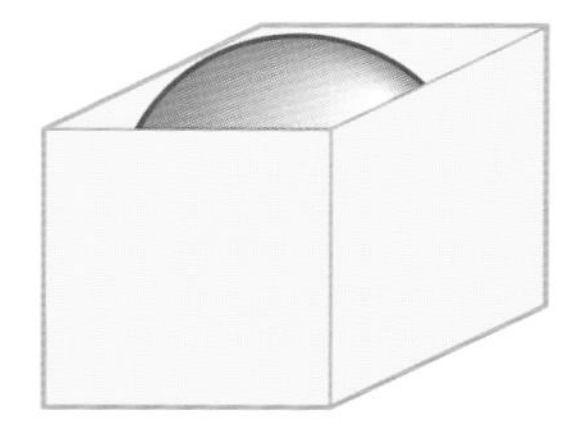

**8** A steel ball is melted down to make eight smaller identical balls. If the radius of the original steel ball was 20 cm, calculate to the nearest millimetre the radius of each of the smaller balls.

**9** A steel ball of volume 600 cm$^3$ is melted down and made into three smaller balls A, B and C. If the volumes of A, B and C are in the ratio 7 : 5 : 3, calculate to one decimal place the radius of each of A, B and C.

**10** The cylinder and sphere shown have the same radius and the same height. Calculate the ratio of their volumes, giving your answer in the form: volume of cylinder : volume of sphere.

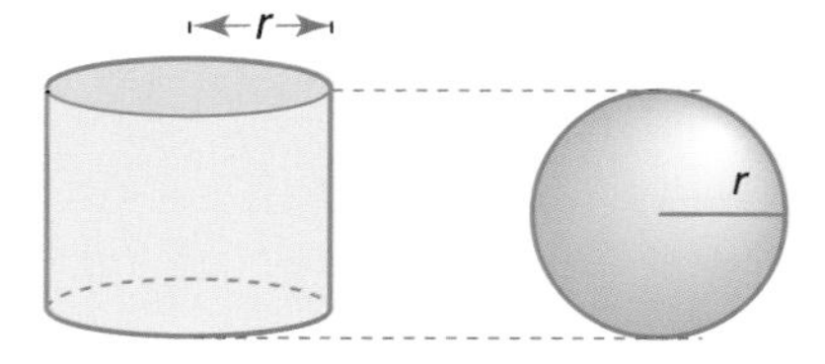


**11** Sphere A has a radius of 8 cm and sphere B has a radius of 16 cm. Calculate the ratio of their surface areas in the form 1 : $n$.

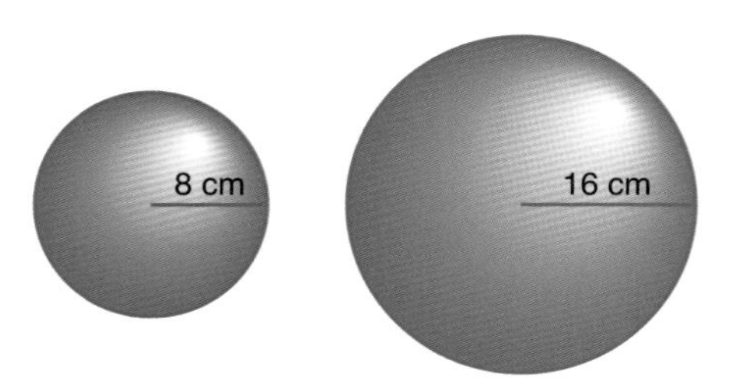

12 A hemisphere of diameter 10 cm is attached to a cylinder of equal diameter as shown.

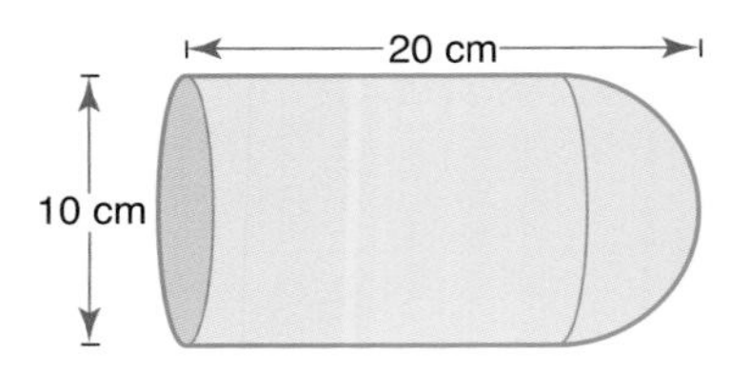


If the total length of the shape is 20 cm, calculate:

a the surface area of the hemisphere
b the length of the cylinder
c the surface area of the whole shape.

13 A sphere and a cylinder both have the same surface area and the same height of 16 cm.

Calculate, to one decimal place:

a the surface area of the sphere
b the radius of the cylinder.

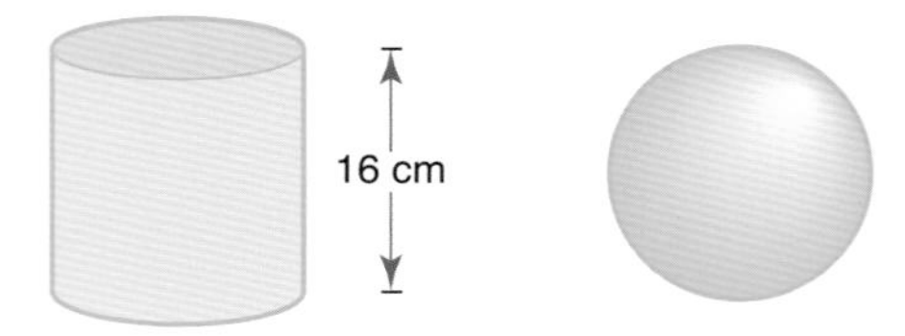


## Volume and surface area of a pyramid

A **pyramid** is a three-dimensional shape. Each of the faces of a pyramid is planar (not curved). A pyramid has a polygon for its base and the other faces are triangles with a common vertex, known as the **apex**. A pyramid's individual name is taken from the shape of the base.

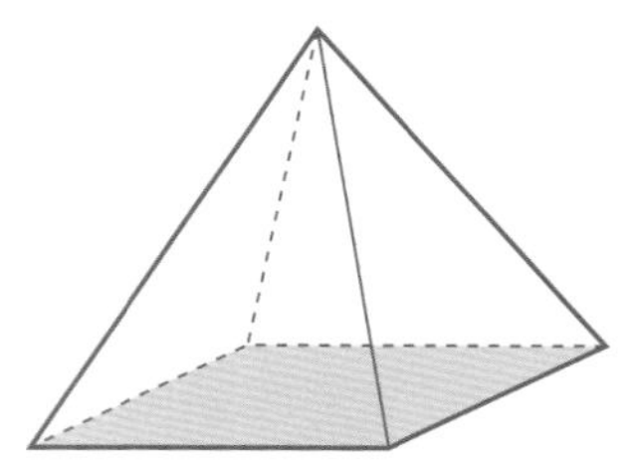

Square-based pyramid

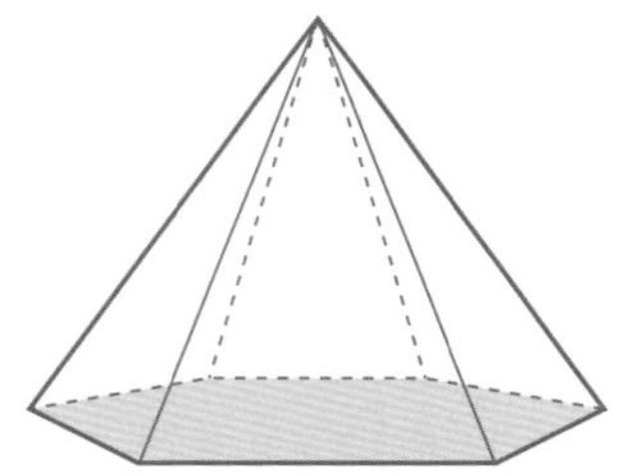

Hexagonal-based pyramid

The volume of any pyramid is given by the following formula:

Volume of a pyramid $= \frac{1}{3} \times$ area of base $\times$ perpendicular height

The surface area of a pyramid is found simply by adding together the areas of all of its faces. You may need to use Pythagoras' theorem to work out the dimensions you need to calculate the volume and surface area.

Before we look at examples of finding the volume and surface area of a cone it is useful to look at how a cone is formed. A cone can be constructed from a sector of a circle.

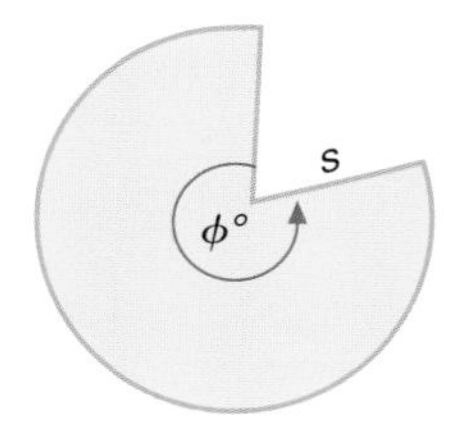


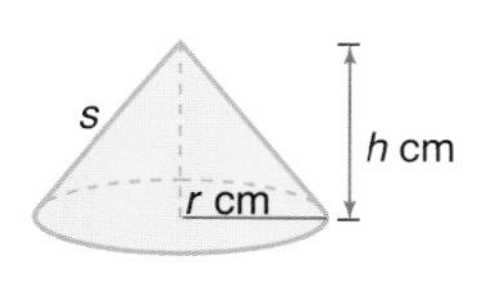


The length of the sloping side of the cone is equal to the radius of the sector. The base circumference is equal to the arc length of the sector. The curved surface area of the cone is equal to the area of the sector.

## Arc length and sector area

An **arc** is part of the circumference of a circle between two radii. Its length is proportional to the size of the angle $\phi$ between the two radii. The length of the arc as a fraction of the circumference of the whole circle is therefore equal to the fraction that $\phi$ is of 360°.

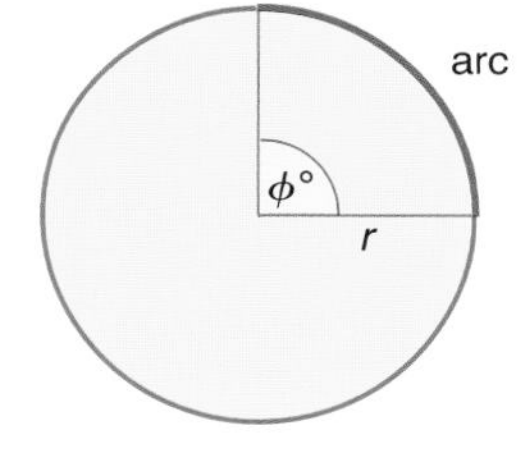


$$\text{Arc length} = \frac{\phi}{360} \times 2\pi r$$

A **sector** is the region of a circle enclosed by two radii and an arc. The area of a sector is proportional to the size of the angle $\phi$ between the two radii. As a fraction of the area of the whole circle, it is therefore equal to the fraction that $\phi$ is of 360°.

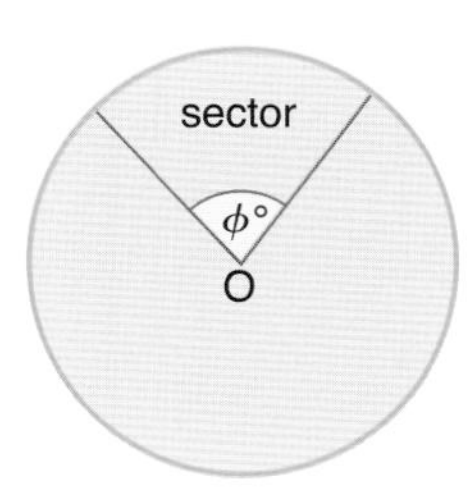


$$\text{Area of sector} = \frac{\phi}{360} \times \pi r^2$$

## Calculating the volume of a cone

As we have seen, the formula for the volume of a cone is:

$$\text{Volume} = \tfrac{1}{3} \times \text{base area} \times \text{height}$$
$$= \tfrac{1}{3}\pi r^2 h$$

**Worked examples**

1 Calculate the volume of the cone.

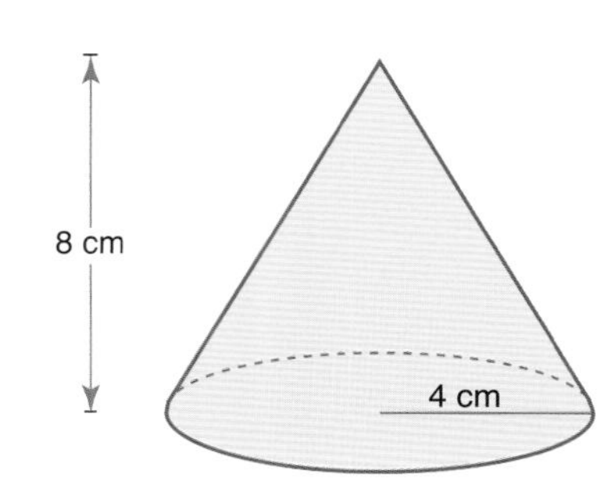


$$\text{Volume} = \tfrac{1}{3}\pi r^2 h$$
$$= \tfrac{1}{3} \times \pi \times 4^2 \times 8$$
$$= 134.0 \text{ (1 d.p.)}$$

The volume is $134\,\text{cm}^3$

**2** The sector below is assembled to form a cone as shown.

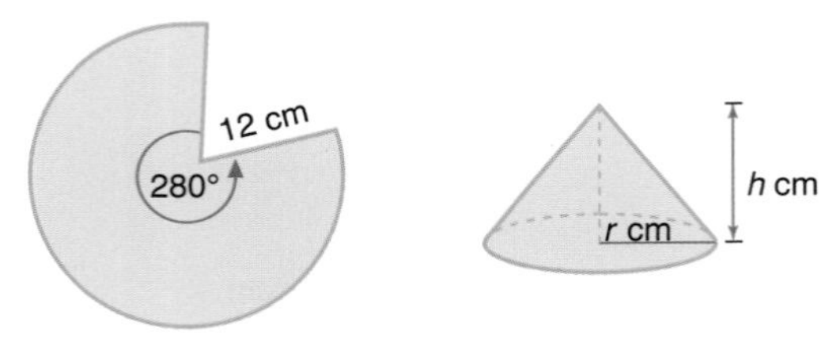


**a** Calculate the base circumference of the cone.
**b** Calculate the base radius of the cone.
**c** Calculate the vertical height of the cone.
**d** Calculate the volume of the cone.

**a** The base circumference of the cone is equal to the arc length of the sector.

$$\text{sector arc length} = \frac{\phi}{360} \times 2\pi r$$
$$= \frac{280}{360} \times 2\pi \times 12 = 58.6$$

So the base circumference is 58.6 cm.

**b** The base of a cone is circular, therefore:

$$C = 2\pi r$$
$$r = \frac{C}{2\pi} = \frac{58.6}{2\pi}$$
$$= 9.33$$

So the radius is 9.33 cm.

**c** The vertical height of the cone can be calculated using Pythagoras' theorem on the right-angled triangle enclosed by the base radius, vertical height and the sloping face, as shown below.

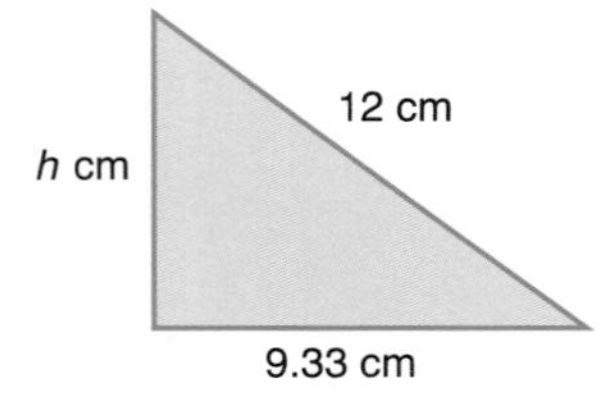


Note that the length of the sloping side is equal to the radius of the sector.

$$12^2 = h^2 + 9.33^2$$
$$h^2 = 12^2 - 9.33^2$$
$$h^2 = 56.9$$
$$h = 7.54$$

So the height is 7.54 cm.

**d** $\text{Volume} = \frac{1}{3} \times \pi r^2 h$

$$= \frac{1}{3} \times \pi \times 9.33^2 \times 7.54 = 688$$

So the volume is 688 cm$^3$.

## Student assessment 6

**Give all your answers to one decimal place.**

1 A sphere has a radius of 6.5 cm. Calculate:
  a its total surface area
  b its volume.

2 A pyramid with a base the shape of a regular hexagon is shown. If the length of each of its sloping edges is 24 cm, calculate:
  a its total surface area
  b its volume.

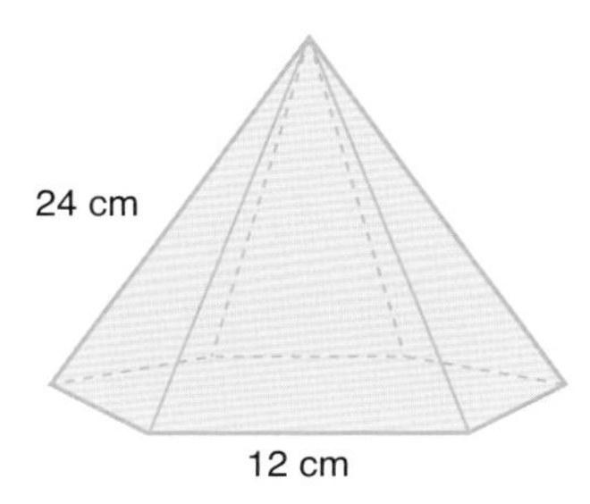


3 The prism here has a cross-sectional area in the shape of a sector.

Calculate:
  a the radius $r$ cm
  b the cross-sectional area of the prism
  c the total surface area of the prism
  d the volume of the prism.

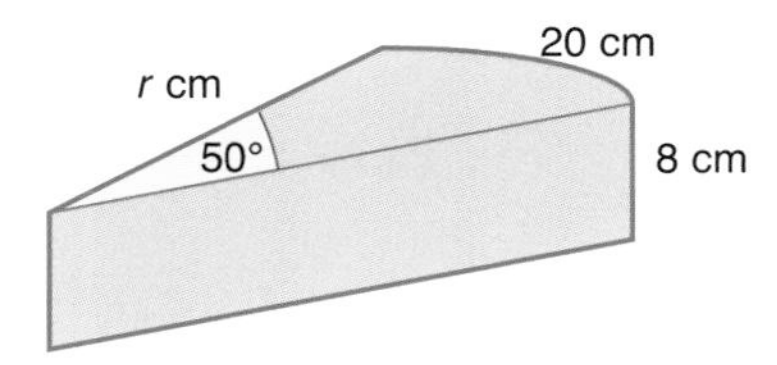


4 The cone and sphere shown here have the same volume.

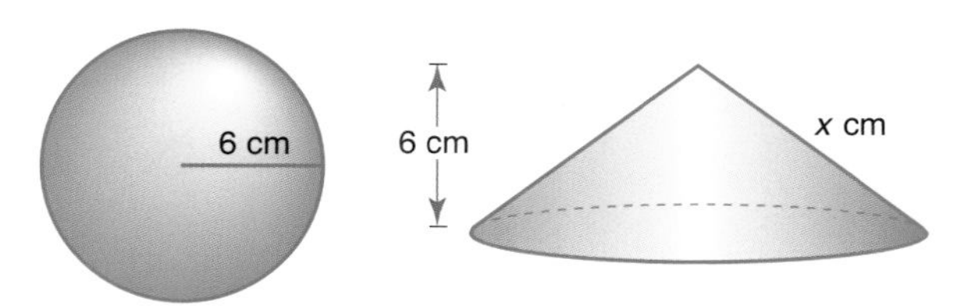


If the radius of the sphere and the height of the cone are both 6 cm, calculate:
  a the volume of the sphere
  b the base radius of the cone
  c the slant height $x$ cm
  d the surface area of the cone.

5 The top of a cone is cut off and a cylindrical hole is drilled out of the remaining truncated cone as shown.

Calculate:
  a the height of the original cone
  b the volume of the original cone
  c the volume of the solid truncated cone
  d the volume of the cylindrical hole
  e the volume of the remaining truncated cone.

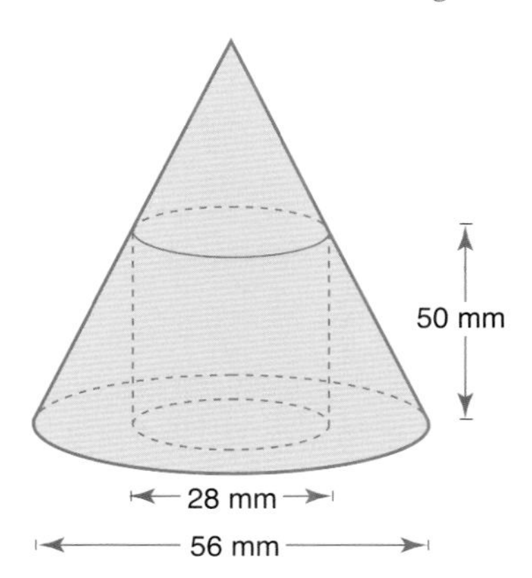

# Examination questions

1 The following diagrams show six lines with equations of the form $y = mx + c$.

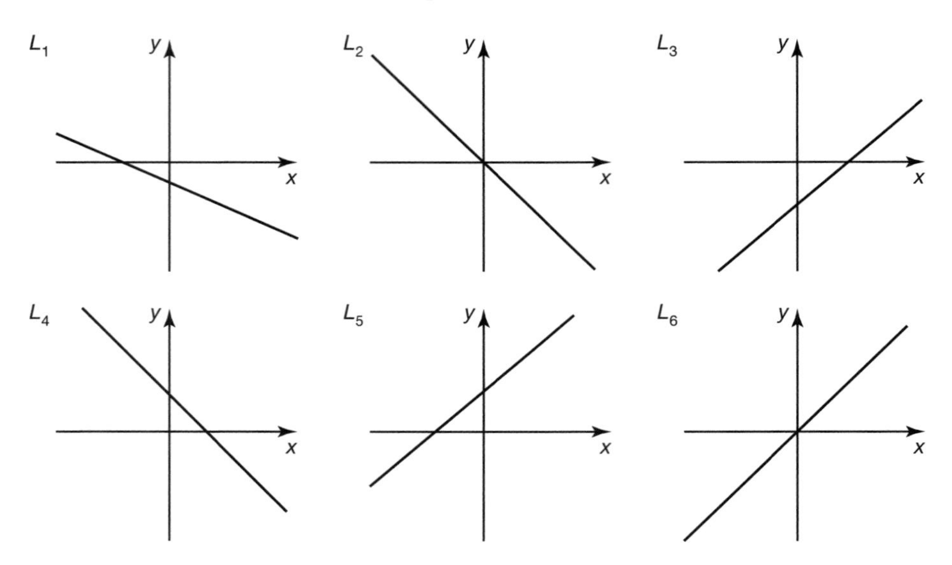


In the table below there are four possible conditions for the pair of values $m$ and $c$.
Match each of the given conditions with one of the lines drawn above.

| Condition | Line |
|---|---|
| $m > 0$ and $c > 0$ | |
| $m < 0$ and $c > 0$ | |
| $m < 0$ and $c < 0$ | |
| $m > 0$ and $c < 0$ | |

[6]

**Paper 1, Nov 10, Q4**

2 The base of a prism is a **regular hexagon**. The centre of the hexagon is O and the length of OA is 15 cm.

a Write down the size of angle AOB. [1]

b Find the area of triangle AOB. [3]

The height of the prism is 20 cm.

c Find the volume of the prism. [2]

**Paper 1, Nov 10, Q12**

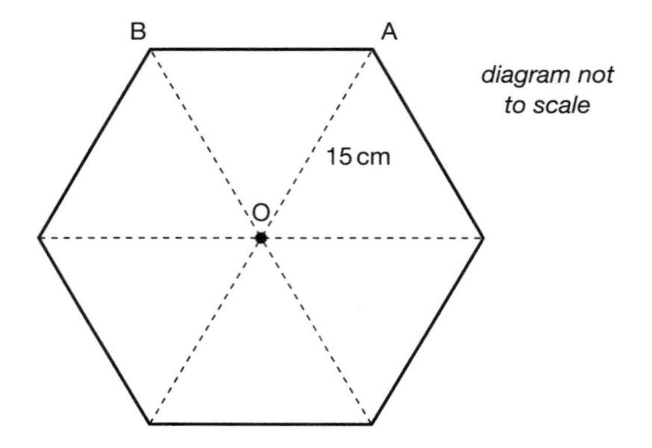

3 A room is in the shape of a cuboid. Its floor measures 7.2 m by 9.6 m and its height is 3.5 m.

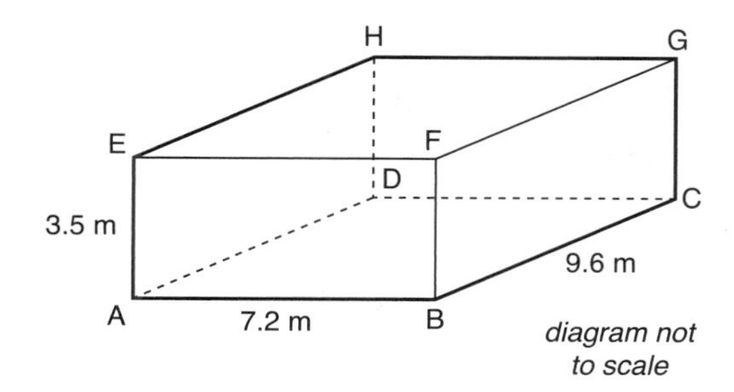


a Calculate the length of AC. [2]
b Calculate the length of AG. [2]
c Calculate the angle that AG makes with the floor. [2]

**Paper 1, May 11, Q7**

4 **Part A**

The diagram below shows a square based right pyramid. ABCD is a square of side 10 cm. VX is the perpendicular height of 8 cm. M is the midpoint of BC.

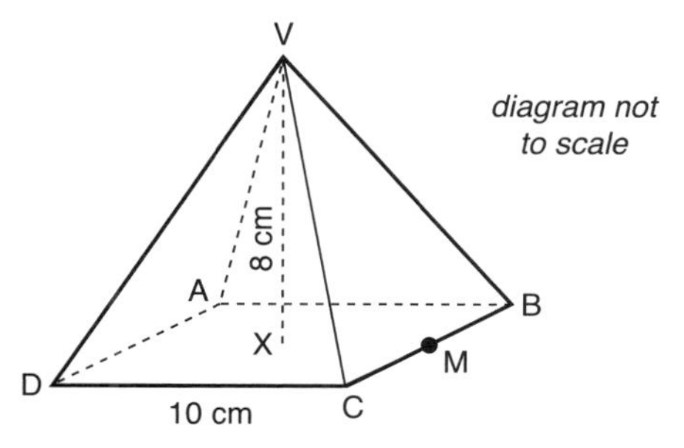


a Write down the length of XM. [1]
b Calculate the angle of VM. [2]
c Calculate the angle between VM and ABCD. [2]

**Part B**

A path goes around a forest so that it forms the three sides of a triangle. The lengths of two sides are 550 m and 290 m. These two sides meet at an angle of 115°. A diagram is shown below.

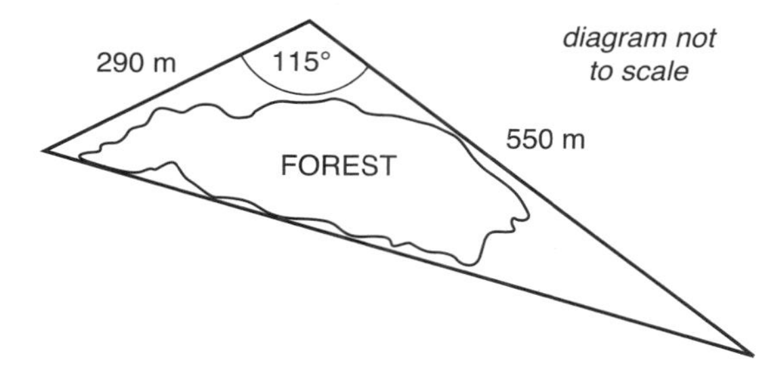


a Calculate the length of the third side of the triangle. Give your answer correct to the nearest 10 m. [4]
b Calculate the area enclosed by the path that goes around the forest. [3]

Inside the forest a second path forms the three sides of another triangle named ABC. Angle BAC is 53°, AC is 180 m and BC is 230 m.

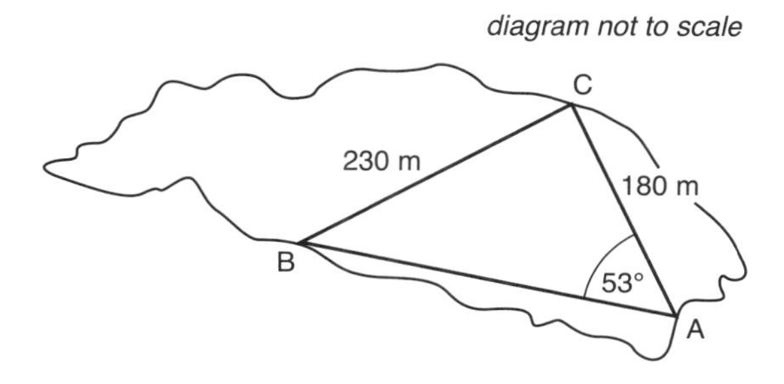


c Calculate the size of angle ACB. [4]

**Paper 2, Nov 09, Q1**

Topic 5

# Applications, project ideas and theory of knowledge

**1** *The human brain has a longitudinal fissure (a long crack), which separates it into two hemispheres. The function of each hemisphere is different. Popular psychology talks of 'left brain, right brain thinking' where one side or another is responsible for logic, creativity, etc. Find out more about this idea and discuss its scientific basis.*

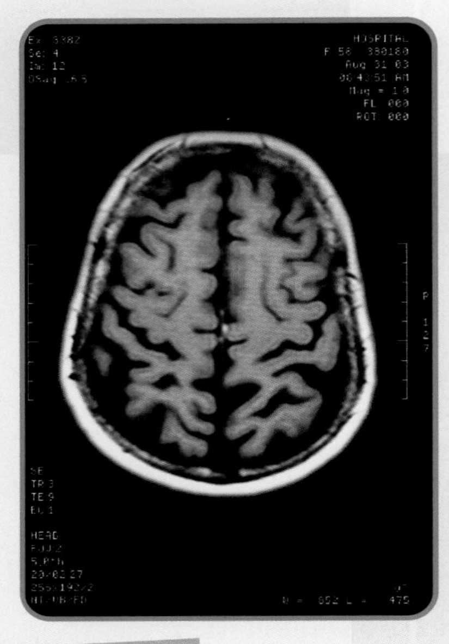

**2** *Pythagoras' theorem has a large number of proofs. These include Garfield's proof, proof using similar triangles, Euclid's proof and others. These types of proof could form the basis of a project. No proof that uses trigonometry is considered valid. Why do you think this is?*

**3** The terms arithmetic, algebra and geometry used to be studied in schools as separate subjects (and still are in parts of the USA). Discuss the statement that these terms are becoming redundant. What you think are now the most important areas of mathematics and should they, or can they, be studied in isolation?

**4** Euclidean geometry is an axiomatic system. Discuss what this means and discover the main axioms of Euclidean geometry. Einstein's Theory of General Relativity maintains that space–time is non-Euclidean. Find out more about non-Euclidean geometry.

**5** Investigate the mathematics of doodles like the one shown. This could be developed as a project.

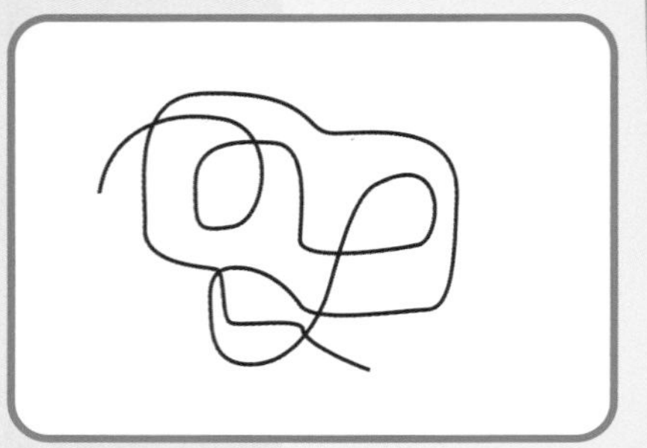

| **Autograph** | |
|---|---|
| Graph the equation $y = x^2 - 4x + 3$.<br>Select the curve then click 'Object' followed by 'Solve $f(x) = 0$'.<br>The points are marked on the graph and their coordinates can be displayed in the results box. This is accessed by selecting . | 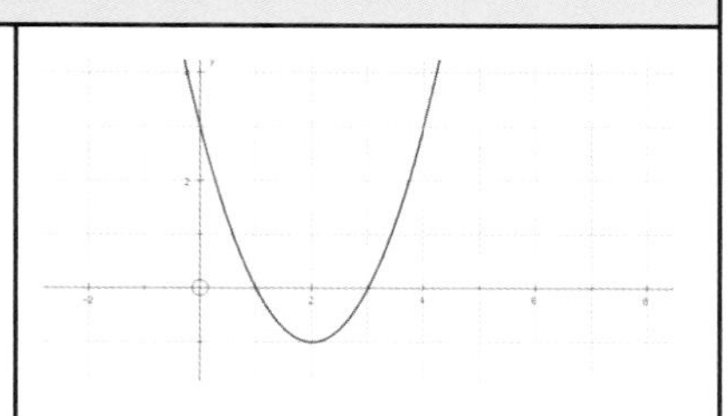 |
| **GeoGebra** | |
| Graph the equation $y = x^2 - 4x + 3$.<br>In the input box type 'Root[f]', this finds where the graph intersects the x-axis. The points are marked on the graph as A and B their coordinates are displayed in the algebra window. | 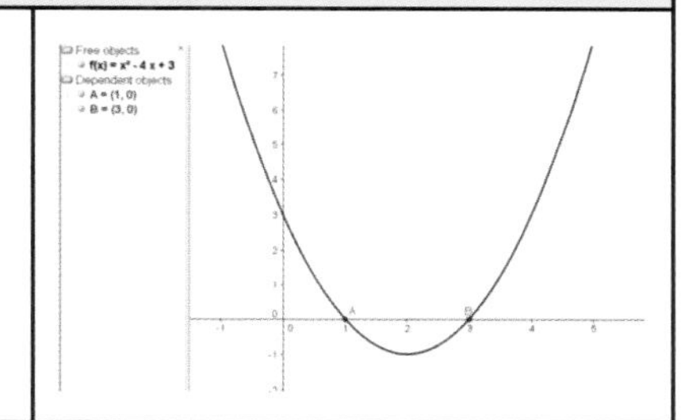 |

The number of roots will depend on the shape and position of the quadratic curve. The maximum number of real roots is two, as a quadratic cannot cross the $x$-axis more than twice. However, if the graph touches the $x$-axis (i.e. the $x$-axis is a tangent to the curve) then there is only one real (repeated) root. If the graph does not cross the $x$-axis then there are no real roots.

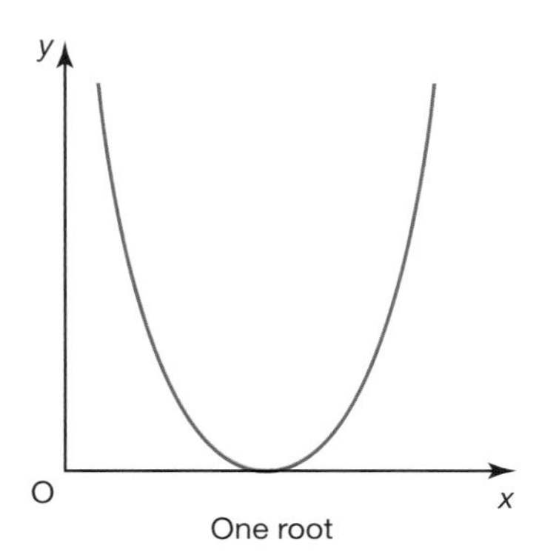


One root

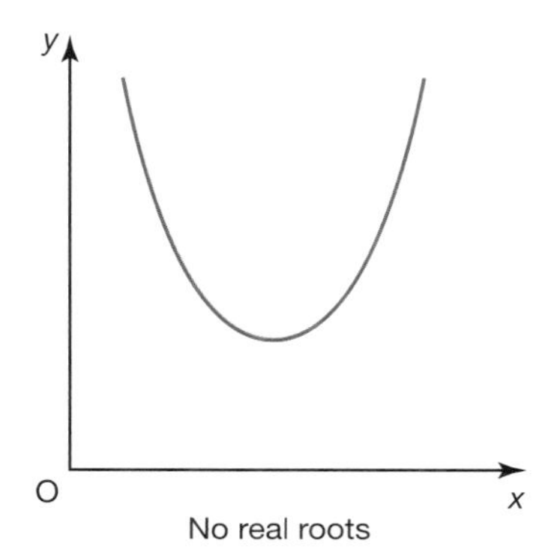


No real roots

## Exercise 6.3.2

Solve each of the quadratic equations below by plotting a graph of the function.

1 $x^2 - x - 6 = 0$

2 $-x^2 + 1 = 0$

3 $x^2 - 6x + 9 = 0$

4 $-x^2 - x - 2 = 0$

5 $x^2 - 4x + 4 = 0$

6 $2x^2 - 7x + 3 = 0$

7 $-2x^2 + 4x - 2 = 0$

8 $3x^2 - 5x - 2 = 0$

In the previous worked example, $y = x^2 - 4x + 3$, a solution could be found to the equation $x^2 - 4x + 3 = 0$ by reading off where the graph crossed the $x$-axis. This graph can, however, also be used to solve other quadratic equations.

**Worked example**

Use the graph of $y = x^2 - 4x + 3$ to solve the equation $x^2 - 4x + 1 = 0$.

$x^2 - 4x + 1 = 0$ can be rearranged to give:
$x^2 - 4x + 3 = 2$

Using the graph of $y = x^2 - 4x + 3$ and plotting the line $y = 2$ on the same graph gives the graph shown.

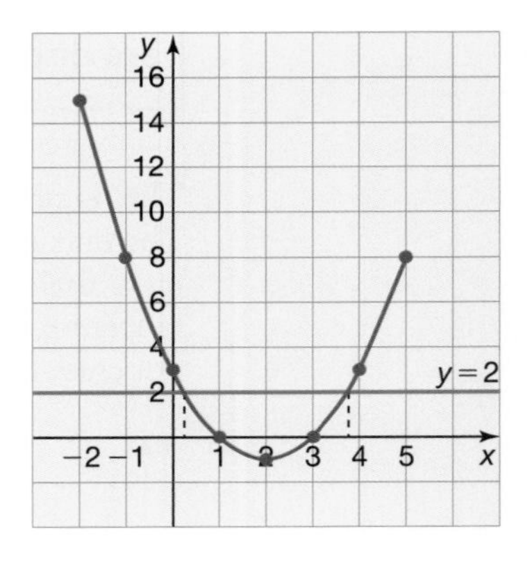


Where the curve and the line cross gives the solution to $x^2 - 4x + 3 = 2$ and hence the solution to $x^2 - 4x + 1 = 0$.

Therefore the solutions to $x^2 - 4x + 1 = 0$ are $x \approx 0.3$ and $3.7$.

## Exercise 6.3.3

Using the graphs that you drew for Exercise 6.3.2, solve the following quadratic equations. Show your method clearly.

1 $x^2 - x - 4 = 0$

2 $-x^2 - 1 = 0$

3 $x^2 - 6x + 8 = 0$

4 $-x^2 - x = 0$

5 $x^2 - 4x + 1 = 0$

6 $2x^2 - 7x = 0$

7 $-2x^2 + 4x = -1$

8 $3x^2 = 2 + 5x$

## Factorizing quadratic expressions

In order to solve quadratic equations algebraically, it is necessary to know how to factorize quadratic expressions.

For example, the quadratic expression $x^2 + 5x + 6$ can be factorized by writing it as a product of two brackets: $(x + 3)(x + 2)$. A method for factorizing quadratics is shown below.

**Worked examples**

1 Factorize $x^2 + 5x + 6$.

On setting up a 2 × 2 grid, some of the information can immediately be entered. As there is only one term in $x^2$, this can be entered, as can the constant +6. The only two values which multiply to give $x^2$ are $x$ and $x$. These too can be entered.

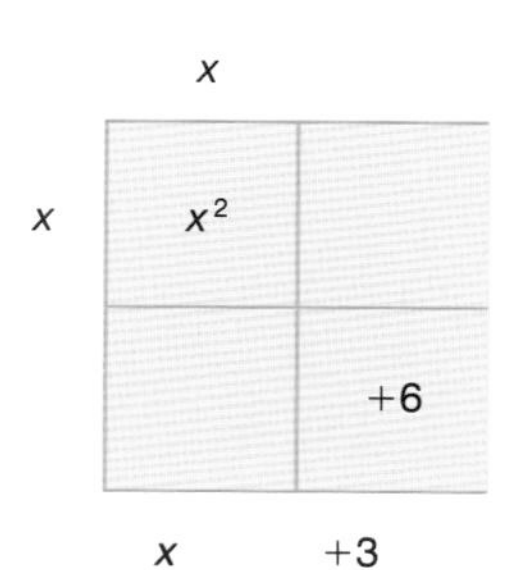


We now need to find two values which multiply to give +6 and which add to give +5. The only two values which satisfy both these conditions are +3 and +2. The grid can then be completed.

| | $x$ | $+3$ |
|---|---|---|
| $x$ | $x^2$ | $3x$ |
| $+2$ | $2x$ | $+6$ |

Therefore $x^2 + 5x + 6 = (x + 3)(x + 2)$.

2 Factorize $x^2 + 2x - 24$.

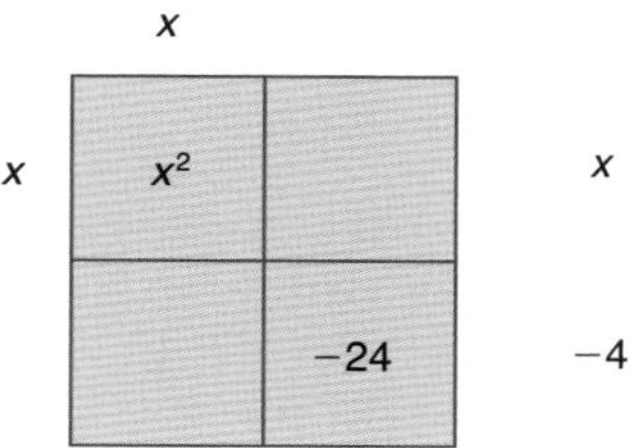


| | $x$ | $+6$ |
|---|---|---|
| $x$ | $x^2$ | $+6x$ |
| $-4$ | $-4x$ | $-24$ |

Therefore $x^2 + 2x - 24 = (x + 6)(x - 4)$.

3 Factorize $2x^2 + 11x + 12$.

| | $2x$ | |
|---|---|---|
| $x$ | $2x^2$ | |
| | | $12$ |

| | $2x$ | $+3$ |
|---|---|---|
| $x$ | $2x^2$ | $3x$ |
| $+4$ | $8x$ | $12$ |

Therefore $2x^2 + 11x + 12 = (2x + 3)(x + 4)$.

4 Factorize $x^2 - 10x + 25$.

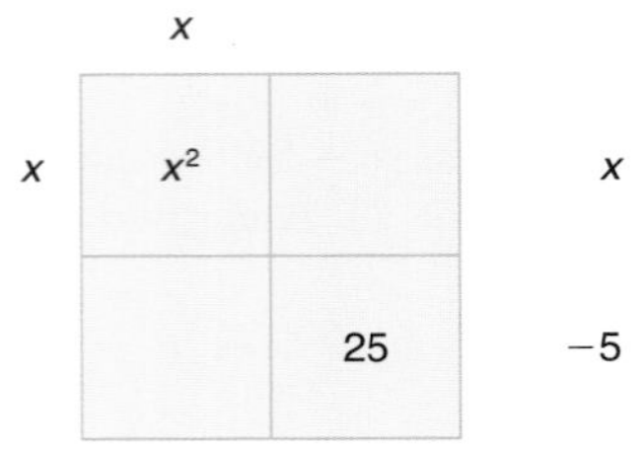


| | $x$ | $-5$ |
|---|---|---|
| $x$ | $x^2$ | $-5x$ |
| $-5$ | $-5x$ | 25 |

Therefore $x^2 - 10x + 25 = (x - 5)(x - 5) = (x - 5)^2$.

## Exercise 6.3.4

Factorize the following quadratic expressions.

1 a $x^2 + 7x + 12$ b $x^2 + 8x + 12$ c $x^2 + 13x + 12$
d $x^2 - 7x + 12$ e $x^2 - 8x + 12$ f $x^2 - 13x + 12$

2 a $x^2 + 6x + 5$ b $x^2 + 6x + 8$ c $x^2 + 6x + 9$
d $x^2 + 10x + 25$ e $x^2 + 22x + 121$ f $x^2 - 13x + 42$

3 a $x^2 + 14x + 24$ b $x^2 + 11x + 24$ c $x^2 - 10x + 24$
d $x^2 + 15x + 36$ e $x^2 + 20x + 36$ f $x^2 - 12x + 36$

4 a $x^2 + 2x - 15$ b $x^2 - 2x - 15$ c $x^2 + x - 12$
d $x^2 - x - 12$ e $x^2 + 4x - 12$ f $x^2 - 15x + 36$

5 a $x^2 - 2x - 8$ b $x^2 - x - 20$ c $x^2 + x - 30$
d $x^2 - x - 42$ e $x^2 - 2x - 63$ f $x^2 + 3x - 54$

6 a $2x^2 + 4x + 2$ b $2x^2 + 7x + 6$ c $2x^2 + x - 6$
d $2x^2 - 7x + 6$ e $3x^2 + 8x + 4$ f $3x^2 + 11x - 4$
g $4x^2 + 12x + 9$ h $9x^2 - 6x + 1$ i $6x^2 - x - 1$

## Solving quadratic equations algebraically

$x^2 - 3x - 10 = 0$ is a quadratic equation which, when factorized, can be written as $(x - 5)(x + 2) = 0$.

Therefore either $(x - 5) = 0$ or $(x + 2) = 0$ since, if two things multiply to make zero, then one of them must be zero.

$x - 5 = 0$ or $x + 2 = 0$
$x = 5$ or $x = -2$

It is important to understand the relationship between a quadratic equation written in factorized form and the graph of the quadratic.

In the example above $x^2 - 3x - 10$ factorized to $(x - 5)(x + 2)$. The equation $x^2 - 3x - 10 = 0$ had solutions $x = 5$ and $x = -2$.

The graph of $y = x^2 - 3x - 10$ is as shown.

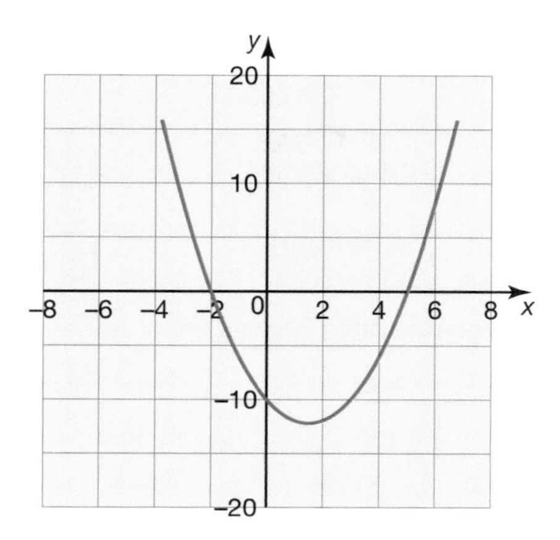


It crosses the $x$-axis at 5 and $-2$ because these are the points where the function has the value 0. These values are directly related to the factorized form.

**Worked example**

Solve each of the following equations to give solutions for $x$.

**a** $x^2 - x - 12 = 0$
**b** $x^2 + 2x = 24$
**c** $x^2 - 6x = 0$
**d** $x^2 - 6x + 9 = 0$

**a**
$$x^2 - x - 12 = 0$$
$$(x - 4)(x + 3) = 0$$
so either $x - 4 = 0$ or $x + 3 = 0$
$$x = 4 \qquad x = -3$$

**b** This becomes $x^2 + 2x - 24 = 0$
$$(x + 6)(x - 4) = 0$$
so either $x + 6 = 0$ or $x - 4 = 0$
$$x = -6 \qquad x = 4$$

**c**
$$x^2 - 6x = 0$$
$$x(x - 6) = 0$$
so either $x = 0$ or $x - 6 = 0$
$$x = 6$$

**d**
$$x^2 - 6x + 9 = 0$$
$$(x - 3)(x - 3) = 0$$
$$(x - 3)^2 = 0$$
so $x = 3$

Note: For a repeated factor, there is only one distinct root.

You can use your GDC to solve a quadratic equation. Although you are expected to be able to solve quadratic equations, your calculator is a useful tool for checking your answers.

**Worked example**

Solve the quadratic equation $x^2 - x - 12 = 0$ using your GDC.

## Casio

Select 'Equation' from the main menu.

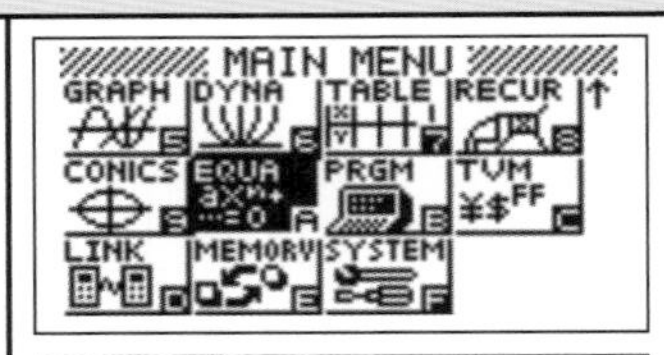

 to select Polynomial, followed by Trace F1 to select Polynomial of Degree 2 (i.e. quadratic).

Enter the coefficients of each of the terms into the matrix, i.e. $a = 1$, $b = -1$ and $c = -12$.

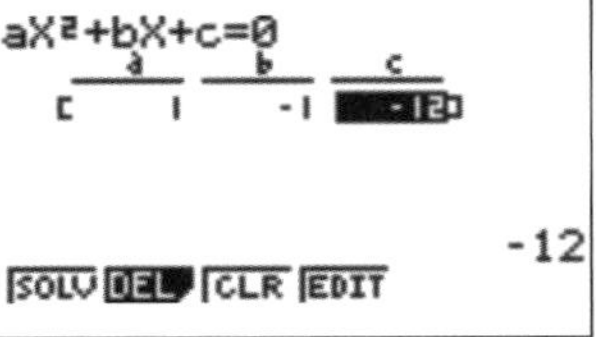

 to solve the equation and display the results on the screen, i.e. $x = 4$ and $-3$.

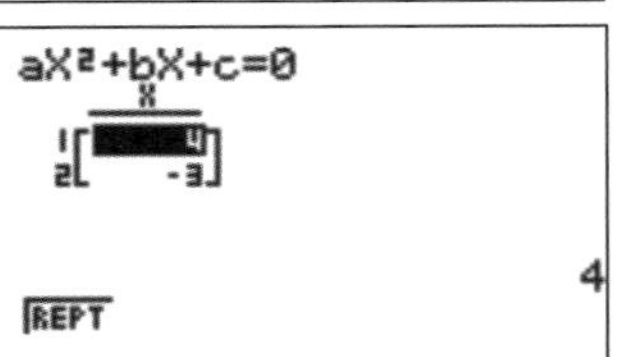


## Texas

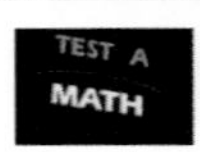  to access the equation solver and enter the equation $0 = x^2 - x - 12$.

EQUATION SOLVER
eqn:0=X²-X-12

Type an initial value for $x$, e.g. 0. Leave the bound at its default setting. The calculator will search for a solution in this range.

X²-X-12=0
X=■
bound={-1E99,1...

Highlight the initial value of $x$.

  to solve and display a solution to the equation, i.e. $x = -2.999...$

X²-X-12=0
▪X=-2.999999999...
bound={-1E99,1...
▪left-rt=0

To find the other solution restrict the bound to include the second solution, e.g. Bound = {0,5} and repeat the above steps.

X²-X-12=0
▪X=3.9999999999...
bound={0,5}
▪left-rt=0

Note: With this calculator, you need to know how many solutions there are and roughly where the solutions lie before using the equation solver.

## Exercise 6.3.5

1 Solve the following quadratic equations by factorizing.

a $x^2 + 7x + 12 = 0$
b $x^2 + 8x + 12 = 0$
c $x^2 + 3x - 10 = 0$
d $x^2 - 3x - 10 = 0$
e $x^2 + 5x = -6$
f $x^2 + 6x = -9$
g $x^2 - 2x = 8$
h $x^2 - x = 20$
i $x^2 + x = 30$
j $x^2 - x = 42$

2 Solve the following quadratic equations.

a $x^2 - 9 = 0$
b $x^2 = 25$
c $x^2 - 144 = 0$
d $4x^2 - 25 = 0$
e $9x^2 - 36 = 0$
f $x^2 - \frac{1}{9} = 0$
g $x^2 + 6x + 8 = 0$
h $x^2 - 6x + 8 = 0$
i $x^2 - 2x - 24 = 0$
j $x^2 - 2x - 48 = 0$

3 Solve the following quadratic equations.

a $x^2 + 5x = 36$
b $x^2 + 2x = -1$
c $x^2 - 8x = 0$
d $x^2 - 7x = 0$
e $2x^2 + 5x + 3 = 0$
f $2x^2 - 3x - 5 = 0$
g $x^2 + 12x = 0$
h $x^2 + 12x + 27 = 0$
i $2x^2 = 72$
j $3x^2 - 12 = 288$

In questions 4–10, construct equations from the information given and then solve them to find the unknown.

4 When a number $x$ is added to its square, the total is 12. Find two possible values for $x$.

5 If the area of the rectangle below is $10\,\text{cm}^2$, calculate the only possible value for $x$.

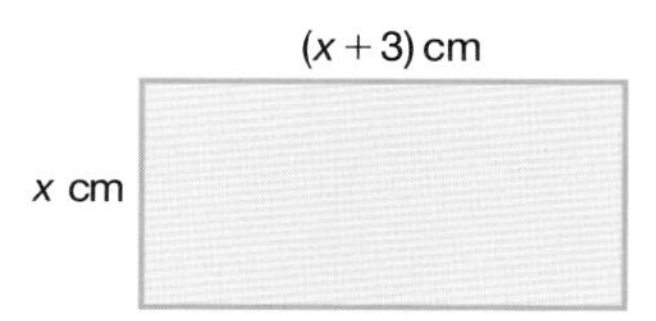


6 If the area of the rectangle below is $52\,\text{cm}^2$, calculate the only possible value for $x$.

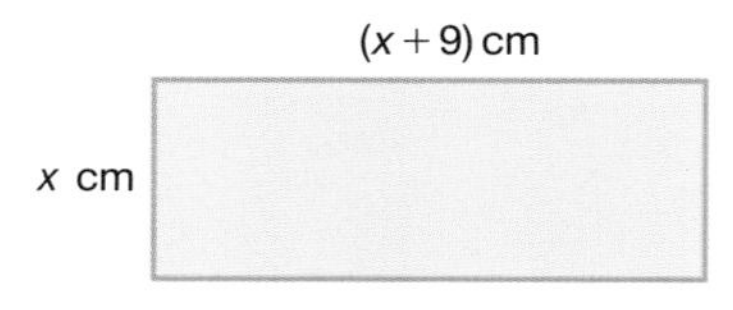


7 A triangle has a base length of $2x$ cm and a height of $(x - 3)$ cm. If its area is $18\,\text{cm}^2$, calculate its height and base length.

8 A triangle has a base length of $(x - 8)$ cm and a height of $2x$ cm. If its area is $20\,\text{cm}^2$, calculate its height and base length.

9 A right-angled triangle has a base length of $x$ cm and a height of $(x - 1)$ cm. If its area is $15\,\text{cm}^2$, calculate its base length and height.

10 A rectangular garden has a square flowerbed of side length $x$ m in one of its corners. The remainder of the garden consists of lawn and has dimensions as shown.

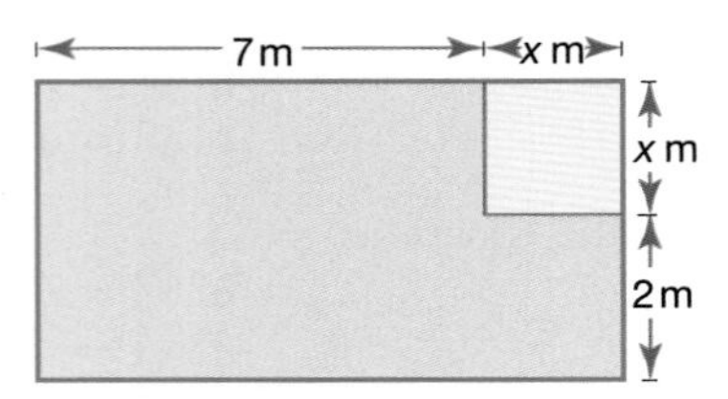


If the total area of the lawn is $50\,\text{m}^2$, calculate the length and width of the whole garden.

## The quadratic formula

In general a quadratic equation takes the form $ax^2 + bx + c = 0$ where $a$, $b$ and $c$ are integers. Quadratic equations can be solved by the use of the quadratic formula, which states that:

$$x = \frac{-b \pm \sqrt{b^2 - 4ac}}{2a}$$

This is particularly useful when the quadratic equation has solutions but does not factorize neatly.

**Worked examples**

1 Solve the quadratic equation $x^2 + 7x + 3 = 0$.

$a = 1$, $b = 7$ and $c = 3$.

Substituting these values into the quadratic formula gives:

$$x = \frac{-7 \pm \sqrt{7^2 - 4 \times 1 \times 3}}{2 \times 1}$$

$$x = \frac{-7 \pm \sqrt{49 - 12}}{2}$$

$$x = \frac{-7 \pm \sqrt{37}}{2}$$

Therefore $x = \dfrac{-7 + 6.08}{2}$ or $x = \dfrac{-7 - 6.08}{2}$

$x = -0.46$ (2 d.p.) or $x = -6.54$ (2 d.p.)

## Texas

This calculator does not have a dynamic graphing facility. However, several of the functions can be graphed simultaneously to observe the family of curves, e.g. $y = \frac{1}{x}$, $y = \frac{2}{x}$, $y = \frac{3}{x}$ etc.

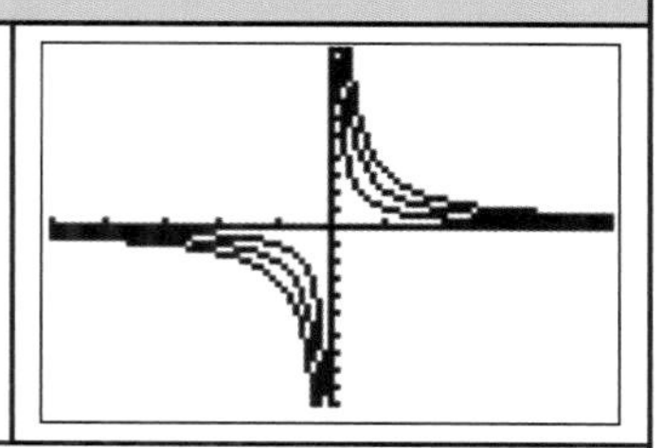

Note: The scale on the axes may need to be changed by selecting WINDOW and entering the following: Xmin = −5, Xmax = 5, Xscl = 1, Ymin = −10, Ymax = 10, Yscl = 1.

## Autograph

Select and enter the equation $y = \frac{a}{x}$.

Select the constant controller .

Change the settings so that $a$ changes in increments of 1.

Use to change the value of $a$ and see the graph move.

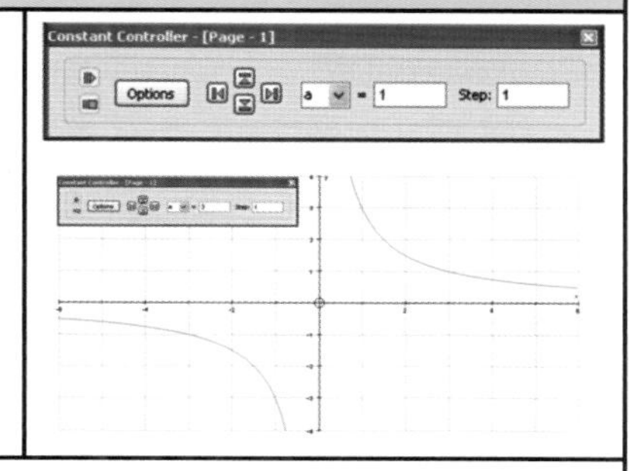


Note: Autograph can display the family of curves or animate the display, by selecting Options and selecting either 'family plot' or 'animation' and entering the parameters as required.

## GeoGebra

Select the 'slider' tool and click on the drawing pad. This will enable you to enter the values for '$a$' and the incremental change as shown.

Apply to place the slider on the drawing pad.

Type $f(x) = \frac{a}{x}$ into the input box.

The effect of '$a$' can be observed by dragging the slider.

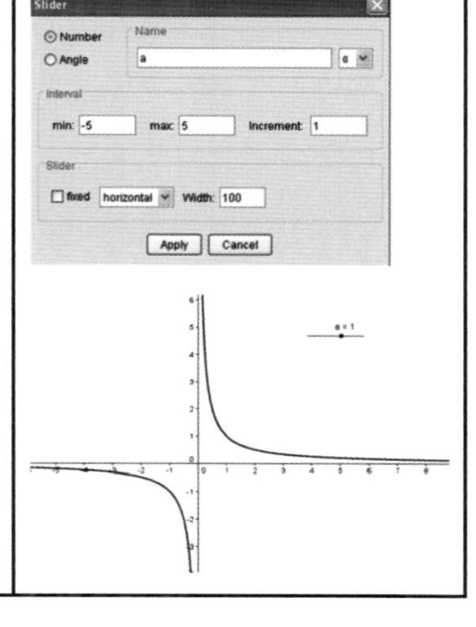

To see the effect of $b$ on the shape of the graph $y = \frac{a}{x + b}$, let $a = 1$ and change the values of $b$ in a similar way to that shown above.

## Exercise 6.5.1

1 Describe the effect that changing $a$ has on the shape of the graph $y = \frac{a}{x + b}$.

2 Describe the effect that changing $b$ has on the shape of the graph $y = \frac{a}{x + b}$.

The graph $y = \frac{1}{x}$ has a particular property. It does not cross the $y$-axis (the line $x = 0$) or the $x$-axis (the line $y = 0$). The graph gets closer and closer to these lines but does not meet or cross them. These lines are known as **asymptotes** to the curve.

Therefore the graph of $y = \frac{1}{x}$ has a vertical asymptote at $x = 0$ and a horizontal asymptote at $y = 0$. Although clear from the graph, they can be calculated as follows.

**Worked examples**

1 Calculate the equations of any asymptotes for the graph of $y = \frac{1}{x}$.

Vertical asymptote: This occurs when the denominator $= 0$ as $\frac{1}{0}$ is undefined. Therefore the vertical asymptote is $x = 0$.

Horizontal asymptote: This can be deduced by looking at the value of $y$ as $x$ tends to infinity (written as $x \to \pm\infty$).

As $x \to +\infty$, $y \to 0$ as $\frac{1}{x} \to 0$ As $x \to -\infty$, $y \to 0$ as $\frac{1}{x} \to 0$

Therefore the horizontal asymptote is $y = 0$.

2 Calculate the equations of any asymptotes for the graph of $y = \frac{1}{x - 3} + 1$ and give the coordinates of any points where the graph crosses either axis.

Vertical asymptote: This occurs when the denominator is 0.

$x - 3 = 0$, so $x = 3$. Therefore the vertical asymptote is $x = 3$.

Horizontal asymptote: Look at the value of $y$ as $x \to \pm\infty$.

As $x \to +\infty$, $y \to 1$ as $\frac{1}{x - 3} \to 0$ As $x \to -\infty$, $y \to 1$ as $\frac{1}{x - 3} \to 0$

Therefore the horizontal asymptote is $y = 1$.

To find where the graph intercepts the $y$-axis, let $x = 0$.

Substituting $x = 0$ into $\frac{1}{x - 3} + 1$ gives $y = -\frac{1}{3} + 1 = \frac{2}{3}$

To find where the graph intercepts the $x$-axis, let $y = 0$.

Substituting $y = 0$ into $\frac{1}{x - 3} + 1$ gives:

$$0 = \frac{1}{x - 3} + 1$$

$$\frac{1}{x - 3} = -1$$

$$x = 2$$

Therefore the intercepts with the axes occur at $\left(0, \frac{2}{3}\right)$ and $(2, 0)$.

The graph of $y = \frac{1}{x-3} + 1$ can now be sketched as shown.

Note: The asymptotes are shown with dashed lines.

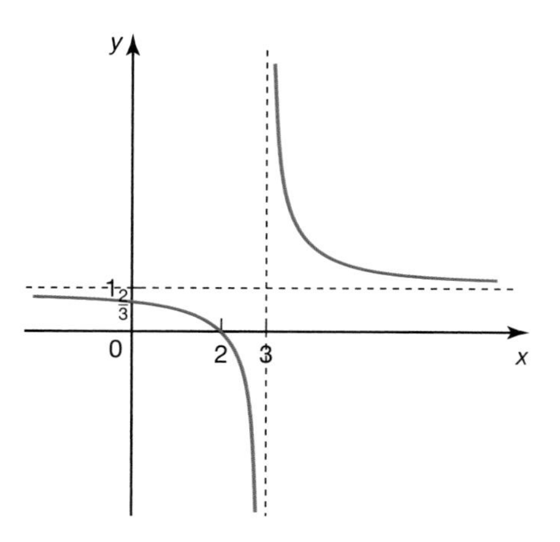

## Exercise 6.5.2

For each of the equations in questions 1–3:

**i)** calculate the equations of the vertical and horizontal asymptotes
**ii)** calculate the coordinates of any points where the graph intercepts the axes
**iii)** sketch the graph
**iv)** check your graph using either a GDC or graphing software.

**1 a** $y = \frac{1}{x+1}$ **b** $y = \frac{1}{x+3}$

**c** $y = \frac{2}{x-4}$ **d** $y = \frac{-1}{x-3}$

**2 a** $y = \frac{1}{x} + 2$ **b** $y = \frac{1}{x} - 3$

**c** $y = \frac{1}{x-1} + 4$ **d** $y = \frac{1}{x+4} - 1$

**3 a** $y = \frac{1}{2x+1}$ **b** $y = \frac{1}{2x-1} + 1$

**c** $y = \frac{2}{3x-1}$ **d** $y = \frac{-1}{4x-1} + 2$

**Worked examples**

1 Use a GDC or graphing software to graph the function $y = \dfrac{1}{(x-2)(x+1)}$ and determine the equations of any asymptotes.

Graphing software will produce the following graph of the function:

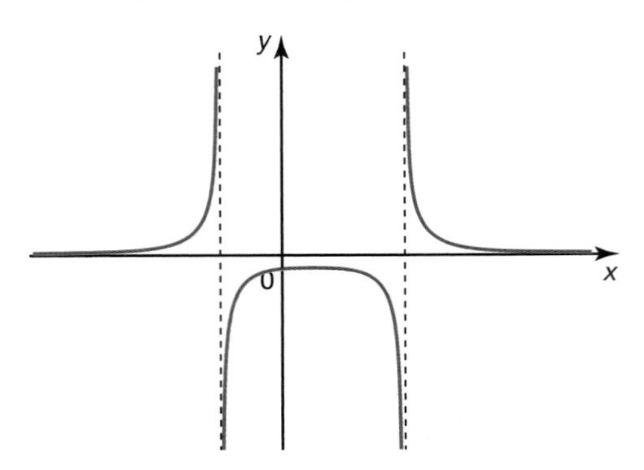


The equations of any asymptotes can be determined as before.

Vertical asymptote: This occurs when the denominator is 0: $(x - 2)(x + 1) = 0$

Therefore the vertical asymptotes are $x = 2$ and $x = -1$.

Horizontal asymptote: Look at the value of $y$ as $x \to \pm\infty$.

As $x \to \pm\infty$,

$y \to 0$ as $\dfrac{1}{(x-2)(x+1)} \to 0$

Therefore the horizontal asymptote is $y = 0$.

Therefore a more informative sketch is:

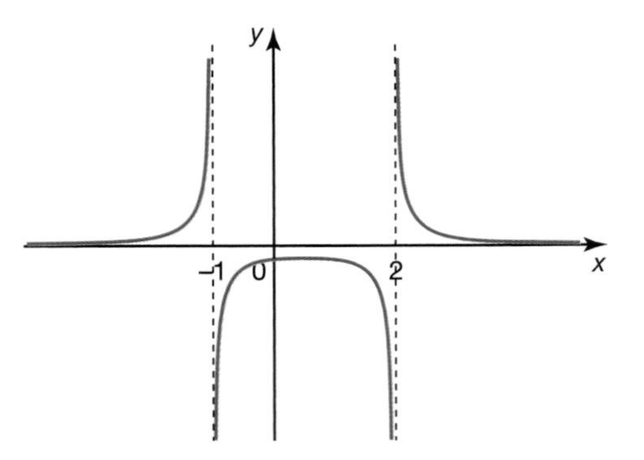

2 Use a GDC or graphing software to graph the function $y = \frac{1}{x^2 + x - 6} + 2$ and determine the equations of any asymptotes.

The graphing software will produce the following graph of the function:

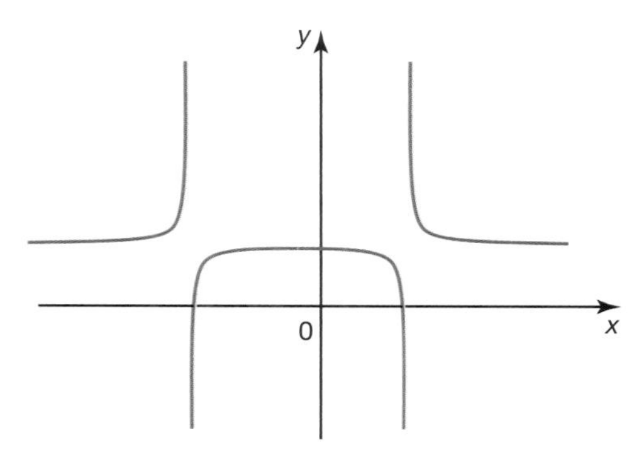


The equations of the asymptotes are determined as before.

Vertical asymptote: This occurs where the denominator is 0.

$x^2 + x - 6 = 0$

Factorising gives $(x - 2)(x + 3) = 0$

Therefore the vertical asymptotes are $x = 2$ and $x = -3$.

Horizontal asymptote: Look at the value of $y$ as $x \to \pm\infty$.

As $x \to \pm\infty$, $y \to 2$ as $\frac{1}{x^2 + x - 6} \to 0$

Therefore the horizontal asymptote is $y = 2$.

Therefore a more informative sketch is:

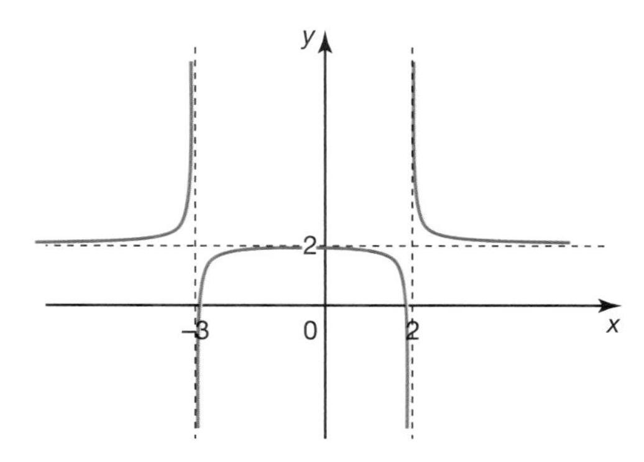


## Exercise 6.5.3

For each of the equations in the following questions:

**i)** calculate the equations of any vertical or horizontal asymptotes

**ii)** calculate the coordinates of any points where the graph intercepts the axes

**iii)** sketch the graph of the function with the aid of a GDC or graphing software.

**1** **a** $y = \frac{1}{(x - 1)(x - 2)}$ **b** $y = \frac{1}{(x - 4)(x + 3)}$

**c** $y = \frac{1}{x(x - 5)} + 1$ **d** $y = \frac{1}{(x + 2)^2} - 3$

2 a $y = \dfrac{1}{x^2 + 3x - 4}$

b $y = \dfrac{1}{x^2 + 3x - 10}$

c $y = \dfrac{1}{x^2 + 2x + 1} + 3$

d $y = \dfrac{1}{2x^2 - 7x - 4} - 1$

## Higher-order polynomials

Earlier in this topic you saw how a quadratic equation can sometimes be factorized, e.g.

$$x^2 - 4x - 5 = (x - 5)(x + 1)$$

This can be used to find where the graph of $y = x^2 - 4x - 5$ intercepts the $x$-axis, i.e. when $y = 0$. $(x - 5)(x + 1) = 0$ gives $x = 5$ and $x = -1$ as shown.

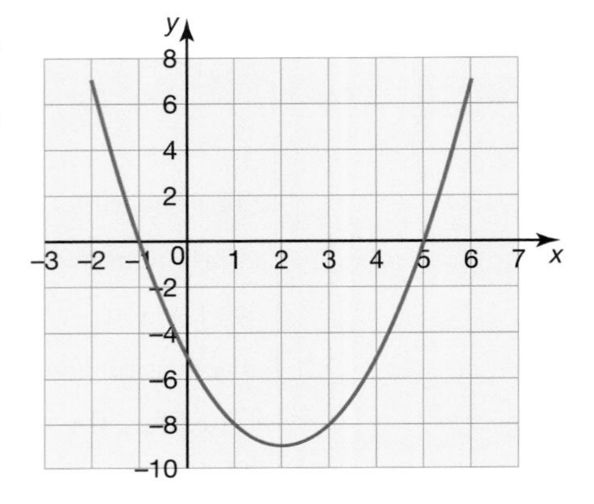


The reverse is therefore also true. A quadratic equation that intercepts the $x$-axis at $x = 3$ and $-6$ can be written as $y = (x - 3)(x + 6) = x^2 + 3x - 18$.

The same is true of higher-order polynomials.

**Worked example**

Using a GDC or graphing software, graph the cubic equation $y = x^3 - 2x^2 - 5x + 6$ and, by finding its roots, rewrite the equation in factorized form.

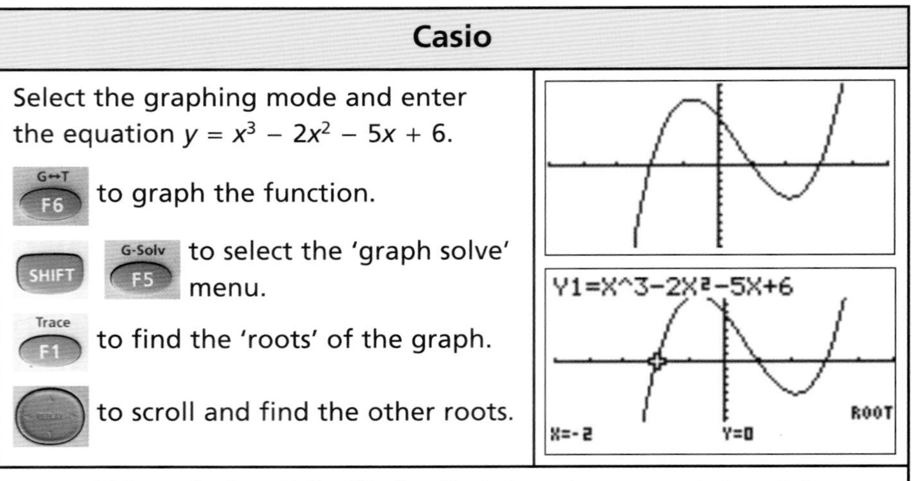


Note: Although the GDC will also find the $y$-intercept, it is quicker simply to substitute $x = 0$ into the equation, i.e. the $y$-intercept is $+6$.

## Texas

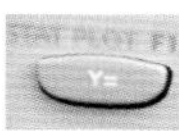

and enter the equation $y = x^3 - 2x^2 - 5x + 6$.

to graph the function.

to select the 'graph calc.' menu.

to find where the graph is zero (i.e. crosses the $x$-axis).

Use the cursor key to select a left bound. Press enter and then use the cursor key to select a right bound. The calculator will search for a solution within this range. Press enter twice to display the answer.

Repeat the process for the other roots.

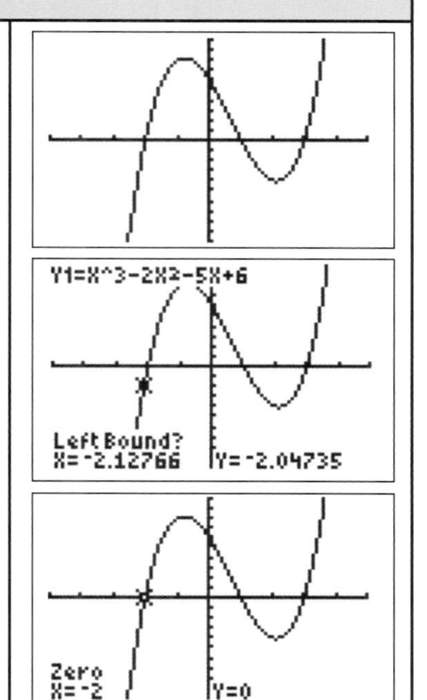


Note: Although the GDC will also find the $y$-intercept by entering a value of $x = 0$, it is quicker simply to substitute $x = 0$ into the equation, i.e. the $y$-intercept is $+6$.

Select and enter the equation $y = x^3 - 2x^2 - 5x + 6$.

Select the graph and choose 'Object' and 'Solve $f(x) = 0$'.

The results are displayed at the bottom of the screen, or by accessing the 'results box' .

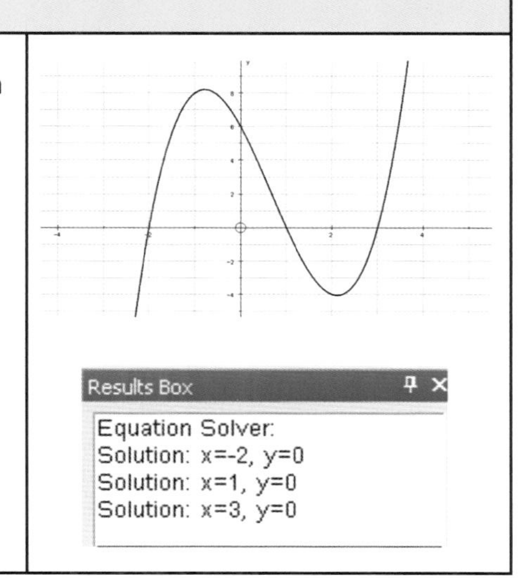

| GeoGebra | |
|---|---|
| Type $f(x) = x \wedge 3 - 2x \wedge 2 - 5x + 6$ into the input box. | 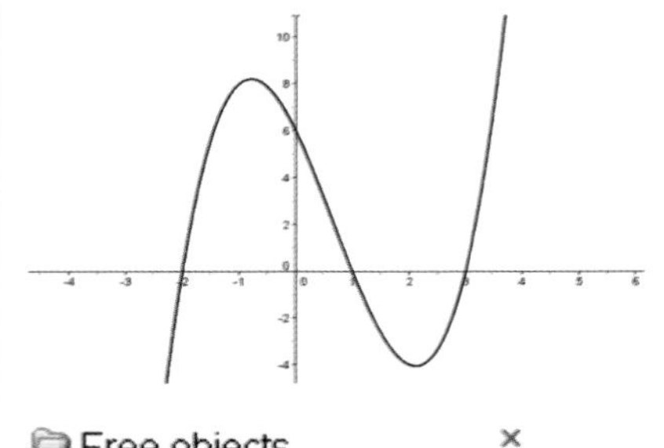 |
| Type: Root[*f*]. This finds the roots for the polynomial *f*. The results are displayed in the algebra window. | Free objects<br>$f(x) = x^3 - 2x^2 - 5x + 6$<br>Dependent objects<br>A = (-2, 0)<br>B = (1, 0)<br>C = (3, 0) |

The roots of the equation are therefore $x = -2$, 1 and 3, so the equation can be written in factorized form as $y = (x + 2)(x - 1)(x - 3)$.

The $y$-intercept is found by substituting $x = 0$ into the equation, giving $y = 6$.

You can also use your GDC in a similar way to find the stationary points of the curve. The algebraic method for finding the stationary points of a curve is covered in Topic 7.

## Exercise 6.5.4

For each of the equations in the following questions:

**i)** Use a GDC or graphing software to sketch the function.
**ii)** Determine the $y$-intercept by using $x = 0$.
**iii)** Rewrite the equation in factorized form by finding the roots of the equation.

1 a $y = x^3 + 6x^2 + 11x + 6$
b $y = x^3 - 3x^2 + 3$
c $y = x^3 - x^2 - 12x$
d $y = x^3 - 4x$

2 a $y = x^3 - 3x + 2$
b $y = x^3 - 4x^2 + 4x$
c $y = -x^4 + 2x^3 + 3x^2$
d $y = -x^3 + x^2 + 22x - 40$

# 6.6 Solving unfamiliar equations graphically

When an equation has to be solved, it can be done graphically or algebraically. The following example shows how equations can be solved graphically.

**Worked example**

Solve $x + 2 = \frac{3}{x}$.

Method 1: Rearrange the equation to form the quadratic equation $x^2 + 2x - 3 = 0$. This produces the graph shown. The roots can be calculated as shown earlier in the topic.

Method 2: Both sides of the original equation can be plotted separately and the $x$-coordinates of their points of intersection calculated.

Solution: $x = 1$ and $x = -3$.

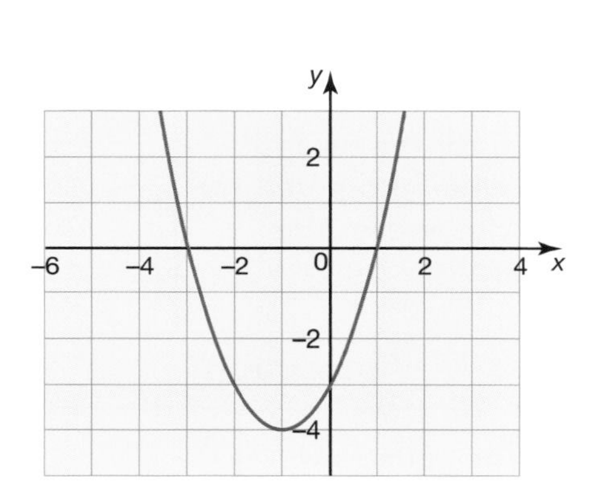


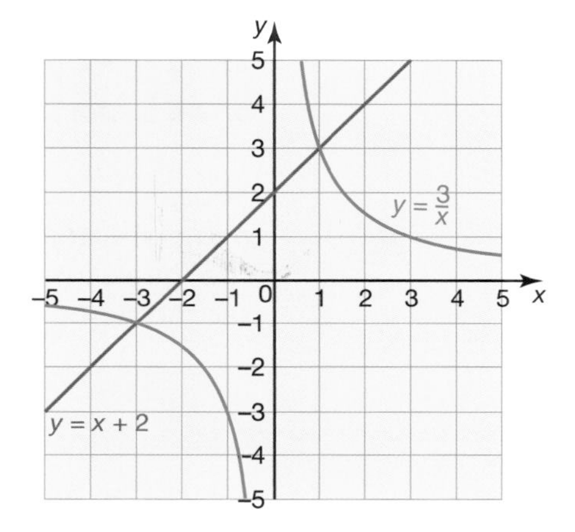


## Casio

Select the graphing mode and enter the equations $y = x + 2$ and $y = \frac{3}{x}$.

to graph the functions.

to select the 'graph solve' menu.

to find the intersection of the graphs.

to scroll and find the other points of intersection.

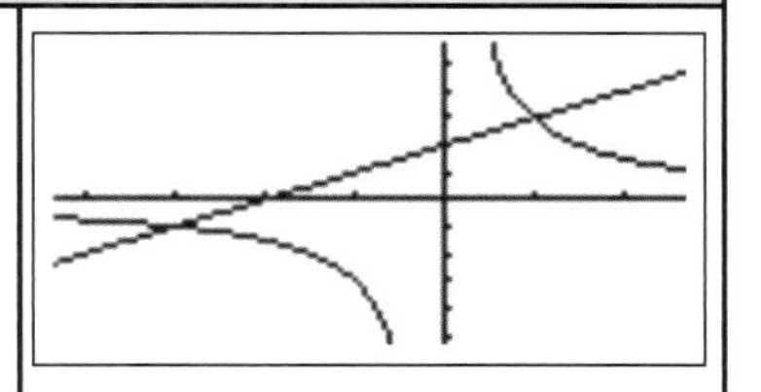

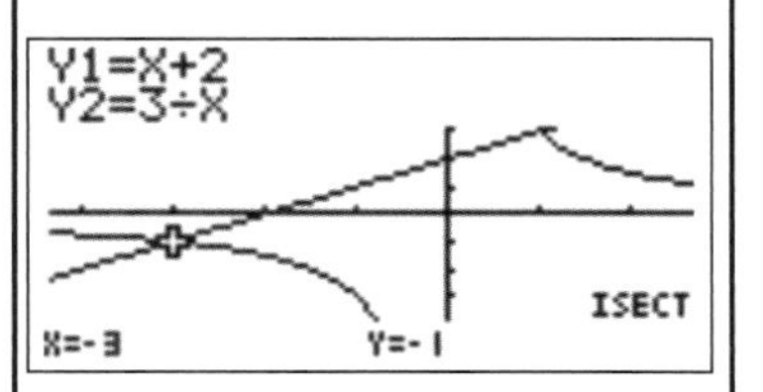


## Texas

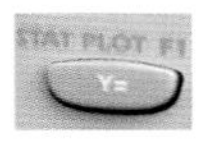

and enter the equations $y = x + 2$ and $y = \frac{3}{x}$.

to graph the functions.

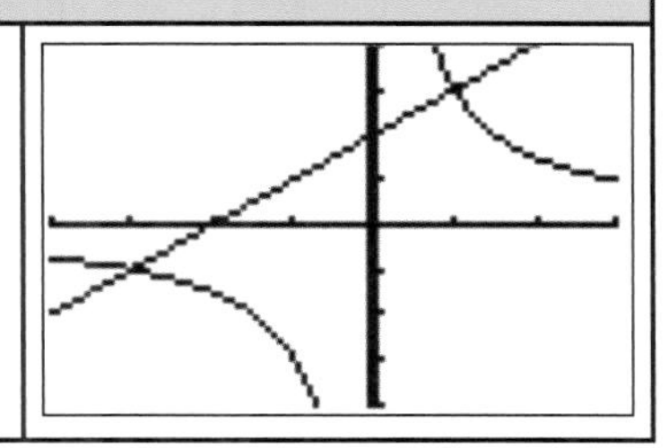

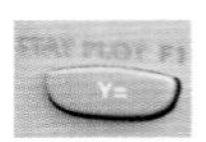 and enter the equations $y = x + 2$ and $y = \frac{3}{x}$.

 to graph the functions.

 to select the 'graph calc.' Menu.

 to find the intersection of the two graphs.

   to select both graphs and confirm the first intersection point.

Repeat the above steps for the second point of intersection. When prompted for the 'guess', move the cursor over the second point of intersection.

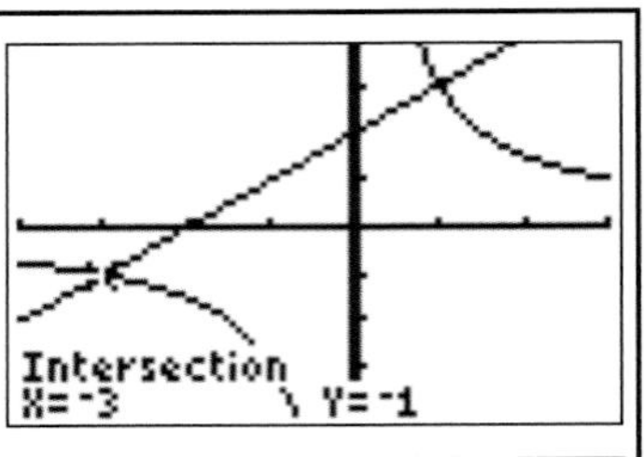


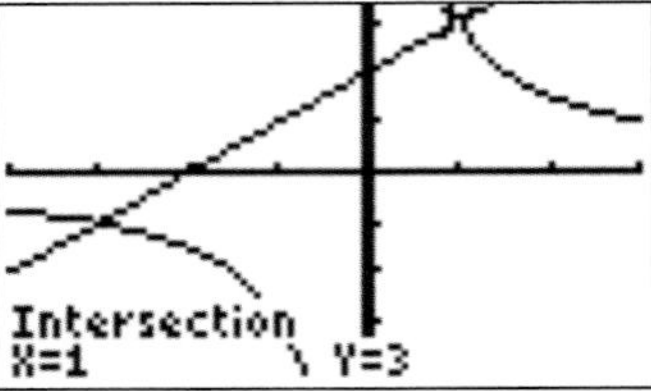


## Autograph

Select and enter the equations $y = x + 2$ and $y = \frac{3}{x}$.

Select the graphs, choose 'Object' and 'Solve $f(x) = g(x)$'.

The results are displayed at the bottom of the screen, or by accessing the 'results box' .

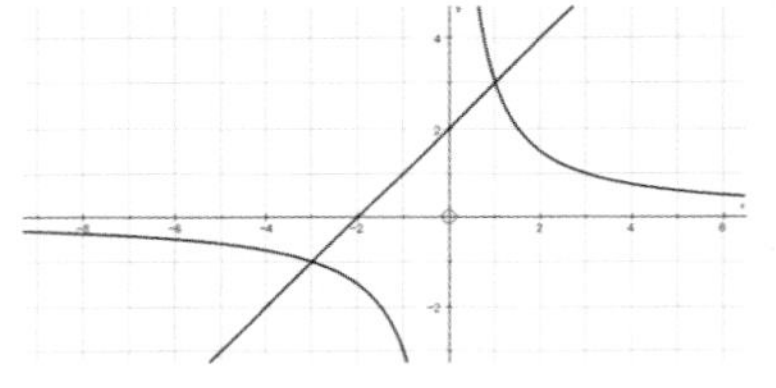

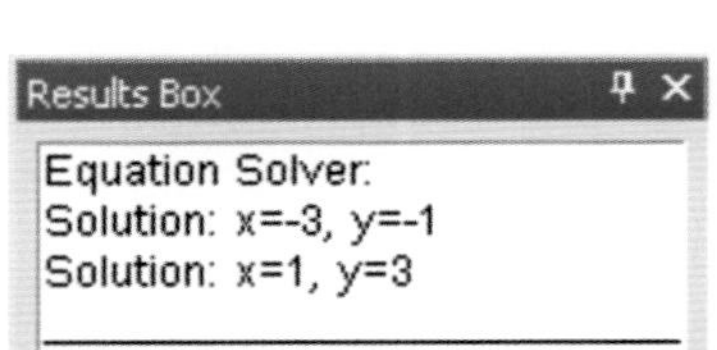


## GeoGebra

Type $f(x) = x + 2$ and $g(x) = \frac{3}{x}$ into the input box.

To find the points of intersection, select the 'intersect two objects' icon.

Click on the curve and the straight line; the intersection point is marked. Its coordinates appear in the algebra window.

Repeat for the second point of intersection.

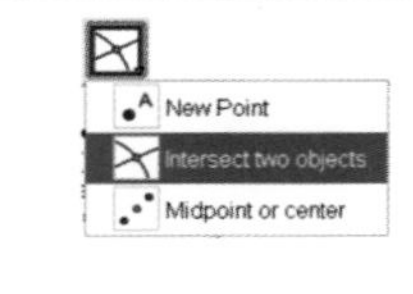


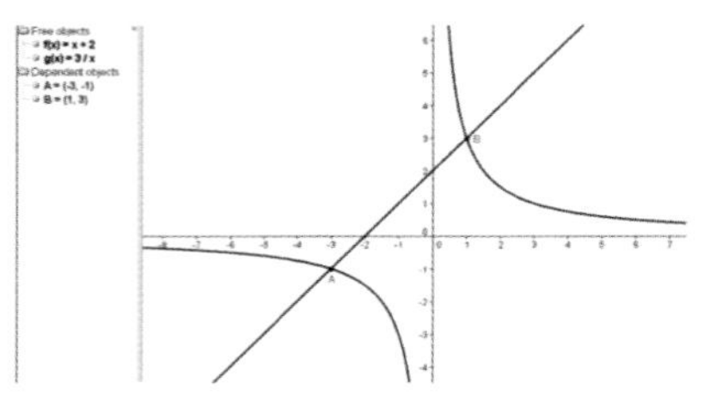

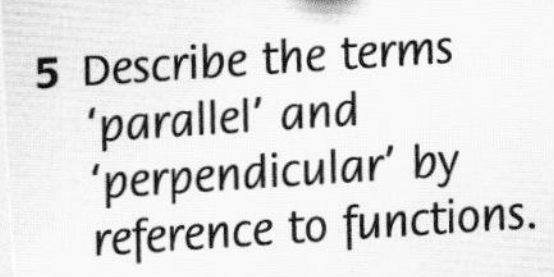

**5** Describe the terms 'parallel' and 'perpendicular' by reference to functions.

**6** Does a graph without labels or a scale have any meaning? Would reversing the position of the $x$ and $y$ axes have any effect upon the study of graphs of functions?

**7** Any business is concerned with cash flow, and cost and revenue functions. The study of these in a company could form the basis of a project.

**8** Discuss the statement that 'to know about Descartes, Newton and Gauss has no relevance to learning mathematics'. Can mathematics be studied without any reference to its historical context?

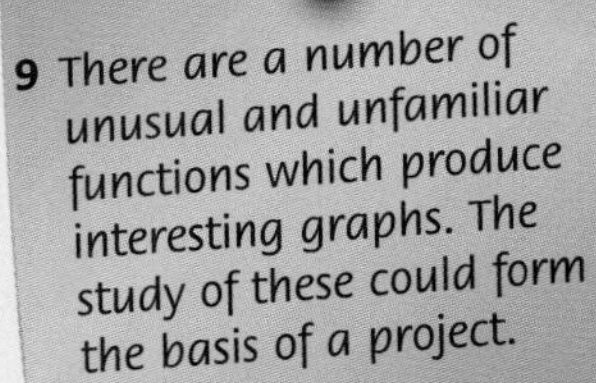

**9** There are a number of unusual and unfamiliar functions which produce interesting graphs. The study of these could form the basis of a project.

**10** Leonhard Euler wrote seventy volumes of work on mathematics. Is a great mathematician one who is prolific like him, or one who makes only one major breakthrough?

Topic 7

# Introduction to differential calculus

## Syllabus content

**7.1** Concept of the derivative as a rate of change.

Tangent to a curve.

**7.2** The principle that $f(x) = ax^n \Rightarrow f'(x) = anx^{n-1}$.

The derivative of functions of the form

$f(x) = ax^n + bx^{n-1} + \ldots$, where all exponents are integers.

**7.3** Gradients of curves for given values of $x$.

Values of $x$ where $f'(x)$ is given.

Equation of the tangent at a given point.

Equation of the line perpendicular to the tangent at a given point (normal).

**7.4** Increasing and decreasing functions.

Graphical interpretation of $f'(x) > 0$, $f'(x) = 0$, $f'(x) < 0$.

**7.5** Values of $x$ where the gradient of a curve is 0 (zero).

Solution of $f'(x) = 0$.

Stationary points.

Local maximum and minimum points.

**7.6** Optimization problems.

## Introduction

Pierre de Fermat was a great French mathematician who proposed and solved many mathematical problems. He is most famous for 'Fermat's Last Theorem', which he proposed in 1637. Although seemingly simple, the theorem was not proved until 358 years later, in 1995.

Fermat's last theorem is an extension of Pythagoras' theorem. It states that: 'If an integer $n$ is greater than 2, then the equation $a^n + b^n = c^n$ has no solutions.' Fermat suggested that he had found a simple proof of this, however it is not accepted by mathematicians today.

Andrew Wiles

Andrew Wiles, a Cambridge mathematician, worked secretly for many years to find a proof. This proof was extremely complex, building on work on elliptical curves by Eves Hellegouach, and required a proof of the Taniyama-Shimura conjecture. However, the final proof by Wiles is accepted as a work of genius because of the innovative way in which he brought together ideas.

Everyone accepts that Andrew Wiles proved Fermat's Last Theorem. However, this is not the case with the discovery of calculus. Sir Isaac Newton (1643–1727), another Cambridge mathematician, is accepted as one of the most influential men in human history. His work on gravitation and the Laws of Motion in his book, *Philosophae Naturalis Principia Mathematica*, influenced mathematics and science for hundreds of years. His work was only taken further by the great mind of Albert Einstein. Newton is credited by many as discovering calculus.

Gottfried Wilhelm Leibniz (1646–1716), the German mathematician and Philosopher, worked on what we now term calculus at the same time as Newton. The suggestion that Leibniz, who wrote poetry in addition to writing on maths, politics, law, theology, history and philology, had stolen Newton's ideas on calculus and merely improved them, caused a bitter argument that went on long after they had both died.

Claims as to who should get the credit for an invention are not unusual. Did Alexander Graham Bell or Antonio Meucci invent the telephone? Did the Scot John Logie Baird, the American Philo Taylor Farnsworth or the Russian Vladimir Kosma Zworkin, discover television? It often depends upon the country in which the book you read was published. It has been suggested that 'there is a time for a discovery' and that if one person had not made the breakthrough, someone else would have. This claim is supported by one of the great modern discoveries, the structure of the DNA molecule. Watson and Crick discovered the double helix structure but other scientists, particularly Rosalind Franklin, were very close to a solution.

Calculus is the cornerstone of much of the mathematics studied at a higher level. Differential calculus deals with finding the formula for the gradient of a function. In this topic, the functions will be of the form $f(x) = ax^n + bx^{n-1} + \ldots$ where $n$ is an integer.

## 7.1 Gradient

You will already be familiar with finding the gradient of a straight line, shown below.

The gradient of the line passing through points $(x_1, y_1)$ and $(x_2, y_2)$ is calculated by $\frac{y_2 - y_1}{x_2 - x_1}$. Therefore the gradient of the line passing through points P and Q is $\frac{10 - 5}{11 - 1} = \frac{5}{10} = \frac{1}{2}$

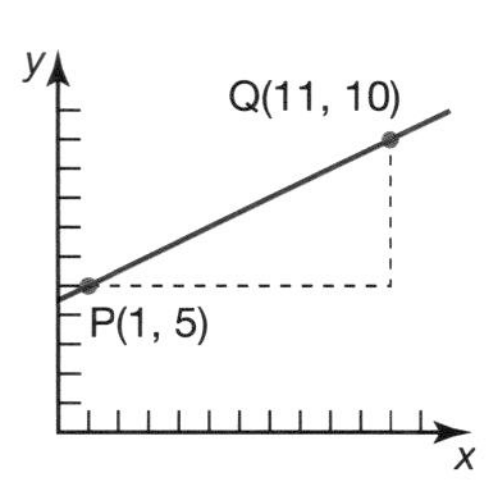


The gradient of a linear function (a straight line) is constant, i.e. the same at any point on the line. However non-linear functions, such as those that produce curves when graphed, are more difficult to work with as the gradient is not constant.

The graph opposite shows the function $f(x) = x^2$. Point P is on the curve at (3, 9). If P moves along the curve to the right, the gradient of the curve becomes steeper. If P moves along the curve towards the origin, the gradient of the curve becomes less steep.

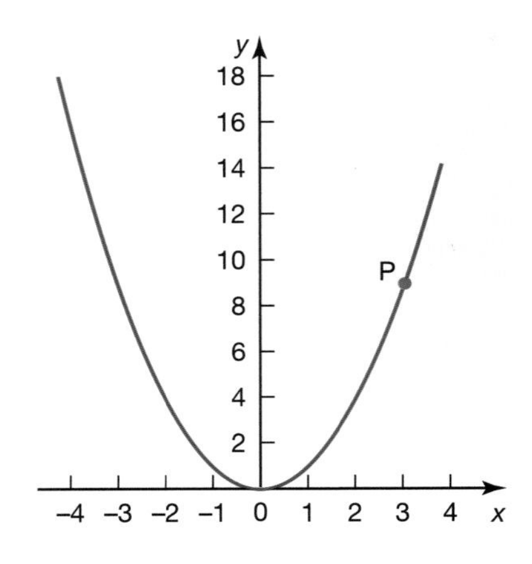


WEB
GEOGEBRA FILES

See GeoGebra file '7.1 Gradient of $y = x^2$'.

The gradient of the function $f(x) = x^2$ at the point P(1, 1) can be calculated as follows.

Mark a point $Q_1(3, 9)$ on the graph and draw the line segment $PQ_1$. The gradient of the line segment $PQ_1$ is only an approximation of the gradient of the curve at P.

$$\text{Gradient of } PQ_1 = \frac{9 - 1}{3 - 1} = 4$$

Mark a point $Q_2$ closer to P, for example (2, 4) and draw the line segment $PQ_2$. The gradient of the line segment $PQ_2$ is still only an approximation of the gradient of the curve at P, but it is a better approximation than the gradient of $PQ_1$.

$$\text{Gradient of } PQ_2 = \frac{4 - 1}{2 - 1} = 3$$

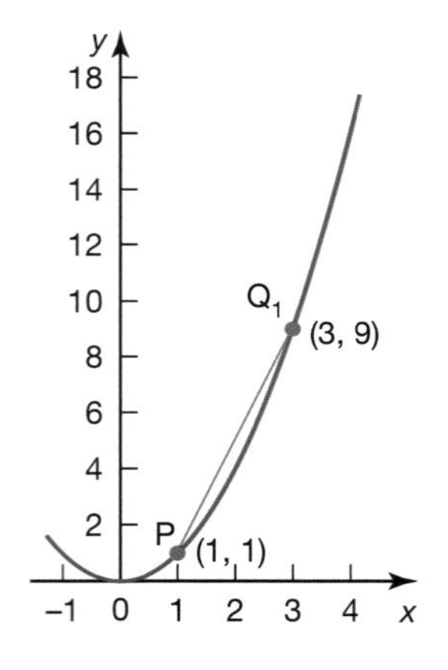


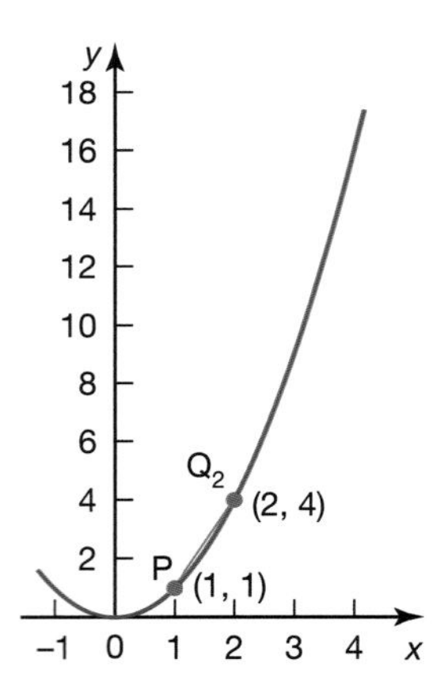


If a point $Q_3$ $(1.5, 1.5^2)$ is chosen, the gradient $PQ_3$ will be an even better approximation.

$$\text{Gradient of } PQ_3 = \frac{1.5^2 - 1}{1.5 - 1} = 2.5$$

For the point $Q_4$ $(1.25, 1.25^2)$, the gradient of $PQ_4 = \dfrac{1.25^2 - 1}{1.25 - 1} = 2.25$

For the point $Q_5$ $(1.1, 1.1^2)$, the gradient of $PQ_5 = \dfrac{1.1^2 - 1}{1.1 - 1} = 2.1$

See GeoGebra file '7.1 Gradient of line segment PQ'.

The results above suggest that, as point Q gets closer to P, the gradient of the line segment PQ gets closer to 2.

It can be proved that the gradient of the function is $f(x) = x^2$ is 2 when $x = 1$.

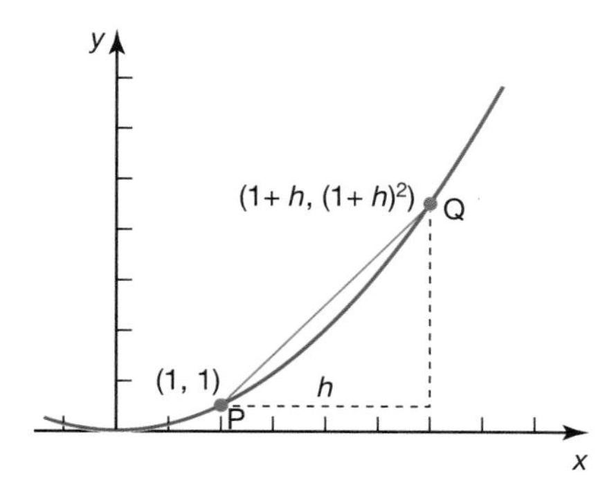


Consider points P and Q on the function $f(x) = x^2$. P is at (1, 1) and Q, $h$ units from P in the $x$-direction has coordinates $(1+h, (1+h)^2)$.

$$\text{Gradient of line segment PQ} = \frac{(1+h)^2 - 1}{1+h-1} = \frac{1+2h+h^2-1}{h} = \frac{h(2+h)}{h} = 2 + h$$

As Q gets closer to P, $h$ gets smaller and smaller (tends to 0), therefore the gradient $(2 + h)$ of the line segment PQ tends to 2.

Therefore the gradient at P(1, 1) is 2.

In general:

**The gradient of a curve at the point P is the same as the gradient of the tangent to the curve at P.**

See GeoGebra file '7.1 Line passing through P and Q'.

## Exercise 7.1.1

1 Using the proof above as a guide, find the gradient of the function $f(x) = x^2$ when:

  **a** $x = 2$ **b** $x = 3$ **c** $x = -1$

  **d** By looking at the pattern in your results, complete this sentence. For the function $f(x) = x^2$, the formula for the gradient is … .

2 Find the gradient of the function $f(x) = 2x^2$ when:

  **a** $x = 1$ **b** $x = 2$ **c** $x = -2$

  **d** By looking at the pattern in your results, complete this sentence. For the function $f(x) = 2x^2$, the formula for the gradient is … .

3 Find the gradient of the function $f(x) = \frac{1}{2}x^2$ when:

  **a** $x = 1$ **b** $x = 2$ **c** $x = 3$

  **d** By looking at the pattern in your results, complete this sentence. For the function $f(x) = \frac{1}{2}x^2$, the formula for the gradient is … .

In each case in Exercise 7.1.1, a rule was found for calculating the gradient at any point on the particular curve. This rule is known as the gradient function $f'(x)$ or $\frac{dy}{dx}$, i.e. the function $f(x) = x^2$ has a gradient function $f'(x) = 2x$

$$\text{or } \frac{dy}{dx} = 2x.$$

Note: $\frac{dy}{dx}$ is also known as the rate of change of $y$ with $x$.

You can check this by graphing a function and its gradient function simultaneously on a computer.

**Autograph**

- Type equation $y = x^2 + x$
- Select curve.
- Click on the gradient function icon .

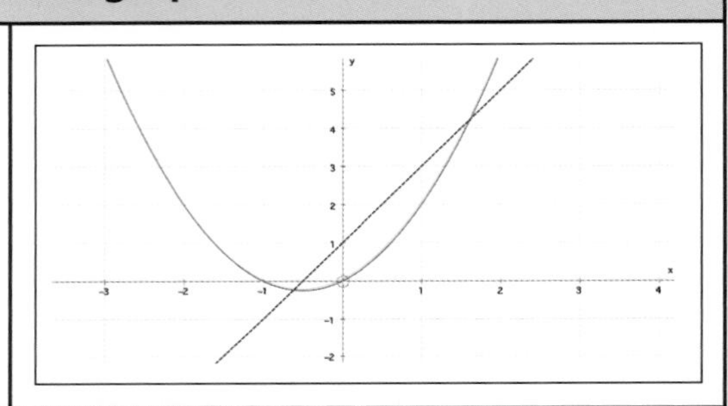

**GeoGebra**

- Type equation $f(x) = x^2 + x$
- Type Derivative [$f$]

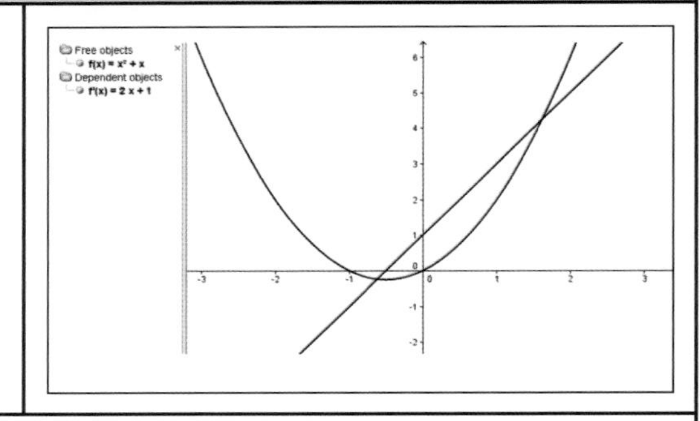


Note: The equation of the gradient function is given in the algebra window.

## Exercise 7.1.2

1 Using graphing software as shown above, find the gradient functions of each of the following functions.

   a $f(x) = x^3$
   b $f(x) = 3x^2$
   c $f(x) = x^2 + 2x$
   d $f(x) = x^2 - 2$
   e $f(x) = 3x - 3$
   f $f(x) = 2x^2 - x + 1$

2 Copy and complete the table by entering your gradient functions from question 1 above and from Exercise 7.1.1.

3 Describe any patterns you notice in your table for question 2, between a function and its gradient function.

| Function $f(x)$ | Gradient function $f'(x)$ |
|---|---|
| $x^2$ | |
| $2x^2$ | |
| $\frac{1}{2}x^2$ | |
| $x^2 + x$ | $2x + 1$ |
| $x^3$ | |
| $3x^2$ | |
| $x^2 + 2x$ | |
| $x^2 - 2$ | |
| $3x - 3$ | |
| $2x^2 - x + 1$ | |

The functions used so far have all been polynomials. There is a relationship between a polynomial function and its gradient function. This is summarized below.

If $f(x) = ax^n$ then $\frac{dy}{dx} = anx^{n-1}$,

i.e. to work out the gradient function of a polynomial, multiply the coefficient of $x$ by the power of $x$ and then subtract 1 from the power.

**Worked examples**

1 Calculate the gradient function of the function $f(x) = 2x^3$.

$$\frac{dy}{dx} = 3 \times 2x^{(3-1)} = 6x^2$$

2 Calculate the gradient function of the function $f(x) = 5x^4$.

$$\frac{dy}{dx} = 4 \times 5x^{(4-1)} = 20x^3$$

## Exercise 7.1.3

1 Calculate the gradient function of each of the following functions.

   a $f(x) = x^4$   b $f(x) = x^5$   c $f(x) = 3x^2$
   d $f(x) = 5x^3$   e $f(x) = 6x^3$   f $f(x) = 8x^7$

2 Calculate the gradient function of each of the following functions.

   a $f(x) = \frac{1}{3}x^3$   b $f(x) = \frac{1}{4}x^4$   c $f(x) = \frac{1}{4}x^2$
   d $f(x) = \frac{1}{2}x^4$   e $f(x) = \frac{2}{5}x^3$   f $f(x) = \frac{2}{9}x^3$

## 7.2 Differentiation

The process of finding the gradient function is known as **differentiation.** Differentiating a function produces the **derivative** or gradient function.

**Worked examples**

1 Differentiate the function $f(x) = 3$ with respect to $x$.

The graph of $f(x) = 3$ is a horizontal line as shown.

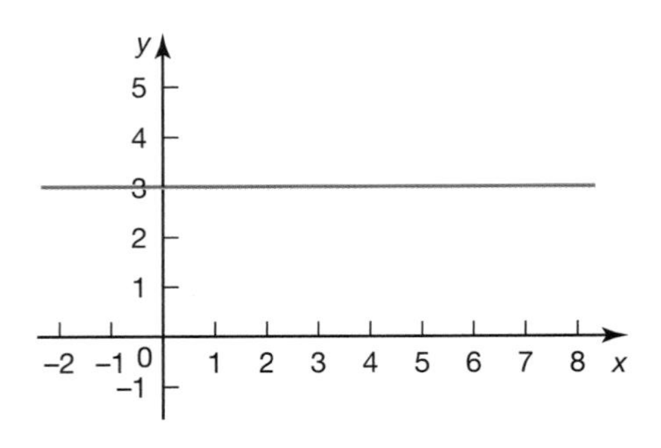


A horizontal line has a gradient of zero. Therefore,

$$f(x) = 3 \Rightarrow \frac{dy}{dx} = 0.$$

This can be calculated using the rule for differentiation:

$f(x) = 3$ can be written as $f(x) = 3x^0$

$$\frac{dy}{dx} = 0 \times 3x^{(0-1)} = 0$$

In general therefore, the derivative of a constant is zero.

$$\text{If } f(x) = c \Rightarrow \frac{dy}{dx} = 0.$$

2 Differentiate the function $f(x) = 2x$ with respect to $x$.

The graph of $f(x) = 2x$ is a straight line as shown.

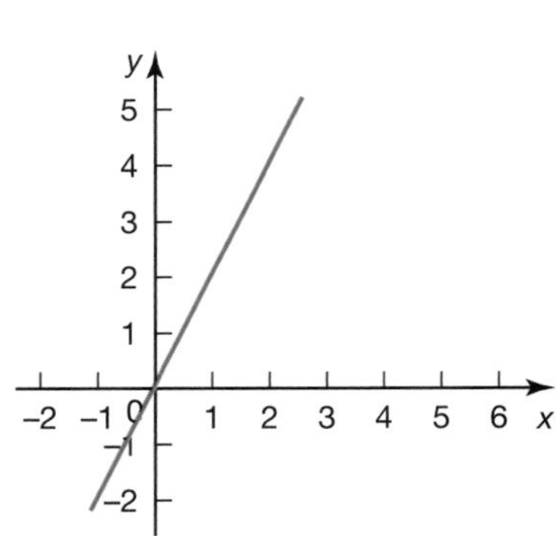


From work done on linear graphs, you know the gradient is 2. Therefore,

$$f(x) = 2x \Rightarrow \frac{dy}{dx} = 2.$$

This too can be calculated using the rule for differentiation:

$f(x) = 2x$ can be written as $f(x) = 2x^1$.

$$\frac{dy}{dx} = 1 \times 2x^{(1-1)} = 2x^0$$

But $x^0 = 1$, therefore $\frac{dy}{dx} = 2$.

In general, therefore, if $f(x) = ax \Rightarrow \frac{dy}{dx} = a$.

3 Differentiate the function $f(x) = \frac{1}{3}x^3 - 2x + 4$ with respect to $x$.

Graphically the function and its derivative are as shown.

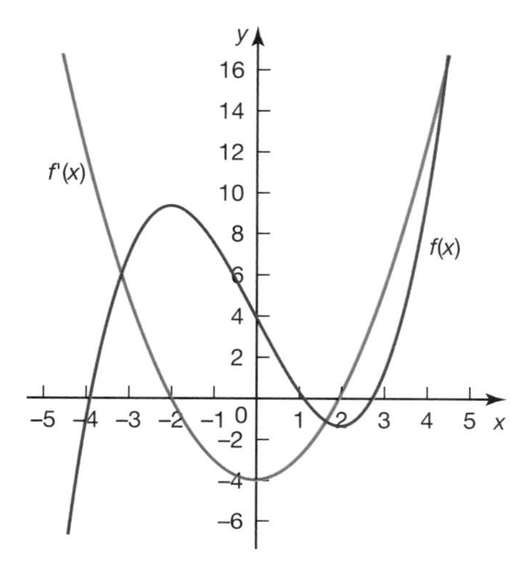


It can be seen that the derivative of the function $f(x)$ is a quadratic. The derivative can be calculated to be $f'(x) = x^2 - 2$.

This suggests that the derivative of a function with several terms can be found by differentiating each of the terms individually, which is indeed the case.

4 Differentiate the function $f(x) = \dfrac{2x^3 + x^2}{x}$ with respect to $x$.

Before differentiating, rewrite functions as sums of terms in powers of $x$.

$\dfrac{2x^3 + x^2}{x}$ can be written as $\dfrac{2x^3}{x} + \dfrac{x^2}{x}$

and simplified to $2x^2 + x$.

Therefore $f(x) = \dfrac{2x^3 + x^2}{x} = 2x^2 + x \Rightarrow \dfrac{dy}{dx} = 4x + 1$.

Note: A common error here is to differentiate each of the terms individually.

The derivative of $\dfrac{2x^3 + x^2}{x}$ is *not* $\dfrac{6x^2 + 2x}{1}$.

## Exercise 7.2.1

1 Differentiate each of the following expressions with respect to $x$.

a $5x^3$ b $7x^2$ c $4x^6$

d $\frac{1}{4}x^2$ e $\frac{2}{3}x^6$ f $\frac{3}{4}x^5$

g $5$ h $6x$ i $\frac{1}{8}$

2 Differentiate the following expressions with respect to $x$.

a $3x^2 + 4x$ b $5x^3 - 2x^2$ c $10x^3 - \frac{1}{2}x^2$

d $6x^3 - 3x^2 + x$ e $12x^4 - 2x^2 + 5$ f $\frac{1}{3}x^3 - \frac{1}{2}x^2 + x - 4$

g $-3x^4 + 4x^2 - 1$ h $-6x^5 + 3x^4 - x + 1$ i $-\frac{3}{4}x^6 + \frac{2}{3}x^3 - 8$

3 Differentiate the following expressions with respect to $x$.

a $\dfrac{x^3 + x^2}{x}$ b $\dfrac{4x^3 - x^2}{x^2}$ c $\dfrac{6x^3 + 2x^2}{2x}$

d $\dfrac{x^3 + 2x^2}{4x}$ e $3x(x + 1)$ f $2x^2(x - 2)$

g $(x + 5)^2$ h $(2x - 1)(x + 4)$ i $(x^2 + x)(x - 3)$

## Negative powers of *x*

So far all the polynomials that have been differentiated have had positive powers of $x$. This section looks at the derivative of polynomials with negative powers of $x$, for example differentiating from first principles the function $f(x) = x^{-1}$ with respect to $x$.

You will know from your work on indices that $x^{-1}$ can be written as $\frac{1}{x}$.

The function $f(x)$ is shown graphically with $P\left(x, \frac{1}{x}\right)$ and $Q\left(x + h, \frac{1}{x + h}\right)$.

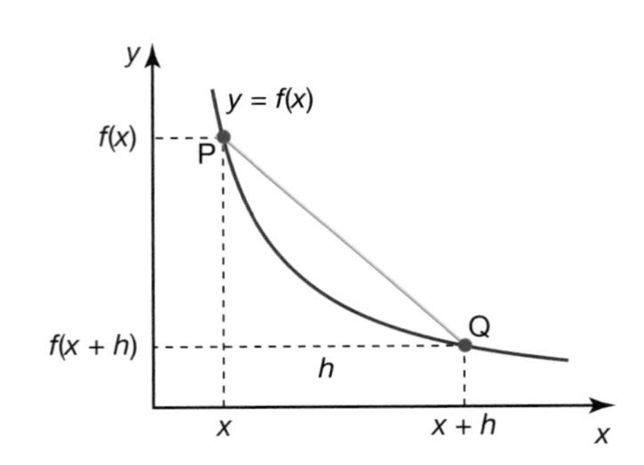


Gradient of line segment PQ

$$= \frac{\frac{1}{x} - \frac{1}{x+h}}{x - (x+h)} = \frac{\frac{x+h}{x(x+h)} - \frac{x}{x(x+h)}}{-h} = \frac{\frac{h}{x(x+h)}}{-h} = -\frac{1}{x(x+h)}$$

As Q gets closer to P, $h$ gets smaller and therefore the gradient of PQ tends to $-\frac{1}{x^2}$.

This tells us that for $f(x) = x^{-1}$, the derivative $f'(x) = -\frac{1}{x^2} = -x^{-2}$.

It can be seen that the original rule for the derivative of polynomials, namely if $f(x) = ax^n$ then $f'(x) = anx^{n-1}$, is still valid when $n$ is negative.

$f(x) = x^{-1}$

$f'(x) = -1 \times x^{(-1-1)} = -x^{-2}$

**Worked examples**

1 Find the derivative of $x^{-2}$.

$$\frac{dy}{dx} = -2 \times x^{(-2-1)}$$
$$= -2x^{-3} = -\frac{2}{x^3}$$

2 Calculate $\frac{dy}{dx}$, when $y = 2x^{-1} + x^{-2} + 2$.

$$\frac{dy}{dx} = -1 \times 2x^{(-1-1)} + -2 \times x^{(-2-1)} + 0$$
$$= -2x^{-2} - 2x^{-3}$$
$$= -2\left(\frac{1}{x^2} + \frac{1}{x^3}\right)$$

3 Differentiate $\frac{2}{x^5}$ with respect to $x$.

First write the expression $\frac{2}{x^5}$ in the form $ax^n$, where $a$ is a constant and $n$ an integer:

$$\frac{2}{x^5} = 2 \times \frac{1}{x^5} = 2x^{-5}$$

$$\frac{dy}{dx} = -5 \times 2x^{(-5-1)}$$
$$= -10x^{-6} = -\frac{10}{x^6}$$

## Exercise 7.2.2

1 Find the derivative of each of the following expressions.

a $x^{-1}$ b $x^{-3}$ c $2x^{-2}$ d $-x^{-2}$ e $-\frac{1}{3}x^{-3}$ f $-\frac{2}{5}x^{-5}$

2 Write down the following expressions in the form $ax^n$, where $a$ is a constant and $n$ an integer.

a $\frac{1}{x}$ b $\frac{2}{x}$ c $\frac{3}{x^2}$ d $\frac{2}{3x^3}$ e $\frac{3}{7x^2}$ f $\frac{2}{9x^3}$

3 Calculate $f'(x)$ for the following curves.

a $f(x) = 3x^{-1} + 2x$
b $f(x) = 2x^2 + x^{-1} + 1$
c $f(x) = 3x^{-1} - x^{-2} + 2x$
d $f(x) = \frac{1}{x^3} + x^3$
e $f(x) = \frac{2}{x^4} - \frac{1}{x^3} + 1$
f $f(x) = -\frac{1}{2x^2} + \frac{1}{3x^3}$

So far we have only used the variables $x$ and $y$ when finding the gradient function. This does not always need to be the case. Sometimes it is more convenient or appropriate to use other variables.

If a stone is thrown vertically upwards from the ground, with a speed of $10\,\mathrm{m\,s^{-1}}$, its distance ($s$) from its point of release is given by the formula $s = 10t - 4.9t^2$, where $t$ is the time in seconds after the stone's release.

This is represented graphically as shown.

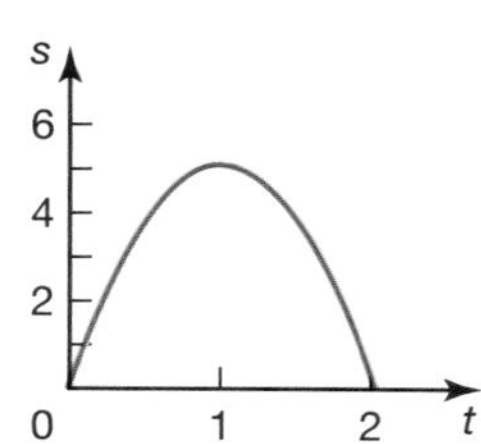

The velocity ($v$) of the stone at any point can be found by calculating the rate of change of distance with respect to time, i.e. $\frac{ds}{dt}$.

Therefore, if $s = 10t - 4.9t^2$ $\qquad v = \frac{ds}{dt} = 10 - 9.8t$

**Worked examples**

1 Calculate $\frac{ds}{dt}$ for the function $s = 6t^2 - 4t + 1$.

$\frac{ds}{dt} = 12t - 4$

2 Calculate $\frac{dr}{dt}$ for the function $r = \frac{6}{t^2} + 2t^2 - 1$.

Rewriting the function as $r = 6t^{-2} + 2t^2 - 1$

$\frac{dr}{dt} = -12t^{-3} + 4t = -\frac{12}{t^3} + 4t$

3 Calculate $\frac{dv}{dt}$ for the function $v = (r^3 + 1)\left(\frac{2}{r^2} - 1\right)$.

Expanding the brackets gives: $v = 2r - r^3 + \frac{2}{r^2} - 1$

Therefore $\frac{dv}{dr} = 2 - 3r^2 - \frac{4}{r^3}$

## Exercise 7.2.3

1 Differentiate each of the following with respect to $t$.

a $y = 3t^2 + t$ b $v = 2t^3 + t^2$ c $m = 5t^3 - t^2$

d $y = 2t^{-1}$ e $r = \frac{1}{2}t^{-2}$ f $s = t^4 - t^{-2}$

2 Calculate the derivative of each of the following functions.

a $y = 3x^{-1} + 4$ b $s = 2t^{-1} - t$ c $v = r^{-2} - \frac{1}{r}$

d $P = \frac{l^{-4}}{2} + 2l$ e $m = \frac{n}{2} - \frac{n^{-3}}{3}$ f $a = \frac{2t^{-2}}{5} - t^3$

3 Calculate the derivative of each of the following functions.

a $y = x(x + 4)$ b $r = t(1 - t)$ c $v = t\left(\frac{1}{t} + t^2\right)$

d $p = r^2\left(\frac{2}{r} - 3\right)$ e $a = x\left(x^{-2} + \frac{x}{2}\right)$ f $y = t^{-1}\left(t - \frac{1}{t^2}\right)$

4 Differentiate each of the following with respect to $t$.

a $y = (t + 1)(t - 1)$ b $r = (t - 1)(2t + 2)$

c $p = \left(\frac{1}{t} + 1\right)\left(\frac{1}{t} - 1\right)$ d $a = (t^{-2} + t)(t^2 - 2)$

e $v = \left(\frac{2t^2}{3} + 1\right)(t - 1)$ f $y = \left(\frac{3}{2t^4} - t\right)\left(2t - \frac{3}{t}\right)$

## 7.3 The gradient of a curve at a given point

You have seen that differentiating the equation of a curve gives the general equation for the gradient of any point on the curve. Using the general equation of the gradient, gradients at specific points on the curve can be calculated.

For the function $f(x) = \frac{1}{2}x^2 - 2x + 4$, the gradient function $f'(x) = x - 2$. The gradient at any point on the curve can be calculated using this.

For example, when $x = 4$, $f'(x) = 4 - 2$
$= 2$

i.e. the gradient of the curve $f(x) = \frac{1}{2}x^2 - 2x + 4$ is 2 when $x = 4$, as shown below.

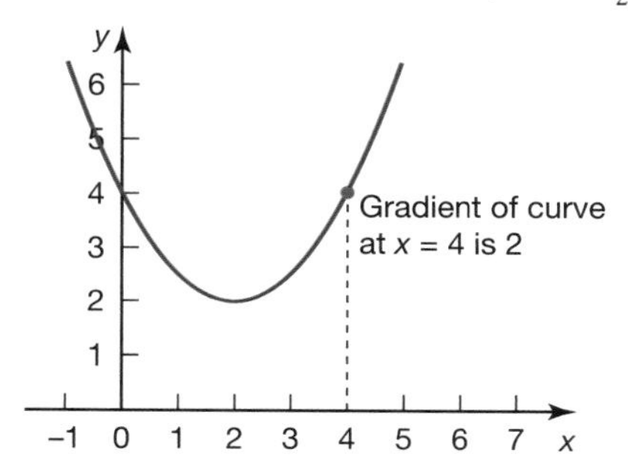


GDCs and graphing software can also help to visualize the question and check the solution.

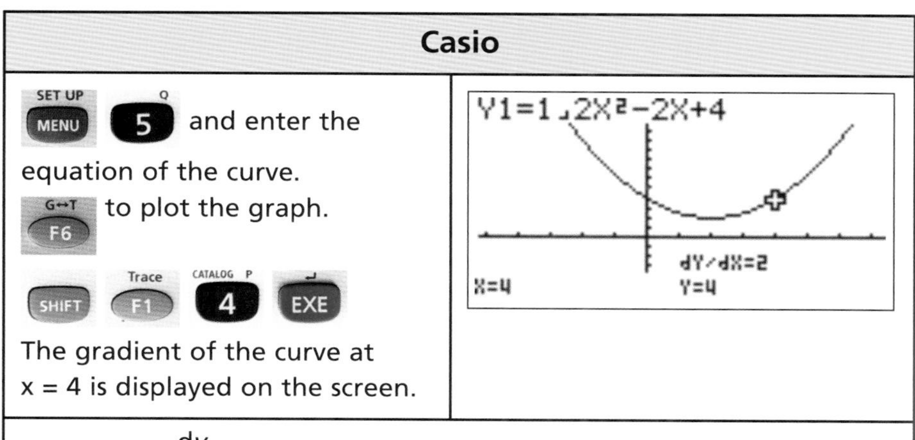


Note: If the $\frac{dy}{dx}$ derivative feature is not displayed on the screen, it can be turned on via the set up menu.

## Texas

 and enter the equation of the curve.

 to plot the graph.

  and select $\frac{dy}{dx}$.

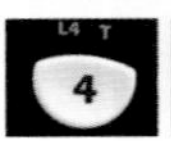  to calculate the gradient when $x = 4$.

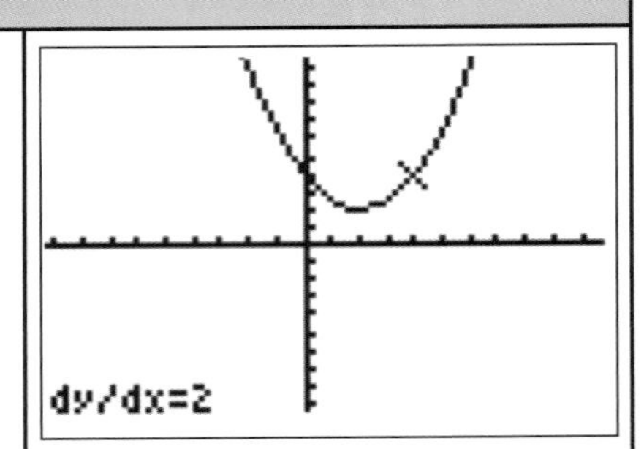


## Autograph

- Type equation $y = \frac{1}{2}x^2 - 2x + 4$
- Select the curve.
  Click on coordinate icon (,) and enter the value $x = 4$.

  A point is plotted on the curve at $x = 4$.
- Click on the point. Select 'object' followed by 'tangent'.

A tangent is drawn to the curve at $x = 4$. Its equation and the coordinate of the point are displayed at the base of the screen.

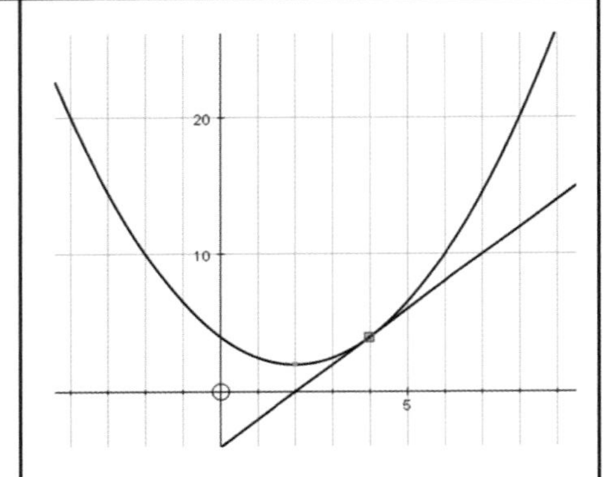


Note: The equation of the tangent is displayed at the base of the screen, i.e. $y = 2x - 4$. The gradient can therefore be deduced from this.

## GeoGebra

- Type equation
  $f(x) = \frac{1}{2}x^\wedge 2 - 2x + 4$
- Type 'Tangent[4, f]'. This draws a tangent to the curve at $x = 4$.

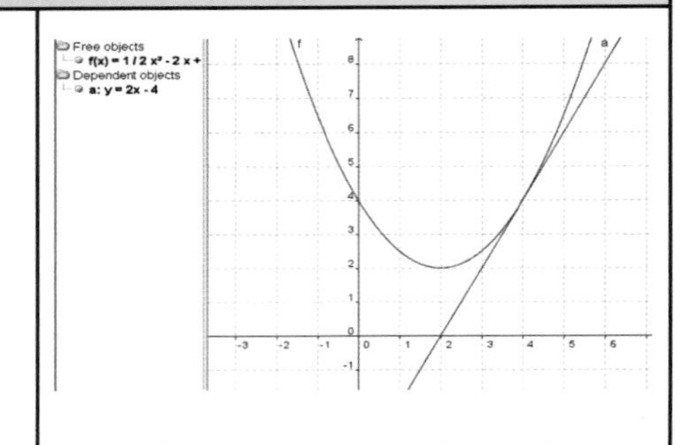


Note: The equation of the tangent is displayed in the algebra window, i.e. $y = 2x - 4$. The gradient can therefore be deduced from this.

**Worked example**

Calculate the gradient of the curve $f(x) = x^3 + x - 6$ when $x = -1$.

The gradient function $f'(x) = 3x^2 + 1$.

When $x = -1$, $f'(-1) = 3(-1)^2 + 1 = 4$,

i.e. the gradient is 4.

## Exercise 7.3.1

1 Find the gradient of each of the following functions at the given values of $x$.

- a $f(x) = x^2$; $x = 3$
- b $f(x) = \frac{1}{2}x^2 - 2$; $x = -3$
- c $f(x) = 3x^3 - 4x^2 - 2$; $x = 0$
- d $f(x) = -x^2 + 2x - 1$; $x = 1$
- e $f(x) = -\frac{1}{2}x^3 + x - 3$; $x = -1$, $x = 2$
- f $f(x) = 6x$; $x = 5$

2 Find the gradient of each of the following functions at the given values of $x$.

- a $f(x) = \frac{1}{x}$; $x = 2$
- b $f(x) = \frac{1}{x^2}$; $x = 1$
- c $f(x) = \frac{1}{x^3} - 3x$; $x = 2$
- d $f(x) = x^2 - \frac{1}{2x^2}$; $x = -1$
- e $f(x) = \frac{1}{6x^3} + x^2 - 1$; $x = 2$
- f $f(x) = \frac{1}{x} - \frac{1}{x^2} + \frac{1}{x^3}$; $x = \frac{1}{2}$, $x = -\frac{1}{2}$

3 The number of people, $N$, newly infected on day $t$ of a stomach bug outbreak is given by $N = 5t^2 - \frac{1}{2}t^3$.

- a Calculate the number of new infections, $N$, when:
  **i)** $t = 1$ **ii)** $t = 3$ **iii)** $t = 6$ **iv)** $t = 10$.
- b Calculate the rate of new infections with respect to $t$, i.e. calculate $\frac{dN}{dt}$.
- c Calculate the rate of new infections when:
  **i)** $t = 1$ **ii)** $t = 3$ **iii)** $t = 6$ **iv)** $t = 10$.
- d Using a GDC, sketch the graph of $N$ against $t$.
- e Using your graph as a reference, explain your answers to part **a**.
- f Using your graph as a reference, explain your answers to part **c**.

4 A weather balloon is released from the ground. Its height $h$ (m) after time $t$ (hours) is given by the formula $h = 30t^2 - t^3$, $t \leq 20$.

   a Calculate the balloon's height when:
      i) $t = 3$  ii) $t = 10$.
   b Calculate the rate at which the balloon is climbing with respect to time $t$.
   c Calculate the rate of ascent when:
      i) $t = 2$  ii) $t = 5$  iii) $t = 20$.
   d Using a GDC, sketch the graph of $h$ against $t$.
   e Using your graph as a reference, explain your answers to part **c**.
   f Use your graph to estimate the time when the balloon was climbing at its fastest rate. Explain your answer.

## Calculating *x*, when the gradient is given

So far we have calculated the gradient of a curve for a given value of $x$. It is possible to work backwards and calculate the value of $x$, when the gradient at a point is given.

Consider the function $f(x) = x^2 - 2x + 1$. It is known that the gradient at a particular point on the curve is 4. It is possible to calculate the coordinate of the point.

The gradient function of the curve is $f'(x) = 2x - 2$.
As the gradient at this particular point is 4 (i.e. $f'(x) = 4$), an equation can be formed:

$$2x - 2 = 4$$
$$2x = 6$$
$$x = 3$$

Therefore, when $x = 3$, the gradient of the curve is 4.

Once again a GDC and graphing software can help solve this type of problem.

**Worked example**

The function $f(x) = x^3 - x^2 - 5$ has a gradient of 8 at a point P on the curve.

Calculate the possible coordinates of point P.

The gradient function $f'(x) = 3x^2 - 2x$

At P, $3x^2 - 2x = 8$.

This can be rearranged into the quadratic $3x^2 - 2x - 8 = 0$ and solved either algebraically or graphically as shown in Section 1.6.

Graphically

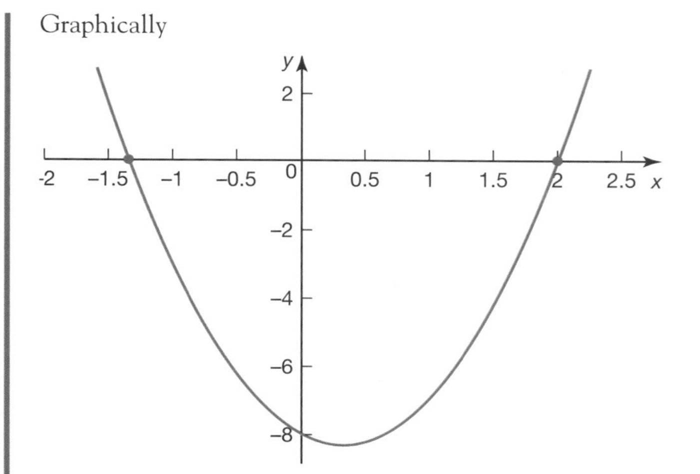


From the graph it can be seen that $x = -1\frac{1}{3}$ and $x = 2$, i.e. there are two possible positions for point P.

By substituting these values of $x$ into the equation of the curve, the $y$-coordinates of P can be calculated.

$f\left(-1\frac{1}{3}\right) = \left(-1\frac{1}{3}\right)^3 - \left(-1\frac{1}{3}\right)^2 - 5 = -\frac{247}{27} = -9\frac{4}{27}$

$f(2) = 2^3 - 2^2 - 5 = -1$

Therefore, the possible coordinates of P are $\left(-1\frac{1}{3}, -9\frac{4}{27}\right)$ and $(2, -1)$.

Algebraically

The quadratic equation $3x^2 - 2x - 8 = 0$ can be solved algebraically by factorizing.

$(3x + 4)(x - 2) = 0$

Therefore, $(3x + 4) = 0 \Rightarrow x = -\frac{4}{3}$ or $(x - 2) = 0 \Rightarrow x = 2$.

Your GDC can also be used to solve (quadratic) equations.

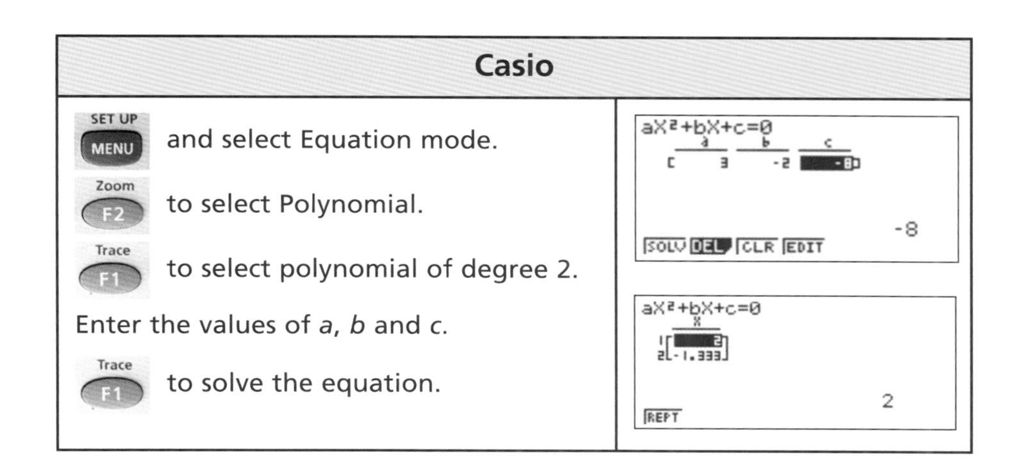

| Texas | |
|---|---|
| 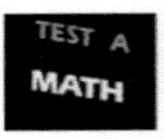  to select equation solver and then enter the equation. | EQUATION SOLVER<br>eqn:0=3X²−2X−8 |
|   to find one solution.<br>Change the value of $x$ to a number closer to the second solution, e.g. $x = 1$. | 3X²−2X−8=0<br>▪X=−1.333333333...<br>bound={−1E99,1...<br>▪left−rt=0 |
|   to find the second solution. | 3X²−2X−8=0<br>▪X=2<br>bound={−1E99,1...<br>▪left−rt=0 |
| Note: To find the second solution, you will already need to know approximately where it is. This can be done by graphing the equation first. | |

Once the values of $x$ have been calculated, the values of $y$ can be calculated as before.

## Exercise 7.3.2

1 Find the coordinate of the point P on each of the following curves, at the given gradient.

a $f(x) = x^2 - 3$, gradient $= 6$

b $f(x) = 3x^2 + 1$, gradient $= 15$

c $f(x) = 2x^2 - x + 4$, gradient $= 7$

d $f(x) = \frac{1}{2}x^2 - 3x - 1$, gradient $= -3$

e $f(x) = \frac{1}{3}x^2 + 4x$, gradient $= 6$

f $f(x) = -\frac{1}{5}x^2 + 2x + 1$, gradient $= 4$

2 Find the coordinate(s) of the point(s) on each of the following curves, at the given gradient.

a $f(x) = \frac{1}{3}x^3 + \frac{1}{2}x^2 + 4x$, gradient $= 6$

b $f(x) = \frac{1}{3}x^3 + 2x^2 + 6x$, gradient $= 3$

c $f(x) = \frac{1}{3}x^3 - 2x^2$, gradient $= -4$

d $f(x) = x^3 - x^2 + 4x$, gradient $= 5$

3 A stone is thrown vertically downwards off a tall cliff. The distance ($s$) it travels in metres is given by the formula $s = 4t + 5t^2$, where $t$ is the time in seconds after the stone's release.

a What is the rate of change of distance with time $\frac{ds}{dt}$? (This represents the velocity.)

b How many seconds after its release is the stone travelling at a velocity of 9 m s$^{-1}$?

c The stone hits the ground travelling at 34 m s$^{-1}$. How many seconds did the stone take to hit the ground?

d Using your answer to part **c**, calculate the distance the stone falls and hence the height of the cliff.

4 The temperature ($T$ °C) inside a pressure cooker is given by the formula $T = 20 + 12t^2 - t^3$; $t \leq 8$, where $t$ is the time in minutes after the cooking started.

a Calculate the temperature at the start.

b What is the rate of temperature increase with time?

c What is the rate of temperature increase when:

**i)** $t = 1$ **ii)** $t = 4$ **iii)** $t = 8$?

d The pressure cooker was switched off when $\frac{dT}{dt} = 36$.

How long after the start could the pressure cooker have been switched off?

e What was the temperature of the pressure cooker if it was switched off at the greater of the two times calculated in part **d**?

## Equation of the tangent at a given point

As was seen in Section 7.1, the gradient of a tangent drawn at a point on a curve is equal to the gradient of the curve at that point.

**Worked example**

What is the equation of the tangent to $f(x) = \frac{1}{2}x^2 + 3x + 1$ in the graph below?

The function $f(x) = \frac{1}{2}x^2 + 3x + 1$ has a gradient function of $f'(x) = x + 3$

At point P, where $x = 1$, the gradient of the curve is 4.

The tangent drawn to the curve at P also has a gradient of 4.

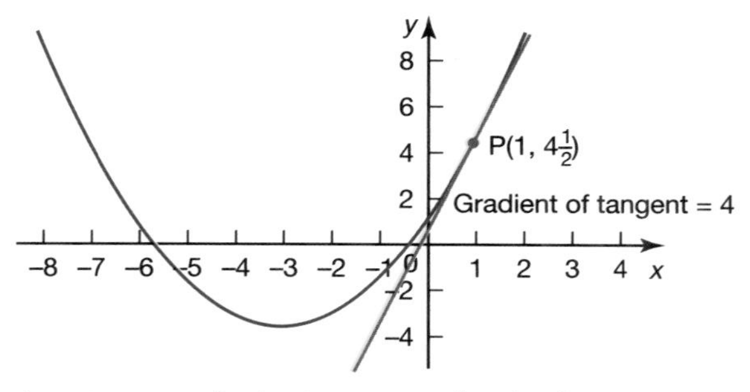


As it is a straight line, it must take the form $y = mx + c$. The gradient $m$ is 4 as shown above.

Therefore, $y = 4x + c$.

As the tangent passes through the point P(1, $4\frac{1}{2}$), these values can be substituted for $x$ and $y$ so that $c$ can be calculated.

$4\frac{1}{2} = 4 + c$

$\Rightarrow c = \frac{1}{2}$

The equation of the tangent is therefore $y = 4x + \frac{1}{2}$.

## Equation of a normal to a curve

You saw in Section 5.1 that there is a relationship between the gradients of lines that are perpendicular (at right angles) to each other. If two lines are perpendicular to each other, the product of their gradients is −1, i.e. $m_1 m_2 = -1$. Therefore the gradient of one line is the negative reciprocal of the other line. i.e. $m_1 = -\frac{1}{m_2}$.

In the previous section, it was possible to work out the equation of a tangent to the curve at a given point. A line drawn at right angles to the tangent, passing through the same point on the curve, is known as the **normal** to the curve at that point.

It is possible to calculate the equation of the normal to the curve in a similar way to calculating the equation of the tangent.

**Worked examples**

In the previous example the equation of the tangent to the curve $f(x) = \frac{1}{2}x^2 + 3x + 1$ at the point P(1, $4\frac{1}{2}$) was calculated as $y = 4x + \frac{1}{2}$. Calculate the equation of the normal to the curve at P.

The curve, the tangent and normal at P are shown below.

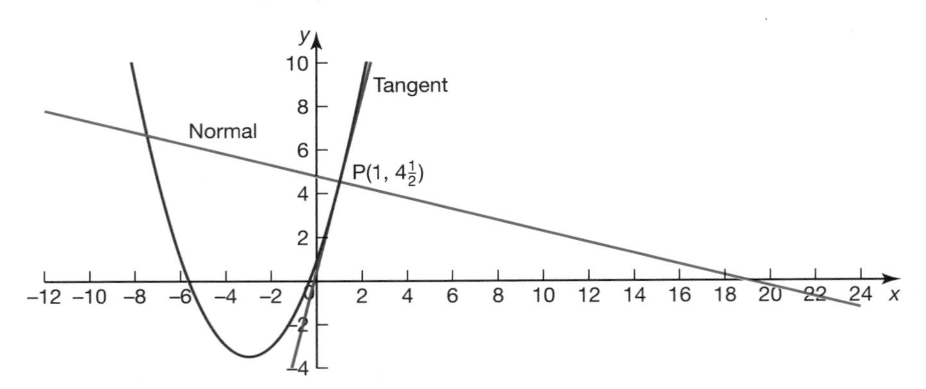


It is already known that the tangent at P has a gradient of 4.

As the product of the gradients of the tangent and normal is −1, the gradient of the normal can be calculated to be $-\frac{1}{4}$.

The normal is a straight line, so its equation must take the form $y = mx + c$.

As the normal also passes through the point P(1, $4\frac{1}{2}$), the values of $x$, $y$ and $m$ can be substituted into $y = mx + c$ so that $c$ can be calculated.

$$4\frac{1}{2} = (-\frac{1}{4} \times 1) + c$$

$$c = 4\frac{3}{4} = \frac{19}{4}$$

The equation of the normal is therefore $y = -\frac{1}{4}x + \frac{19}{4}$.

In the form $ax + by + c = 0$ the equation of the normal is as follows:

$$4y = -x + 19$$
$$x + 4y - 19 = 0$$

## Exercise 7.3.3

1 For the function $f(x) = x^2 - 3x + 1$:
  a calculate the gradient function
  b calculate the gradient of the curve at the point A(2, 1).
 A tangent is drawn to the curve at A.
  c What is the gradient of the tangent?
  d Calculate the equation of the tangent in the form $y = mx + c$.
  e What is the gradient of the normal to the curve at A?
  f Calculate the equation of the normal in the form $y = mx + c$.

2 For the function $f(x) = 2x^2 - 4x - 2$:
  a calculate the gradient of the curve where $x = 2$.
 A tangent is drawn to the curve at the point (2, −2).
  b Calculate the equation of the tangent in the form $y = mx + c$.
  c What is the gradient of the normal to the curve at the point (2, −2)?
  d Calculate the equation of the normal in the form $ax + by + c = 0$.

3 A tangent is drawn to the curve $f(x) = \frac{1}{2}x^2 - 4x - 2$ at the point P(0, −2).
  a Calculate the gradient of the tangent at P.
  b Calculate the equation of the tangent in the form $y = mx + c$.
  c Calculate the equation of the normal to the curve at P. Give your answer in the form $y = mx + c$.

4 A tangent $T_1$ is drawn to the curve $f(x) = -\frac{1}{4}x^2 - 3x + 1$ at the point P(−2, 6).
  a Calculate the equation of $T_1$.
 Another tangent, $T_2$, with equation $y = 10$, is also drawn to the curve at a point Q.
  b Calculate the coordinates of point Q.
  c $T_1$ and $T_2$ are extended so that they intersect. Calculate the coordinates of the point of intersection.

5 A tangent $T_1$ is drawn to the curve $f(x) = -x^2 + 4x + 1$ at the point A(4, 1).
  a Calculate the gradient of the tangent at A.
  b Calculate the equation of the tangent in the form $y = mx + c$.
 Another tangent $T_2$ is drawn at the point B(2, 5).
  c Calculate the equation of $T_2$.
  d Calculate the equations of the normals $N_1$ and $N_2$, where $N_1$ is the normal to $T_1$ at point A and $N_2$ is the normal to $T_2$ at point B. Give your answers in the form $y = mx + c$.
  e $N_1$ and $N_2$ are extended so that they intersect. Calculate the coordinates of the point of intersection.

6 A tangent T, drawn on the curve $f(x) = -\frac{1}{2}x^2 - x - 4$ at P, has an equation $y = -3x - 6$.
   a Calculate the gradient function of the curve.
   b What is the gradient of the tangent T?
   c What are the coordinates of the point P?

## 7.4 Increasing and decreasing functions

The graph shows the heart rate of an adult male over a period of time.

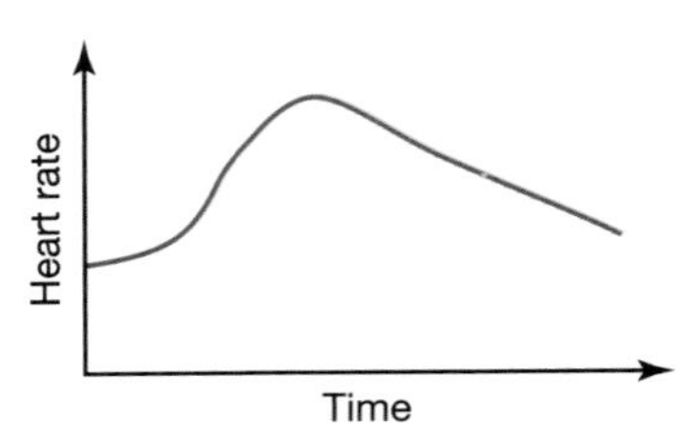


By simply looking at the graph, it is easy to see when his heart rate is increasing, decreasing and when it is at a maximum.

By comparing the shape of the graph with its gradient, we can see that when the gradient is positive, the heart rate is increasing. When the gradient is negative, the heart rate is decreasing. When the gradient is zero, the heart rate is at its maximum.

This section will look at the properties of increasing and decreasing functions.

For the function $f(x) = x^2$ the following observations can be made.

When $x > 0$ the gradient is positive, therefore $f(x)$ is an increasing function for this range of values of $x$.

When $x < 0$ the gradient is negative, therefore $f(x)$ is a decreasing function for this range of values of $x$.

When $x = 0$ the gradient is zero, therefore $f(x)$ is stationary at this value of $x$.

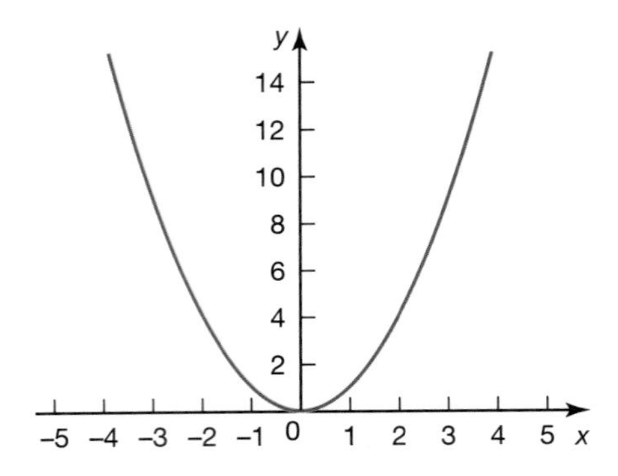

For questions 3 and 4:

i) calculate the gradient function
ii) calculate the coordinates of any stationary points
iii) determine the type of stationary point
iv) calculate the value of the $y$-intercept
v) sketch the graph of the function.

**3 a** $f(x) = 1 - 4x - x^2$ **b** $f(x) = \frac{1}{3}x^3 - 4x^2 + 12x - 3$
**c** $f(x) = -\frac{2}{3}x^3 + 3x^2 - 4x$ **d** $f(x) = x^3 - \frac{9}{2}x^2 - 30x + 4$

**4 a** $f(x) = x^3 - 9x^2 + 27x - 30$ **b** $f(x) = x^4 - 4x^3 + 16x$

## 7.6 Optimization

You will come across this in the Physics Diploma course.

You will have seen earlier in this topic that the process of calculating a rate of change, known as differentiation, is useful for calculating maxima and minima. This has many applications in the real world, in particular with optimization problems. Optimization problems are often concerned with minimising costs, maximising profits, minimising waste, etc. and therefore differentiation is a useful tool.

**Worked example**

A cardboard box manufacturer uses sheets of cardboard, with dimensions 50 cm × 75 cm, to make the main part of the box (i.e. the box without the lid).

The box is made by cutting squares out of each corner and then folding up the sides as shown below:

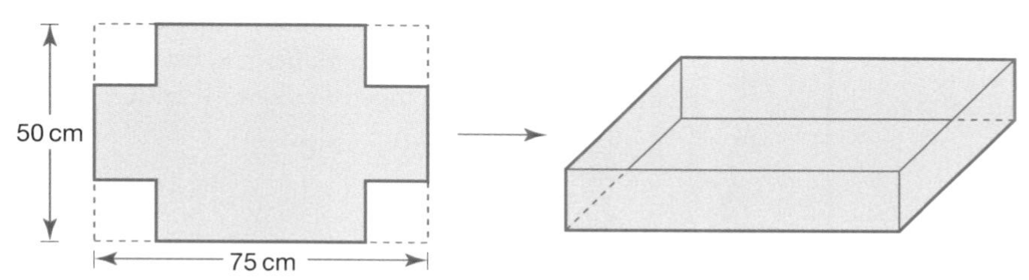


Calculate the maximum volume possible for the box.

The shape of the box depends on the size of the square removed from each corner of the sheet of cardboard. Let each square removed from each corner have a side length of $x$ cm.

The dimensions of the box will therefore be as shown:

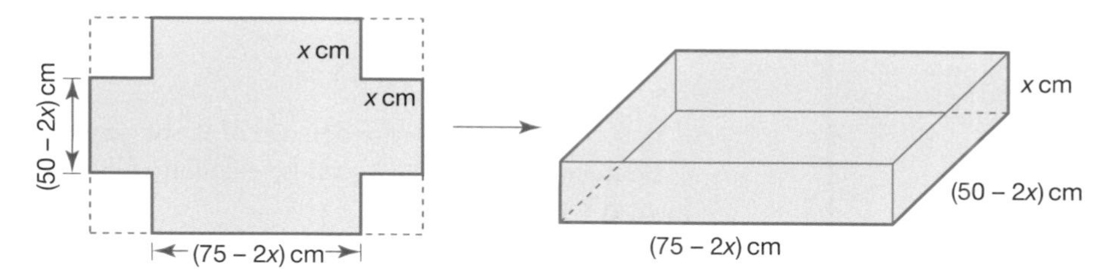

The volume of a cuboid is calculated by Length × Width × Height
In this example $V = (75 - 2x)(50 - 2x)(x)$
Expanded this gives $V = 4x^3 - 250x^2 + 3750x$
The above equation implies that the volume of the box is also dependent on the value of $x$ (i.e. the size of the square removed from each corner).
A graph can be plotted of $V$ against $x$ to see how the volume changes with $x$.

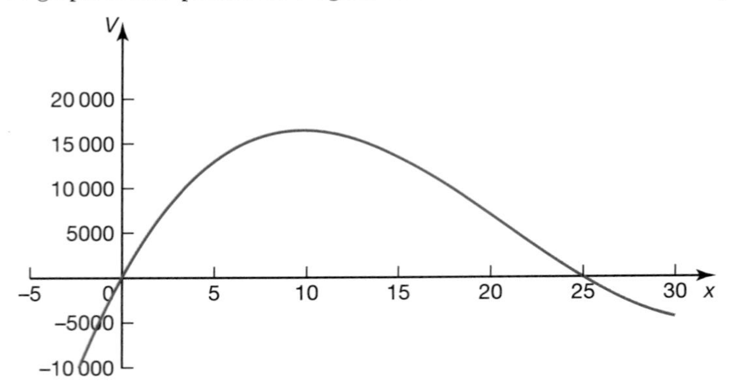


As volume cannot be negative, the values of $x$ must lie within $0 \le x \le 25$. This can also be deduced from the size of the card. As the width is only 50 cm, the maximum sized square that can be removed from each corner is one with a side length of 25 cm.

From the graph we can see that the volume changes as $x$ changes. The volume appears to have a maximum value when $x$ is approximately 10 cm.

To calculate the optimum value of $x$, we can use differentiation to calculate the rate of change of volume $V$ with $x$.

$$\frac{dV}{dx} = 12x^2 - 500x + 3750$$

At a maxima, $\frac{dV}{dx} = 0$

Therefore we need to solve the equation $12x^2 - 500x + 3750 = 0$

$$6x^2 - 250x + 1875 = 0$$

Using the quadratic formula to calculate $x$:

$$x = \frac{-b \pm \sqrt{b^2 - 4ac}}{2a}$$

$$x = \frac{250 \pm \sqrt{250^2 - 4 \times 6 \times 1875}}{2 \times 6}$$

$x = 9.8$ or $31.9$ (1 d.p.)

As 31.9 cm falls outside the domain of $x$, the optimum value of $x = 9.8$ cm.

The maximum volume $V$ can be calculated by substituting $x = 9.8$ into the equation for $V$.

$V = 4 \times 9.8^3 - 250 \times 9.8^2 + 3750 \times 9.8$
$V = 16\ 505\ \text{cm}^3$

## Exercise 7.6.1

1 A farmer has 500 m of fencing. He wants to make a rectangular enclosure for some of his cattle. One side of the enclosure runs along the side of an existing wall. The fencing will be used to make the other three sides of the enclosure as shown.

The farmer wants to build an enclosure with the greatest possible area.

- **a** If the length of the shorter side of the enclosure is $x$ metres, write down, in terms of $x$, an expression for the length of the longer side.
- **b** Write down a formula for the area A of the enclosure in terms of $x$.
- **c** Find the rate of change of A with $x$ (i.e. $\frac{dA}{dx}$).
- **d** Deduce the value of $x$ that produces the maximum area.
- **e** Calculate the maximum area of the enclosure.
- **f** Using a GDC or graphing software to help, sketch the graph of $x$ against A and find the coordinates of the maximum point.

2 A shop sells a type of mobile phone for $100. At this price the shop sells on average 150 of the phones each month. The owner of the store wishes to increase her profits. She knows that each $1 increase in the price of the phone will lead to two fewer phones being sold each month. The shop buys each phone for $60.

- **a** What is the profit currently made on each phone?
- **b** If the price of the phones is increased by $\$x$, write down, in terms of $x$, the profit now made on each phone sold.
- **c** Comment on why the new number of phones sold each month is now given by the expression $150 - 2x$.
- **d** Write down, in terms of $x$, a formula for the new profit $P$.
- **e** Calculate the rate of change of $P$ with $x$ (i.e. $\frac{dP}{dx}$).
- **f** What is the value of $x$ that maximises the profit?
- **g** Assuming that $x$ must be an integer value:
  - **i)** What price should each phone be sold for?
  - **ii)** What is the maximum profit?

3 The diagram shows a metal tin in the shape of a cuboid. The base of the tin is $x$ cm by $4x$ cm, the height of the tin is $h$ cm and its volume is $6400\,\text{cm}^3$.

a Find an expression for $h$ in terms of $x$.

b Show that the formula for the area ($A$ $\text{cm}^2$) of metal sheet used to make the tin is $A = 8x^2 + \dfrac{16000}{x}$

c Calculate $\dfrac{dA}{dx}$.

d Find the value of $x$ for which $A$ is a minimum.

e Prove that your value of $A$ is a minimum.

4 The diagram shows a closed cylinder used as a water tank. The radius of the base of the tank is $r$ cm, the height of the tank is $h$ cm. The total surface area of the tank is $50000\,\text{cm}^2$.

a Show that the volume ($V\,\text{cm}^3$) of the cylinder is given by $V = 25000r - \pi r^3$.

b Show that the maximum volume of the cylinder is $858400\,\text{cm}^3$ correct to four significant figures.

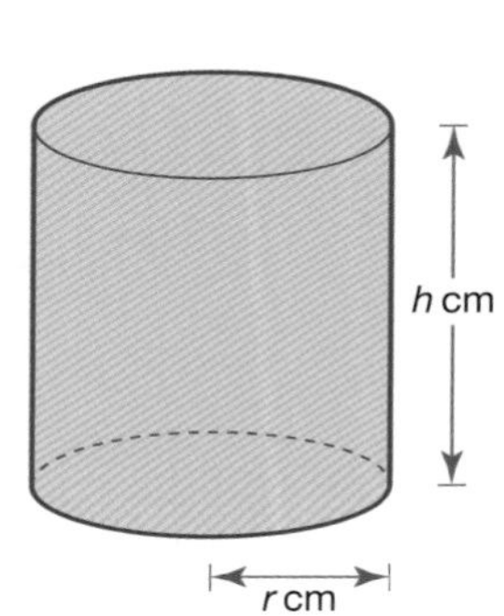


## Student assessment 1

1 Find the gradient function of each the following.

a $y = x^3$

b $y = 2x^2 - x$

c $y = -\frac{1}{2}x^2 + 2x$

d $y = \frac{2}{3}x^3 + 4x^2 - x$

2 Differentiate the following functions with respect to $x$.

a $f(x) = x(x + 2)$

b $f(x) = (x + 2)(x - 3)$

c $f(x) = \dfrac{x^3 - x}{x}$

d $f(x) = \dfrac{x^3 + 2x^2}{2x}$

e $f(x) = \dfrac{3}{x}$

f $f(x) = \dfrac{x^2 + 2}{x}$

3 Find the gradient of the following curves at the given values of $x$.

a $f(x) = \frac{1}{2}x^2 + x$; $\quad x = 1$

b $f(x) = -x^3 + 2x^2 + x$; $\quad x = 0$

c $f(x) = \dfrac{1}{2x^2} + x$; $\quad x = -\frac{1}{2}$

d $f(x) = (x - 3)(x + 8)$; $\quad x = \frac{1}{4}$

4 Determine the gradient of the normal to each of the curves in question 3 at the given values of $x$.

5 A stone is dropped from the top of a cliff. The distance it falls ($s$) is given by the equation $s = 5t^2$, where $s$ is the distance in metres and $t$ the time in seconds.

a Calculate the velocity $v$, by differentiating the distance $s$ with respect to time $t$.

b Calculate the stone's velocity after 3 seconds.

c The stone hits the ground travelling at $42\,\text{m s}^{-1}$. Calculate:

i) how long it took for the stone to hit the ground

ii) the height of the cliff.

## Student assessment 2

1 The function $f(x) = x^3 + x^2 - 1$ has a gradient of zero at points P and Q, where the $x$-coordinate of P is less than that of Q.
  a Calculate the gradient function $f'(x)$.
  b Calculate the coordinates of P.
  c Calculate the coordinates of Q.
  d Determine which of the points P or Q is a maximum. Explain your method clearly.

2 a Explain why the point A(1, 1) lies on the curve $y = x^3 - x^2 + x$.
  b Calculate the gradient of the curve at A.
  c Calculate the equation of the tangent to the curve at A.
  d Calculate the equation of the normal to the curve at A. Give your answer in the form $ax + by + c = 0$.

3 $f(x) = (x - 2)^2 + 3$
  a Calculate $f'(x)$.
  b Determine the range of values of $x$ for which $f(x)$ is a decreasing function.

4 $f(x) = x^4 - 2x^2$
  a Calculate $f'(x)$.
  b Determine the coordinates of any stationary points.
  c Determine the nature of any stationary point.
  d Find where the graph intersects or touches:
    i) the $y$-axis
    ii) the $x$-axis.
  e Sketch the graph of $f(x)$.

# Examination questions

1 Let $f(x) = 2x^2 + x - 6$.
  a Find $f'(x)$. [3]
  b Find the value of $f'(-3)$. [1]
  c Find the value of $x$ for which $f'(x) = 0$. [2]

**Paper 1, Nov 09, Q6**

2 Consider the function $f(x) = x^3 + \frac{48}{x}, x \neq 0$.
  a Calculate $f(2)$. [2]
  b Sketch the graph of the function $y = f(x)$ for $-5 \leq x \leq 5$ and $-200 \leq y \leq 200$. [4]
  c Find $f'(x)$. [3]
  d Find $f'(2)$. [2]
  e Write down the coordinates of the local maximum point on the graph of $f$. [2]
  f Find the range of $f$. [3]
  g Find the gradient of the tangent to the graph of $f$ at $x = 1$. [2]

There is a second point on the graph of $f$ at which the tangent is parallel to the tangent at $x = 1$.

  h Find the $x$-coordinate of this point. [2]

**Paper 2, May 11, Q3**

3 **Part A**
  a Sketch the graph of $y = 2^x$ for $-2 \leq x \leq 3$. Indicate clearly where the curve intersects the $y$-axis. [3]
  b Write down the equation of the asymptote of the graph $y = 2^x$. [2]
  c On the same axes sketch the graph of $y = 3 + 2x - x^2$. Indicate clearly where this curve intersects the $x$ and $y$ axes. [3]
  d Using your graphic display calculator, solve the equation $3 + 2x - x^2 = 2^x$. [2]
  e Write down the maximum value of the function $f(x) = 3 + 2x - x^2$. [1]
  f Use Differential Calculus to verify that your answer to part **e** is correct. [5]

**Part B**

The curve $y = px^2 + qx - 4$ passes through the point (2, –10).

  a Use the above information to write down an equation in $p$ and $q$. [2]

The gradient of the curve $y = px^2 + qx - 4$ at the point (2, –10) is 1.

  b i) Find $\frac{dy}{dx}$. [2]
    ii) Hence, find a second equation in $p$ and $q$. [3]
  c Solve the equations to find the value of $p$ and of $q$. [3]

**Paper 2, Nov 09, Q5**

Topic 7

# Applications, project ideas and theory of knowledge

1 If you look up the history of calculus, you will find that Leibniz and Newton both discovered calculus at the same time. Their work followed on from the earlier work of the Iraqi mathematician Ibn al Haytham. Is there a 'readiness' for major discoveries to be made? Discuss with reference to the telephone, television, radioactivity and the structure of DNA.

2 Terms used in calculus, such as 'differential', 'derivative' and 'integration', have other meanings elsewhere. How did these words become mathematical terms?

3 Discuss the statement that calculus is easier to do than to understand.

4 How far was China's 'one child' policy influenced by calculations of the probable future population of China? Did these predictions use calculus?

5 'Calculus is the point at which mathematics and real life part company'. Discuss with reference to calculus and other advanced mathematics.

6 Was calculus invented or discovered? Discuss with reference to point 1 and to the history of mathematics.

7 A possible project would be to extend your knowledge of calculus beyond the syllabus requirements of Mathematical Studies. Consult your teacher for some possible areas of study.

**8** 'A good GDC means everyone is capable of doing higher maths.' Discuss with particular reference to calculus.

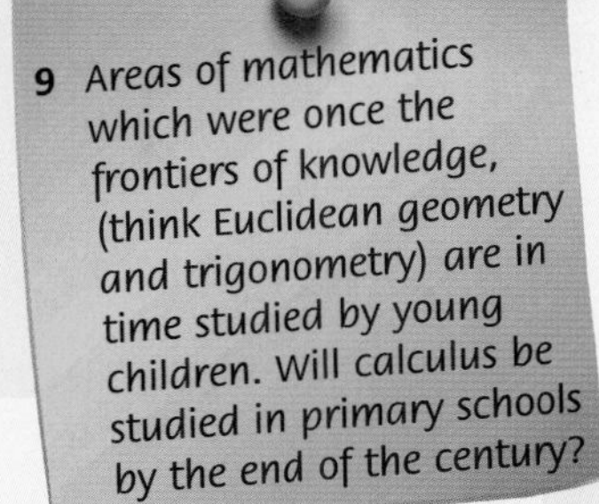

**9** Areas of mathematics which were once the frontiers of knowledge, (think Euclidean geometry and trigonometry) are in time studied by young children. Will calculus be studied in primary schools by the end of the century?

**10** The spread of a pandemic can be calculated using calculus. This could form the basis of a project.

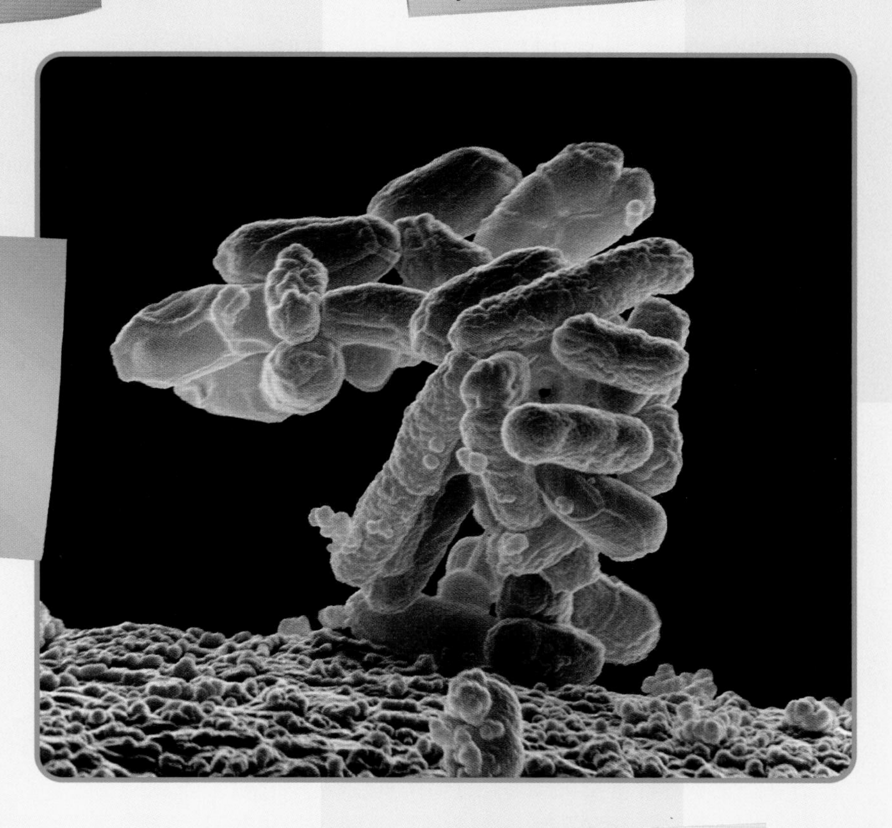

**11** The introduction to this topic suggests that a genius is one who 'brings ideas together in a new way'. Do you agree with this definition or do you have another?

**12** Look up Fermat's Last Theorem. The BBC Horizon series made a programme on Andrew Wiles' work to prove the theorem. It may be available and is a brilliant example of making a very difficult proof understandable. Watch it if you can.

# Revision exercises

## 1.2 Approximation

Approximation: decimal places; significant figures. Percentage errors. Estimation.

**1** Round each of the following numbers to the nearest:
**i** 1000 **ii** 100 **iii** 10
**a** 2842 **b** 12 938 **c** 9581 **d** 496

**2** Round each of the following numbers to:
**i** one decimal place **ii** two decimal places **iii** three decimal places.
**a** 2.1827 **b** 0.9181 **c** 9.9631 **d** 0.0386

**3** Round each of the following numbers to:
**i** one significant figure **ii** two significant figures **iii** three significant figures.
**a** 3.9467 **b** 20.36 **c** 0.015 48 **d** 0.9752

**4** Without using a calculator estimate the answer to the following calculations. Show your workings clearly.
**a** $305 \times 9$ **b** $19.2^2$ **c** $\frac{26.1 \times 3.8}{11}$
**d** $408 \div 18.8$ **e** $\frac{8.7^2 \times 1.9^2}{21.3}$ **f** $(32.2 \times 3.1)^2$

**5** Estimate the area of each of the following shapes. Show your workings clearly.

**a**

**b**
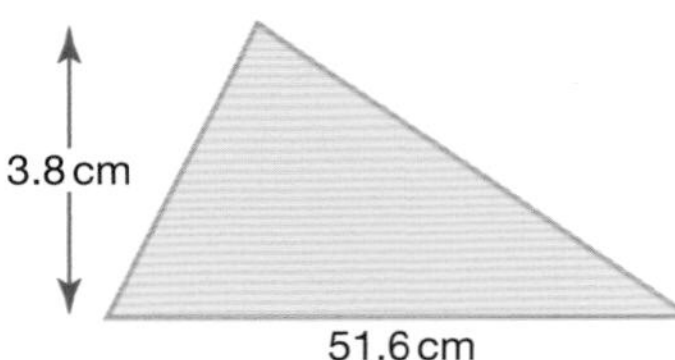


**c**
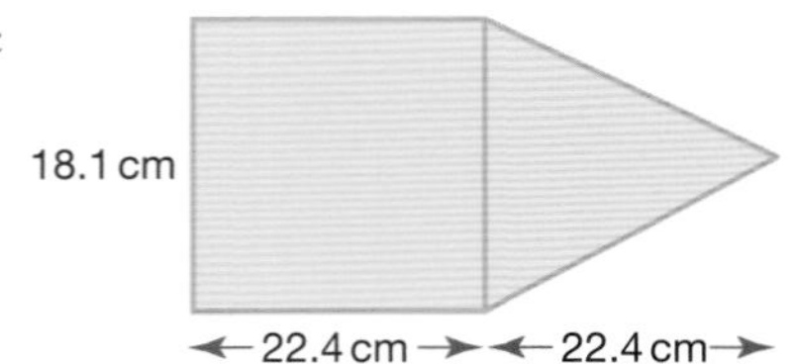


**d**
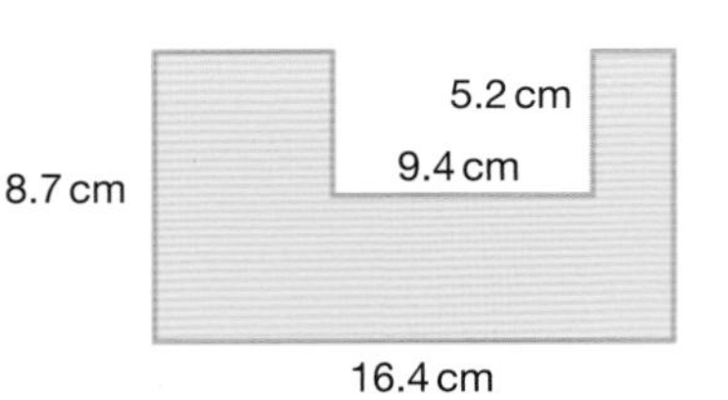


**6** **a** Calculate the actual area of each of the shapes in question 5.
**b** Calculate the percentage error in your estimates for each of the areas.

*continued on next page ...*

7 A certain brand of weighing scales claim that they are accurate to within ±3% of the actual mass being weighed.
A suitcase is weighed and the scales indicate a mass of 18.5 kg. Calculate
a the maximum possible mass of the case
b the minimum possible mass of the case.

8 The value $\pi$ is sometimes approximated to either 3 or $\frac{22}{7}$.
A circle has a radius of 8 cm.
a Using the $\pi$ button on your calculator, calculate the area of the circle, giving your answer correct to five decimal places.
b Calculate the percentage error in the area if $\pi$ is approximated to 3.
c Calculate the percentage error in the area if $\pi$ is approximated to $\frac{22}{7}$.

9 The formula for converting temperatures given in degrees Celsius (C) to temperatures in degrees Fahrenheit ($F$) is $F = \frac{9}{5}C + 32$.
a The temperature in a classroom is recorded as 18 °C. Convert this to degrees Fahrenheit using the formula above.
An approximate conversion is to use the formula $F = 2C + 30$.
b Calculate the temperature of the classroom in degrees Fahrenheit using the approximation above.
c Calculate the percentage error in using this approximation for a temperature of 18 °C.
d What would the percentage error be for a temperature of 30 °C?
e At what temperature would the percentage error be zero?

10 The formula for calculating the velocity of a stone dropped from rest off a cliff is given by $v = gt$, where $v$ is the velocity in $\text{m s}^{-1}$, $g$ the acceleration in $\text{m s}^{-2}$ and $t$ the time in seconds.
a Taking $g$ as $9.81\,\text{m s}^{-2}$, calculate the velocity of the stone after 6 seconds.
b Calculate the velocity of the stone if $g$ is approximated to $10\,\text{m s}^{-2}$.
c Calculate the percentage error in the approximation.

# 1.3 Standard form

Expressing numbers in the form $a \times 10^k$ where $1 \le a < 10$ and $k$ is an integer.
Operations with numbers in this form.

1 Which of the following numbers are not in the form $a \times 10^k$ where $1 \le a < 10$ and $k \in \mathbb{Z}$?

a $7.3 \times 10^3$ b $60.4 \times 10^2$ c $1.0 \times 10^{-2}$
d $0.5 \times 10^3$ e $3.874 \times 10^5$ f $8 \times 10^{-6}$

2 Write down the following numbers in the form $a \times 10^k$ where $1 \le a < 10$ and $k \in \mathbb{Z}$.

a 32 000 b 620 c 777 000 000
d 90 000 e 8 million f 48.5 million

3 The distance (in kilometres) from London to five other cities in the world is given below.

| | |
|---|---|
| London to Tokyo | 9567 km |
| London to Paris | 343 km |
| London to Wellington | 18 831 km |
| London to Cambridge | 78 km |
| London to Cairo | 3514 km |

Write down each of the distances in the form $a \times 10^k$ where $1 \le a < 10$ and $k \in \mathbb{Z}$ correct to two significant figures.

4 Calculate each of the following, giving your answers in the form $a \times 10^k$ where $1 \le a < 10$ and $k \in \mathbb{Z}$.

a $500 \times 6000$ b $20 \times 450\,000$
c 3 million $\times$ 26 d 5 million $\times$ 8 million

5 Write down the following in the form $a \times 10^k$ where $1 \le a < 10$ and $k \in \mathbb{Z}$.

a 0.04 b 0.0076 c 0.000 005 d 0.030 40

6 Write down the following numbers in ascending order of magnitude.

$3.6 \times 10^{-3}$ $2.5 \times 10^{-2}$ $7.4 \times 10^{-2}$
$9.8 \times 10^{-1}$ $8.7 \times 10^{-4}$ $1.4 \times 10^{-2}$

7 Calculate each of the following, giving your answer in the form $a \times 10^k$ where $1 \le a < 10$ and $k \in \mathbb{Z}$.

a $6.3 \times 10^2 \div 8.4 \times 10^5$ b $400 \div 800\,000$
c $7 \times 10^4 \div 4.2 \times 10^8$ d $\dfrac{1.5 \times 10^2}{9 \times 10^{10}}$

8 Deduce the value of $n$ in each of the following.

a $0.0003 = 3 \times 10^n$ b $0.000046 = 4.6 \times 10^n$
c $0.005^2 = 2.5 \times 10^n$ d $0.0006^n = 2.16 \times 10^{-10}$

9 A boy walks 40 km at a constant rate of $2\,\text{m}\,\text{s}^{-1}$.
Calculate how long, in seconds, the boy takes to walk the 40 km.
Give your answer in the form $a \times 10^k$ where $1 \le a < 10$ and $k \in \mathbb{Z}$.

10 The Earth's radius is approximately 6370 km.
Calculate the Earth's circumference in metres, giving your answer in the form $a \times 10^k$ where $1 \le a < 10$ and $k \in \mathbb{Z}$ correct to three significant figures.

# 1.4 SI units of measurement

SI (*Système International*) and other basic units of measurement: for example, kilogram (kg), metre (m), second (s), litre (l), metre per second (m $s^{-1}$), Celsius scale.

1 Write down an estimate for the following using the appropriate unit.
  a The mass of a large suitcase
  b The length of a basketball court
  c The height of a two-storey building
  d The capacity of a car's fuel tank
  e The distance from the North Pole to the South Pole
  f The mass of a table-tennis ball

2 Convert the following distances.
  a 20 cm into millimetres
  b 35 km into metres
  c 46 mm into centimetres
  d 60 m into kilometres
  e 320 m into millimetres
  f 95 mm into kilometres

3 Convert the following masses.
  a 100 kg into tonnes
  b 60 g into kilograms
  c 3.6 tonnes into kilograms
  d 14 g into milligrams
  e 8.67 kg into milligrams
  f 2560 g into tonnes

4 Convert the following capacities.
  a 2600 ml into litres
  b 80 ml into litres
  c 1.65 litres into millilitres
  d 0.085 litres into millilitres

5 The masses of four containers are as follows:

  25 kg   0.35 t   650 g   0.27 kg

  Calculate the total mass of the four containers in kilograms.

6 The lengths of five objects are as follows:

  56 mm   24 cm   0.672 m   1030 mm   1.5 cm

  Calculate the length, in metres, of the five objects if they are laid end to end.

7 The liquid contents of four containers are emptied into a tank with a capacity of 30 litres. The capacities of the four containers are as follows:

  3250 ml   1.05 litres   26000 ml   762 ml

  Calculate the overspill, in litres, after the liquids have been poured in.

# 1.6 Graphical solution of equations

Use of a GDC to solve: pairs of linear equations in two variables; quadratic equations.

*For these questions, use of a GDC or graphing software is expected.*

1 Sketch the following straight-line graphs on the same axes, labelling each clearly. Write the coordinate of the point at which they intercept the $y$-axis.

a $y = x - 5$ b $y = 2x - 5$ c $y = -x - 5$

2 The diagram shows four straight-line graphs. The line $y = x + 2$ is marked. Write down possible equations for the other three graphs.

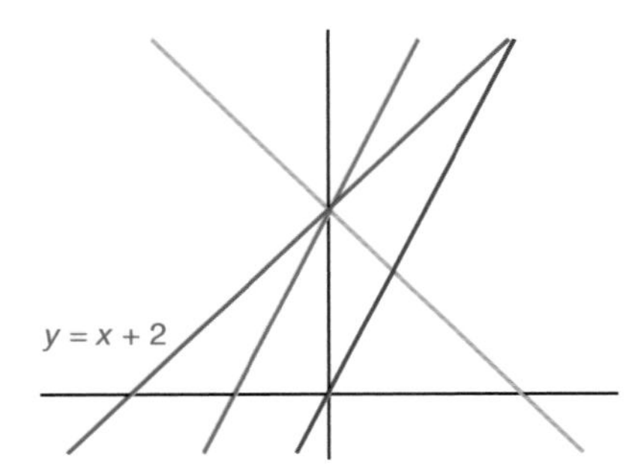


3 Find the coordinates of the points of intersection of the following pairs of linear graphs.

a $y = 8 - x$, $y = 2x - 1$
b $3x + 2y = 2$, $3y = x + 14$
c $y = 3 - 4x$, $3y + 10x = 16$
d $y = \frac{1}{2}x - 3$, $x - 4y = 6$

4 a Sketch the following linear graphs on the same axes.
$y = \frac{1}{2}x + 3$ $x = 2y + 8$
b Explain why there are no points of intersection.

5 Sketch each of the following quadratic equations on separate axes.
a $y = x^2 + 6x + 8$ b $y = x^2 - 16$
c $y = 9 - x^2$ d $y = -(x - 3)(x - 5)$

6 For each of the following quadratic equations:
i sketch the graph
ii find the coordinates of any roots.
a $y = x^2 - 10x + 21$ b $y = 12 + 4x - x^2$
c $y = -x^2 + 10x - 25$ d $y = 2x^2 - 12x + 20$
e $y = 8x^2 - 2x - 1$

# 1.7 Arithmetic sequences and series

Arithmetic sequences and series, and their applications.
Use of the formulae for the $n$th term and the sum of the first $n$ terms of a sequence.

**1** In each of the following sequences, the recurrence relation and $u_1$ are given.
  **i** Calculate and $u_2$, $u_3$ and $u_4$.
  **ii** State whether the sequence is arithmetic or not.
  **a** $u_{n+1} = u_n - 6, u_1 = 15$
  **b** $u_{n+1} = 12 - u_n, u_1 = 15$
  **c** $u_{n+1} = 3u_n + 2, u_1 = \frac{1}{2}$
  **d** $u_{n+1} = \frac{2u_n - 5}{2}, u_1 = 5$

**2** In each of the following arithmetic sequences
  **i** deduce the formula for the $n$th term
  **ii** calculate the 20th term.
  **a** 4, 9, 14, 19, 24
  **b** 3, −5, −13, −21, −29
  **c** −4.5, −2, 0.5, 3, 5.5
  **d** 3.5, 3.25, 3, 2.75, 2.5

**3** In the following arithmetic sequences:
  **i** deduce the common difference $d$
  **ii** deduce the formula for the $n$th term.
  **a** −25, ..., ..., ..., −1
  **b** 7, ..., ..., ..., ..., ..., −14
  **c** $u_4 = -12, u_{20} = 100$
  **d** $u_7 = 19, u_{42} = -128$

**4** Write down in full the terms of the following series.
  **a** $\sum_{1}^{5} 2n - 1$
  **b** $\sum_{3}^{7} -n + 6$
  **c** $\sum_{2}^{8} 6 - \frac{1}{2}n$
  **d** $\sum_{1}^{4} 3(-n + 2)$

**5** Write down the following arithmetic series using the $\sum$ notation. Each series starts at $n = 1$.
  **a** 2 + 6 + 10 + 14 + 18
  **b** 5 + 3 + 1 + −1 + −3 + −5
  **c** $-\frac{1}{2} + 1 + 2\frac{1}{2} + 4 + 5\frac{1}{2} + 7$
  **d** −4.1 + −4.2 + −4.3 + −4.4

**6** Solve the following.
  **a** $\sum_{1}^{50} 3n$
  **b** $\sum_{1}^{25} 20 - n$
  **c** $\sum_{20}^{30} n + 1$
  **d** $\sum_{4}^{32} -2n + 50$

**7** The 5th and 15th terms of an arithmetic series are −10 and 10 respectively.
  Calculate:
  **a** the common difference $d$
  **b** the first term
  **c** the 20th term
  **d** $S_{20}$.

**8** The 11th term of an arithmetic series is 65. If $S_{11} = 495$, calculate:
  **a** the first term
  **b** the common difference
  **c** $S_{20}$.

**9** The 7th term of an arithmetic series is 2.5 times the 2nd term, $x$. If the 10th term is 34, calculate:
  **a** the common difference in terms of $x$
  **b** the first term
  **c** the sum of the first 10 terms.

**10** The first term of an arithmetic series is 24. The last term is −12. If the sum of the series is 150, calculate the number of terms in the series.

# 1.8 Geometric sequences and series

Geometric sequences and series.
Use of the formula for the $n$th term and the sum of the first $n$ terms of the sequence.

1 Describe in words the difference between an arithmetic and a geometric sequence.

2 In each of the following geometric sequences, the recurrence relation and $u_1$ are given.
 i Calculate the values of $u_2$, $u_3$ and $u_4$.
 ii State whether the sequence is geometric or not.
 a $u_{n+1} = 4u_n + 2,\ u_1 = 0$
 b $u_{n+1} = -3u_n,\ u_1 = 1$
 c $u_{n+1} = \frac{5}{2}u_n,\ u_1 = 6$
 d $u_{n+1} = \dfrac{6 - 2u_n}{2},\ u_1 = 4$

3 For the geometric sequences below calculate:
 i the common ratio $r$
 ii the next two terms
 iii the formula for the $n$th term.
 a 5, 15, 45, 135
 b 1296, 216, 36, 6
 c 36, 24, 16, $10\frac{2}{3}$
 d 4, –10, 25, $-62\frac{1}{2}$

4 The $n$th term of a geometric sequence is given by the formula $u_n = -3 \times 4^{n-1}$.
Calculate:
 a $u_1$, $u_2$ and $u_3$
 b the value of $n$ if $u_n = -12\,288$.

5 Part of a geometric sequence is given as …, 27, …, …, 1, … where $u_2$ and $u_5$ are 27 and 1 respectively. Calculate:
 a the common ratio $r$
 b $u_1$
 c $u_{10}$.

6 A homebuyer takes out a loan with a mortgage company for \$300 000. The interest rate is fixed at 5.5% per year. If she is unable to repay the loan during the first four years, calculate the amount extra she will have to pay by the end of the fourth year, due to interest.

7 Solve the following sums.
 a $\sum_{1}^{5} 3^n$
 b $\sum_{1}^{6} -2(3)^{n-1}$
 c $\sum_{1}^{10} \frac{1}{2}(4)^n$
 d $\sum_{2}^{7} 9\left(-\frac{1}{3}\right)^n$

8 In a geometric series, $u_3 = 10$ and $u_6 = \frac{16}{25}$. Calculate:
 a the common ratio $r$
 b the first term
 c $S_7$.

9 Four consecutive terms of a geometric series are $(p - 5)$, $(p)$, $(2p)$ and $(3p + 10)$.
 a Calculate the value of $p$.
 b Calculate the two terms before $(p - 5)$.
 c If $u_3 = (p - 5)$, calculate $S_{10}$.

10 In a geometric series $u_1 + u_2 = 5$. If $r = \frac{2}{3}$, find the sum of the infinite series.

# 1.9 Simple interest and compound interest

Financial applications of geometric sequences and series: compound interest; annual depreciation.

1 Find the simple interest paid on the following capital sums C, deposited in a savings account for $n$ years at a fixed rate of interest of $r$%.
   a C = £550  $n$ = 5 years  $r$ = 3%
   b C = \$8000  $n$ = 10 years  $r$ = 6%
   c C = €12 500  $n$ = 7 years  $r$ = 2.5%

2 A capital sum of £25 000 is deposited in a bank. After 8 years, the simple interest gained was £7000. Calculate the annual rate of interest on the account assuming it remained constant over the 8 years.

3 A bank lends a business \$250 000. The annual rate of interest is 8.4%. When paying back the loan, the business pays an amount of \$105 000 in simple interest. Calculate the number of years the business took out the loan for.

4 \$15 000 is deposited in a savings account. The following arithmetic sequence represents the total amount of money in the savings account each year. Assume that no further money is either deposited or taken out of the account.

| **Number of years** | 0 | 1 | 2 | 3 | 4 | 5 | | $n$ |
|---|---|---|---|---|---|---|---|---|
| **Total savings in account (\$)** | 15 000 | 15 375 | 15 750 | 16 125 | 16 500 | 16 875 | | |

   a Explain, giving reasons, whether the sequence above simulates simple interest or compound interest.
   b Calculate the interest rate.
   c State the formula for calculating the total amount of money ($T$) in the account after $n$ years.
   d State the formula for calculating the total amount of interest ($I$) gained after $n$ years.

*continued on next page ...*

5 \$15 000 is deposited in a savings account. The following geometric sequence represents the total amount of money in the savings account each year. Assume that no further money is either deposited or taken out of the account.

| Number of years | Total savings in account (\$) |
|---|---|
| 0 | 15 000 |
| 1 | 16 500 |
| 2 | 18 150 |
| 3 | 19 965 |
| 4 | 21 961.50 |
| 5 | 24 157.65 |
| | |
| $n$ | |

a Explain, giving reasons, whether the sequence above simulates simple interest or compound interest.
b Calculate the interest rate.
c State the formula for calculating the total amount of money ($T$) in the account after $n$ years.
d State the formula for calculating the total amount of interest ($I$) gained after $n$ years.

6 Find the compound interest paid on the following capital sums C, deposited in a savings account for $n$ years at a fixed rate of interest of $r$% per year.

a C = £400 $n = 2$ years $r = 3\%$
b C = \$5000 $n = 8$ years $r = 6\%$
c C = €18 000 $n = 10$ years $r = 4.5\%$

7 A car is bought for €12 500. Its value depreciates by 15% per year.
a Calculate its value after:
i 1 year ii 2 years.
b After how many years will the car be worth less than €1000?

8 €4000 is invested for three years at 6% per year. What is the interest paid if the interest rate is compounded:
a yearly?
b half-yearly?
c quarterly?
d monthly?

# 2.3 and 2.5 Grouped discrete or continuous data and Measures of central tendency

Grouped discrete or continuous data: frequency tables; mid-interval values; upper and lower boundaries.
Frequency histograms.
Measures of central tendency.
For simple discrete data: mean; median; mode.
For grouped discrete and continuous data: estimate of a mean; modal class.

1 The amount of milk (in litres) drunk by a group of students in a week is given in the table below.

| **Number of litres** | 0 | 1 | 2 | 3 | 4 | 5 | 6 | 7 | 8 |
|---|---|---|---|---|---|---|---|---|---|
| **Frequency** | 6 | 1 | 4 | 9 | 22 | 16 | 2 | 4 | 1 |

a Draw a frequency histogram for this data.
b State the modal value.
c Calculate the mean number of litres drunk per student.
d Calculate the median number of litres drunk per student.

2 The masses $M$ kg of suitcases being checked-in for a flight at an airport are recorded. The results are shown below.

| **Mass (kg)** | $0 \le M < 5$ | $5 \le M < 10$ | $10 \le M < 15$ | $15 \le M < 20$ | $20 \le M < 25$ | $25 \le M < 30$ |
|---|---|---|---|---|---|---|
| **Frequency** | 6 | 18 | 64 | 105 | 94 | 18 |

a State the modal group.
b Estimate the mean mass of the suitcases.
c State, giving reasons, which group the median mass belongs to.

3 The mass $M$ kg of football players in a team is recorded.
For the team of 11 players, the total of their masses is 836 kg.
a Calculate the mean mass, $\overline{M}$, of the 11 players.
b The mean of 11 players and 1 substitute is 76.75 kg. Calculate the mass of the substitute.

continued on next page ...

4 The cost (C euros) of a litre of unleaded petrol at different petrol stations is shown in the frequency histogram below.

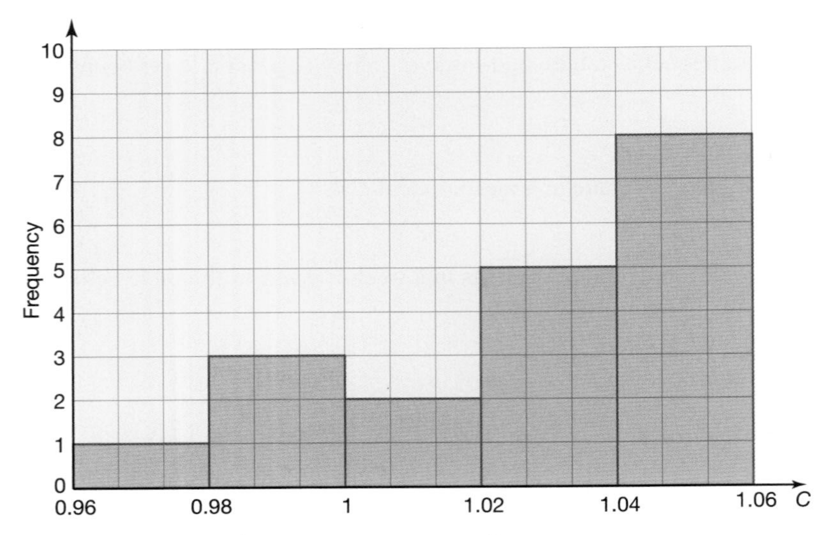


a How many petrol stations were surveyed?

b Estimate the mean price of unleaded petrol.

# 2.4 Cumulative frequency

Cumulative frequency tables for grouped discrete data and for grouped continuous data; cumulative frequency curves; median and quartiles.
Box and whisker diagrams.

1 The percentage test result ($x$) for 20 students in a class are given in the grouped frequency table below.

| **Percentage** | $0 \le x < 20$ | $20 \le x < 40$ | $40 \le x < 60$ | $60 \le x < 80$ | $80 \le x < 100$ |
|---|---|---|---|---|---|
| **Frequency** | 2 | 4 | 6 | 7 | 1 |

a Calculate the cumulative frequency for the data.
b Draw a cumulative frequency curve for the results.
c From your graph estimate the median test result.

2 A business surveys its employees to find how long it takes them to travel to work each morning ($t$ minutes). The results are displayed below.

| **Time (min)** | $0 \le t < 15$ | $15 \le t < 30$ | $30 \le t < 45$ | $45 \le t < 60$ | $60 \le t < 75$ |
|---|---|---|---|---|---|
| **Frequency** | 1 | 3 | 12 | 22 | 12 |

a Calculate the cumulative frequency for the data.
b Draw a cumulative frequency curve for the results.
c From your graph estimate the median time taken to travel to work.
d From your graph estimate the time taken to travel to work by the middle 50% of the employees.
e Draw a box plot to summarize the data.

3 150 students enter an international Maths competition. The scores are out of a maximum 300 points. The scores ($x$) for the 150 students are summarized in the table below.

| **Score** | $0 \le x < 50$ | $50 \le x < 100$ | $100 \le x < 150$ | $150 \le x < 200$ | $200 \le x < 250$ | $250 \le x < 300$ |
|---|---|---|---|---|---|---|
| **Frequency** | 7 | 25 | 56 | 32 | 21 | 9 |

a Draw a cumulative frequency curve of the scores.
b From your graph estimate the median score.
c Students in the top 25% are invited to take part in the next round of the competition. From your graph estimate the score needed for a student to be in the top 25%.

*continued on next page …*

4 The mass ($m$ grams) of forty apples of the same variety was recorded at a market and at a supermarket. The results are given in the table below.

| **Mass (g)** | $60 \le m < 100$ | $100 \le m < 140$ | $140 \le m < 180$ | $180 \le m < 220$ | $220 \le m < 260$ |
|---|---|---|---|---|---|
| **Frequency (market)** | 5 | 8 | 12 | 9 | 6 |
| **Frequency (supermarket)** | 0 | 3 | 28 | 9 | 0 |

a On the same axes draw two cumulative frequency graphs, one for the mass of the apples at the market and the other for the mass of apples at the supermarket.

b Using your graph complete the table below.

| | **Minimum value** | **Lower quartile** | **Median** | **Upper quartile** | **Maximum value** |
|---|---|---|---|---|---|
| **Market** | | | | | |
| **Supermarket** | | | | | |

c Use your results table above to draw a box and whisker diagram for each set of apple data.

# 2.6 Measures of dispersion

Measures of dispersion: range; interquartile range; standard deviation.

1 For the following lists of numbers calculate:
 i the range  ii the interquartile range.
 a 2 6 12 14 15 15 17 21 22 22
 b 26 27 1 14 18 7 19 3 12

2 The number of goals conceded by a football team is recorded. The results are given in the frequency table below.

| Number of goals conceded | 0 | 1 | 2 | 3 | 4 | 5 |
|---|---|---|---|---|---|---|
| Frequency | 7 | 6 | 2 | 5 | 2 | 1 |

Calculate:
 a the range in the number of goals scored
 b the interquartile range of the number of goals scored.

3 Calculate the standard deviation using the formula $s_n = \sqrt{\frac{\sum (x - \bar{x})^2}{n}}$ of the following lists of numbers.
 a 3 4 7 7 7 8 8 8 8 9
 b 1 4 6 8 10 12 15 18 24 30

4 The temperatures (in °C) at two holiday resorts are recorded every other day during the month of June. The results are given in the table below.

| Day | 1 | 3 | 5 | 7 | 9 | 11 | 13 | 15 | 17 | 19 | 21 | 23 | 25 | 27 | 29 |
|---|---|---|---|---|---|---|---|---|---|---|---|---|---|---|---|
| Temperature Resort A | 23 | 24 | 22 | 24 | 25 | 26 | 25 | 23 | 23 | 24 | 22 | 24 | 24 | 26 | 27 |
| Temperature Resort B | 15 | 17 | 24 | 28 | 33 | 33 | 26 | 22 | 22 | 19 | 16 | 16 | 15 | 26 | 31 |

 a Calculate the mean temperature for resort A and B.
 b Calculate the range of temperatures at both resorts.
 c Calculate the interquartile range of temperatures at both resorts.
 d Calculate the standard deviation of the temperature at both resorts.
 e Explain the meaning of your answer to part **d**.

5 The times (in seconds) taken by two sprinters to run 100 m during their training sessions are recorded and given below.

Sprinter A 11.2 10.9 11.0 10.8 10.9 11.0 11.1 11.1 10.9
Sprinter B 10.2 9.9 10.1 11.8 11.2 10.1 10.1 10.3 10.4

 a Calculate the mean sprint time for each runner.
 b Calculate the standard deviation of the sprint times for each runner.
 c Which runner is faster? Justify your answer.
 d Which runner is more consistent? Justify your answer.

*continued on next page ...*

6 Two classes sit the same Maths exam. One of the classes has students of similar mathematical ability, the other has students of different abilities. A summary of their percentage scores is presented below.

| | Mean | Standard deviation |
|---|---|---|
| Class A | 65 | 8 |
| Class B | 50 | 2 |

From the results table deduce which class is likely to have students of similar mathematical ability. Give reasons for your answer, which refer to the table above.

# 3.1 and 3.2 Logic and Sets and logical reasoning

Basic concepts of symbolic logic: definition of a proposition; symbolic notation of propositions.
Compound statements: implication, $\Rightarrow$; equivalence, $\Leftrightarrow$; negation, $\neg$; conjunction, $\wedge$; disjunction, $\vee$; exclusive disjunction, $\veebar$.
Translation between verbal statements and symbolic form.

1 State whether the following are propositions.
For each proposition state whether it is true, false or indeterminate.
  a Five squared is twenty-five.
  b Linear equations can include values of $x^2$.
  c $y$ equals plus nine.
  d No dogs can talk.
  e It is snowing today.
  f How many students are there in your class?

2 Write down the following compound propositions using the symbols for conjunction (and), disjunction (or, or both) and exclusive conjunction (or but not both).
*p*: Anna has a brother.
*q*: Petra has a sister.

  a Anna has a brother and Petra has a sister.
  b Anna has a brother or Petra has a sister or both are true.
  c Anna has a brother or Petra has a sister but not both are true.

# 3.3 and 3.4 Truth tables and Implication: converse; inverse; contrapositive and logical equivalence

Truth tables: concepts of logical contradiction and tautology.
Converse; inverse; contrapositive.
Logical equivalence.
Testing the validity of simple arguments through the use of truth tables.

1 Draw a truth table for the three propositions $p$, $q$ and $r$.
Compare it with the sample space for the result of tossing three coins.

2 Why is $p \wedge q$ and $p \veebar q$ a contradiction?

3 A truth table for the propositions $p$ and $q$ is given below.
Copy and complete the table.

| $p$ | $q$ | $\neg p$ | $p \vee q$ | $p \wedge q$ | $p \veebar q$ |
|---|---|---|---|---|---|
| T | T | | | | |
| | | | | | |
| | | | | | |
| | | | | | |

4 Construct a truth table to show that $\neg(p \vee q)$ is logically equivalent to $(\neg p) \wedge (\neg q)$.

5 For the following propositions,
i rewrite the propositions using 'if … then'
ii state the converse, inverse and contrapositive of the propositions
iii state whether the propositions are true or false.
a An odd number is divisible by two.
b An octagon has eight sides.
c An icosahedron has twelve faces.
d Congruent triangles are also similar.

6 a Draw a truth table for $(p \Rightarrow q) \veebar (q \Rightarrow p)$.
b Comment on the meaning of $(p \Rightarrow q) \veebar (q \Rightarrow p)$ by referring to your table.

7 a Illustrate the proposition $(p \Rightarrow q) \veebar (q \Rightarrow p)$ on a Venn diagram by shading the correct region(s).
b Describe the shaded region(s) using set notation.

# 3.7b Venn diagrams

Use of tree diagrams, Venn diagrams, sample space diagrams and tables of outcomes.

1 a Copy the Venn diagram below.

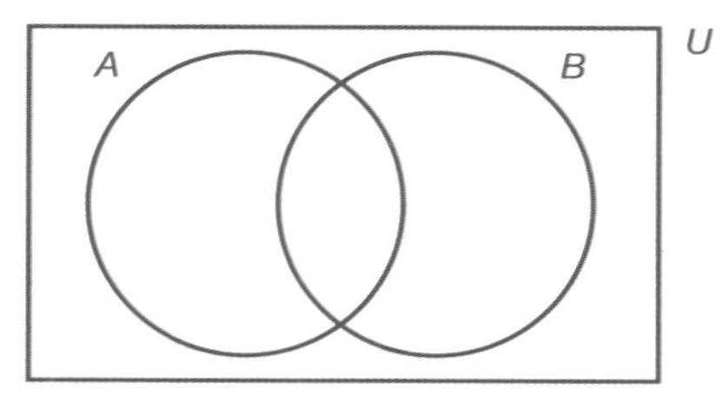


$U$ = {Integers from 1 to 20},
$A$ = {1, 3, 5, 7, 9, 11, 13, 15, 17, 19}
$B$ = {2, 3, 5, 7, 11, 13, 17, 19}

b Enter the information above in the Venn diagram.

2 The Venn diagram shows three sets of numbers.

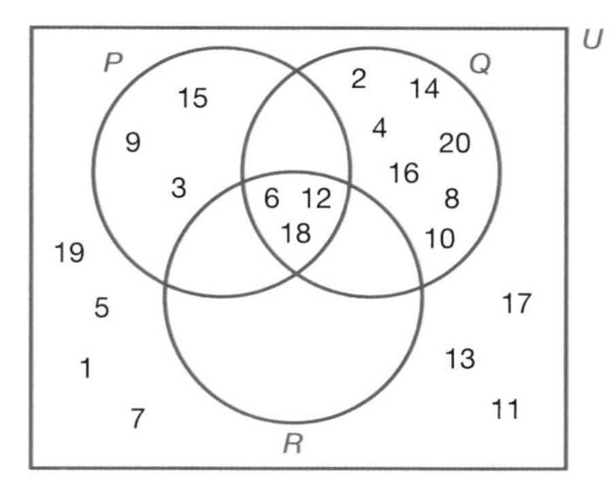


Complete the following.

a $P = \{...\}$ b $R = \{...\}$ c $P \cap Q = \{...\}$
d $P \cup Q = \{...\}$ e $P' \cap Q = \{...\}$

3 The Venn diagram shows three sets of numbers.
Complete the following.

a $L \cap N = \{...\}$
b $N \cup M = \{...\}$
c $L \cap M \cap N = \{...\}$
d $N' \cap L = \{...\}$
e $N' \cup L' = \{...\}$
f $M' \cup L \cap N = \{...\}$

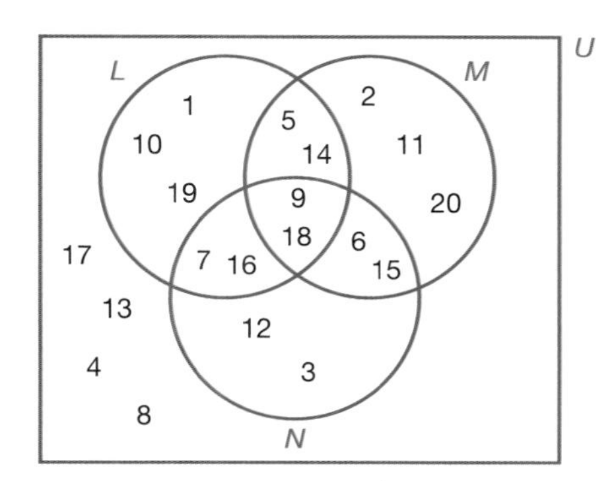


continued on next page ...

4 In the Venn diagram, the numbers shown represent the number of members in each set. For example, $n(E) = 3$.

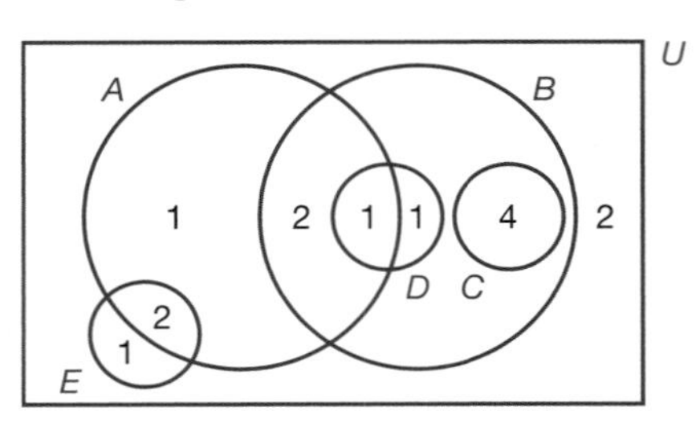


a State whether the following statements are true or false.
   i $C \subset B$   ii $E \subset A$   iii $D \cap C = \varnothing$

b Write down the number of members in each of the following statements.
   i $n(B)$   ii $n(A)$   iii $n(A \cap B)$
   iv $n(A \cup E)$   v $n(A \cap B \cap D)$   vi $n(A' \cap D)$

5 Draw a Venn diagram to represent the following sets.
$A = \{3, 4, 7, 8\}$   $B = \{1, 2, 4, 5, 6, 7, 9\}$   $C = \{1, 2, 6\}$

6 In a college 60% of students study Mathematics and 40% study Science. 75% of students study either Maths or Science or both.
Draw a Venn diagram to represent this information.

7 A language school offers three languages for its students to study: English, Spanish and Chinese. Each student is required to study at least two languages. 85% study English, 50% study Spanish and 20% study all three.
Copy and complete the following Venn diagram for the information above.

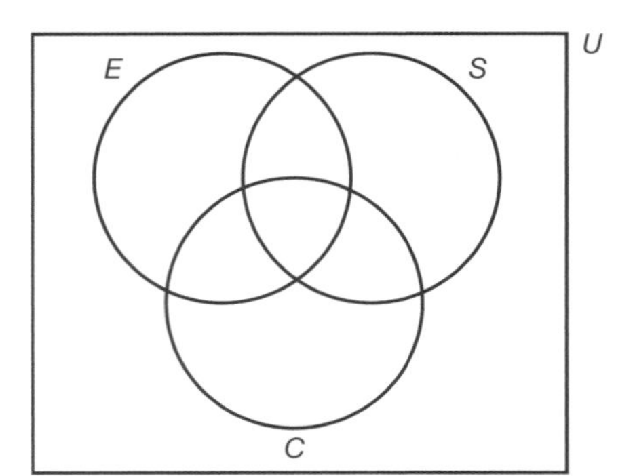

# 3.7c Laws of probability

Probability using 'with replacement' and 'without replacement'.
Conditional probability.

1 In a 100 m sprint, the record holder has a probability of 0.85 of winning. He has a 0.08 probability of coming second.
  a What is his probability of finishing in the first two?
  b Given that he hasn't come first, what is the probability that he has come second?

2 I spin a coin and throw a dice. What is the probability of getting:
  a a tail and a multiple of 2
  b a tail or a multiple of 2
  c a tail or a multiple of 2, but not both?

3 Three friends have a birthday in the same week. Assuming that they are independent events, calculate the probability that they are all on different days.

4 Raul takes a bus followed by a train to work. On a particular day the probability of him catching the bus is 0.65 and the probability of catching the train is 0.6.
  The probability of catching neither is 0.2. $A$ represents catching the bus and $B$ the train.
  a State $P(A \cup B)'$
  b Find $P(A \cup B)$
  c Given that $P(A \cup B) = P(A) + P(B) - P(A \cap B)$ calculate $P(A \cap B)$.
  d Calculate $P(B \mid A)$, the probability of catching the train given that he caught the bus.

5 Julie revised for a multiple choice Science exam. Unfortunately she only managed to revise 60% of the facts necessary. During the exam, if there is a question on any of the topics she revised she gets the answer correct.
  If there is a question on any of the topics she hasn't revised, she has a $\frac{1}{5}$ chance of getting it right.
  a A question is chosen at random. What is the probability that she got the answer correct?
  b If she got a question correct, what is the probability that it was on one of the topics she had revised?

6 Miguel has a driving test on one day and a Drama exam the next. The probability of him passing the driving test is 0.82. The probability of him passing the Drama exam is 0.95. The probability of failing both is 0.01. Given that he has passed the driving test, what is the probability that he passed his Drama exam too?

# 4.1 The normal distribution

The normal distribution.
The concept of a random variable; of the parameters $\mu$ and $\sigma$ ; of the bell shape; the symmetry about $x = \mu$.
Diagrammatic representation.
Normal probability calculations.
Expected value.
Inverse normal calculations.

1 The diagrams below show two normal distribution curves.

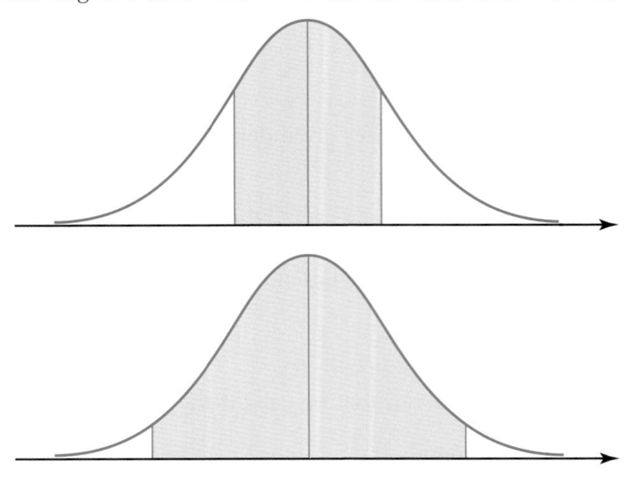

The shaded area in the first graph represents data that is within one standard deviation of the mean. In the second, the shaded area represents data that is within two standard deviations of the mean.

a Approximately what percentage of data is shaded in the first distribution?
b Approximately what percentage of data is shaded in the second distribution?

2 The graph below shows two normal distribution curves, P and Q.

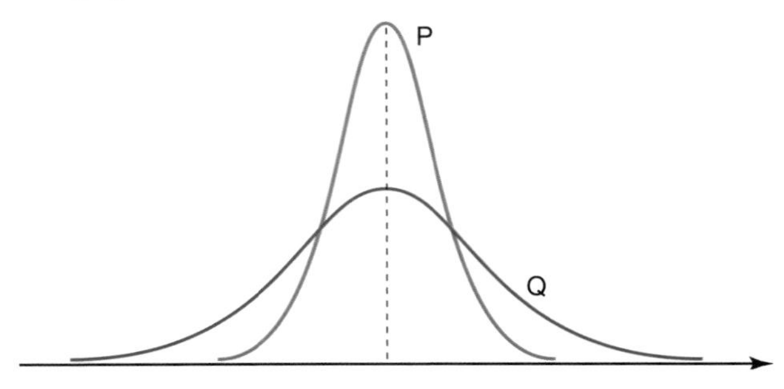


a What statistical measures are the same for both distributions? Justify your answer.
b What statistical measure is different between the two distributions? Justify your answer.

*continued on next page ...*

# 6.3 Quadratic functions and their graphs

Quadratic models.
Quadratic functions and their graphs (parabolas): $f(x) = ax^2 + bx + c, a \neq 0$.
Properties of a parabola: symmetry; vertex; intercepts on the $x$-axis and $y$-axis.
Equation of the axis of symmetry, $x = -\frac{b}{2a}$.

1 For each of the quadratic functions below:
   i use a GDC to sketch the function
   ii write down the coordinates of the points where the graph intercepts the $x$-axis
   iii write down the value of the $y$-intercept.
   a $f(x) = x^2 - 9x + 20$
   b $f(x) = x^2 - 3x - 18$
   c $f(x) = (x - 4)^2$
   d $f(x) = x^2 + 10x + 27$

2 Write down the equation of the axis of symmetry of each of the following quadratic functions.
   a $y = x^2 - 2x$
   b $y = -x^2 - 4x$
   c $y = x(5 - x)$
   d $y = -x^2 + 3x - 10$

3 Write down a possible equation of a quadratic function with each of the following axes of symmetry.
   a $x = 6$
   b $x = -5$

4 Factorize the following quadratic functions.
   a $f(x) = x^2 + 11x + 30$
   b $f(x) = x^2 + 4x - 12$
   c $f(x) = -x^2 + 8x - 15$
   d $f(x) = x^2 - 36$

5 Solve the following quadratic equations by factorizing.
   a $x^2 - 3x - 4 = 0$
   b $x^2 - 2x - 24 = 0$
   c $-x^2 + 10x - 16 = 0$
   d $x^2 = 11x - 28$

6 The following quadratic equations are of the form $ax^2 + bx + c = 0$.

   Solve them by using the quadratic formula $x =$ .

   a $x^2 + 5x - 25 = 0$
   b $x^2 + 9x - 24 = 0$
   c $4x^2 + 8x + 3 = 0$
   d $-x^2 + 9x - 15 = 0$

7 For each of the following:
   i form an equation in $x$
   ii solve the equation to find the possible value(s) of $x$.

   a

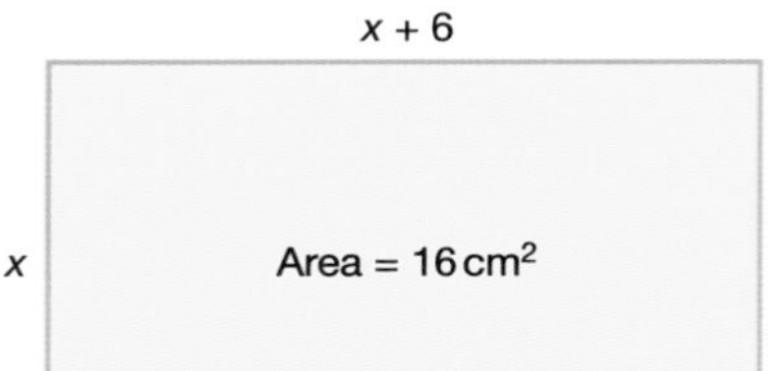


   b

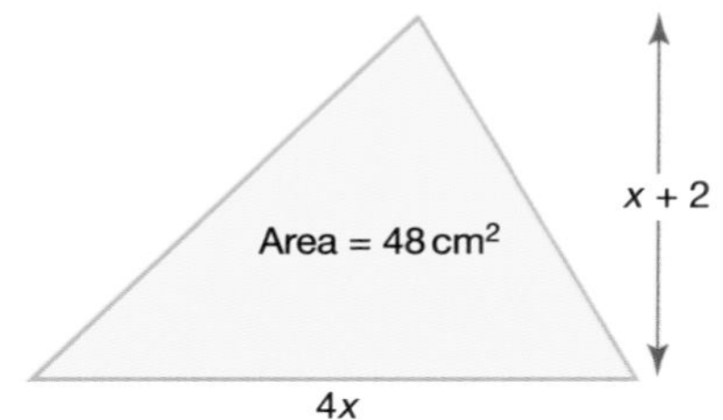

# 6.4 Exponential functions and their graphs

Exponential models.
Exponential functions and their graphs: $f(x) = ka^x + c, a \in \mathbb{Q}^+, a \neq 1, k \neq 0$;
$f(x) = ka^{-x} + c, a \in \mathbb{Q}^+, a \neq 1, k \neq 0$
Concept and equation of a horizontal asymptote.

**1** **i** Plot the following exponential functions.
**ii** State the equation of any asymptotes.
**a** $f(x) = 2^x + 1$ **b** $f(x) = -2^x + 2$ **c** $f(x) = 3^x - 3$

**2** A tap is dripping at a constant rate into a container. The level ($l$ cm) of water in the container is given by the equation $l = 3^t + 5$, where $t$ is the time in hours.
**a** Calculate the level of water in the container at the start.
**b** Calculate the level of water in the container after 4 hours.
**c** Calculate the time taken for the level of the water to reach 248 cm.
**d** Plot a graph to show the level of water over the first 6 hours.
**e** Use your graph to estimate the time taken for the water to reach a level of 1 m.

**3** **a** Plot a graph of $y = 5^x$ for values of $x$ between $-1$ and 3.
**b** Use your graph to find approximate solutions to the following equations.
**i** $5^x = 100$ **ii** $5^x = 50$

**4** The graph below shows a graph of the function $f(x) = 2^x$.

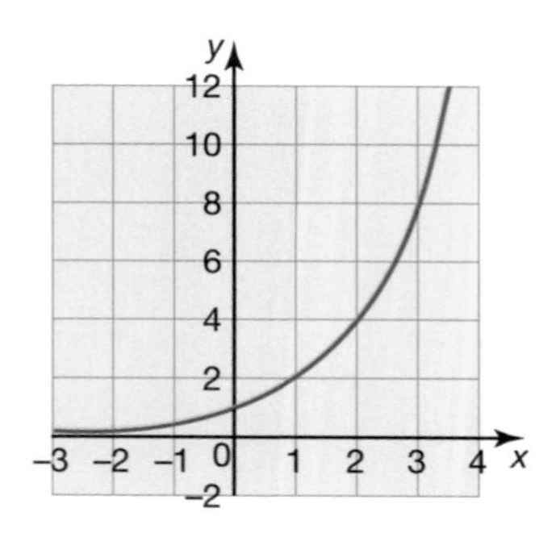


**a** Copy the graph and on the same axes sketch the graph of $f(x) = 2^x + 3$. Label it clearly.
**b** On the same axes sketch the graph of $f(x) = -2^x + 6$. Label it clearly.

**5** The half-life of plutonium 239 is 24 000 years. How long will 2 g of plutonium 239 take to decay to 62.5 mg?

## 6.5, 6.6 and 6.7 Sketching and drawing graphs

Models using functions of the form $f(x) = ax^m + bx^n + \ldots,\ m, n \in \mathbb{Z}$.
Functions of this type and their graphs.
The $y$-axis as a vertical asymptote.
Drawing accurate graphs.
Creating a sketch from information given.
Transferring a graph from GDC to paper.
Reading, interpreting and making predictions using graphs.
Included all the functions above and additions and subtractions.
Use of a GDC to solve equations involving combinations of the functions above.

For the equations given in questions 1 and 2:

**a** Calculate the equation of any asymptote.
**b** Calculate the coordinates of any points where the graph intercepts the axes.
**c** Sketch the graph.
**d** Check your graph using a GDC.

1 $y = \dfrac{1}{x - 5}$

2 $y = -\dfrac{1}{3x - 1} + 2$

For the equations given in questions 3 and 4:

**a** Calculate the equations of any vertical or horizontal asymptotes.
**b** Calculate the coordinates of any points where the graph intercepts the axes.
**c** Sketch the graph of the function with the aid of a GDC.

3 $y = \dfrac{1}{(x + 2)(x - 3)}$

4 $y = -\dfrac{1}{2x^2 + x - 6}.$

5 For the equation $y = -2x^3 - 13x^2 - 13x + 10$:
  **a** Use a GDC to sketch the function.
  **b** Determine where the graph intersects both the $x$ and $y$ axes.
  **c** Rewrite the equation in factorised form.

6 Using a GDC, solve the equation $2x^2 - 11 = \dfrac{4}{x} - 5x$.

# 7.2 and 7.3 Differentiation and The gradient of a curve at a given point

The principle that $f(x) = ax^n \Rightarrow f'(x) = anx^{n-1}$.
The derivative of functions of the form $f(x) = ax^n + bx^{n-1} + \dots$, where all exponents are integers.
Gradients of curves for given values of $x$.
Values of $x$ where $f'(x)$ is given.
Equation of the tangent at a given point.
Equation of the line perpendicular to the tangent at a given point (normal).

1 Differentiate the following functions with respect to $x$.
a $f(x) = x^2 + 3x - 4$
b $f(x) = \frac{1}{2}x^2 - 5x + 4$
c $f(x) = 2x^3 - 4x^2$
d $f(x) = \frac{1}{3}x^6 - \frac{1}{2}x^4 - 1$

2 Find the derivative of the following expressions.
a $x^{-1}$
b $2x^{-3}$
c $x^{-2} + 2x^{-1} - 3$
d $\frac{3}{x^2}$

3 For each of the following functions, find the derivative $f'(x)$ with respect to $x$.
a $f(x) = x(x - 3)$
b $f(x) = 2x^2(x + 2)$
c $f(x) = (x - 2)(x + 3)$
d $f(x) = (x^2 - 3x)(x + 4)$

4 Find the derivative of each of the following expressions.
a $\frac{2x^3 - x^2}{x}$
b $\frac{2x^5 - x^3}{3x^2}$
c $\frac{x^3 - 2x^2}{x^4}$
d $\frac{(x - 6)(2x^2 - 1)}{x}$

5 The graph of the function $f(x) = x^2 - 4x + 1$ is shown below.

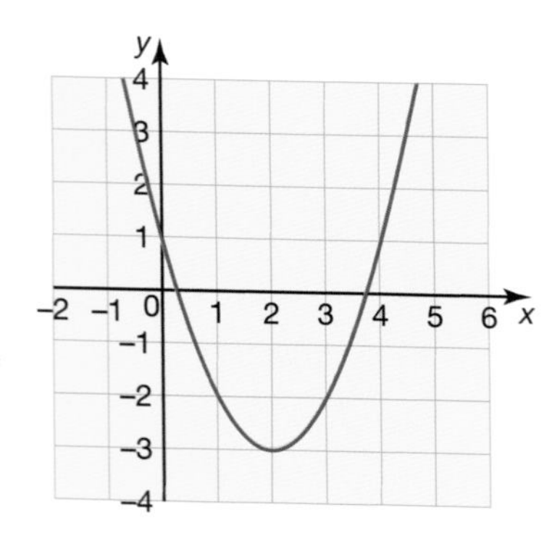


a Calculate the gradient function $f'(x)$.
b Calculate the gradient of the graph when:
i $x = 3$
ii $x = 2$
iii $x = 0$.

*continued on next page ...*

6 The graph of the function $f(x) = \frac{1}{2}x^2 - 4x + 2$ is shown below.

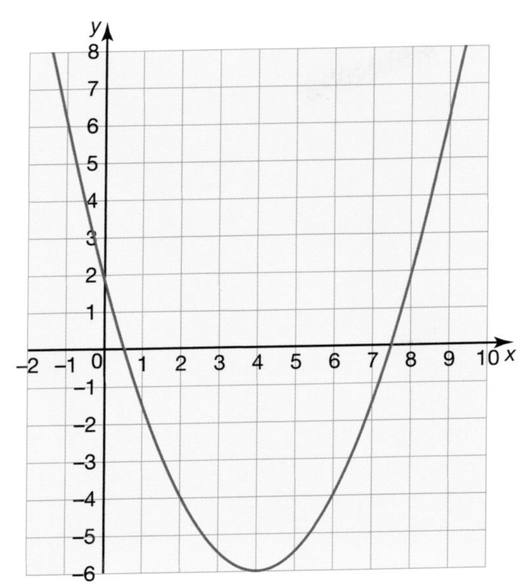


a Calculate the gradient function $f'(x)$.

b Calculate the values of $x$ at the points on the graph where the gradient is:

i 0 ii 2 iii −5.

7 A curve has the equation $y = x^3 + 4x + 2$.

a Find $\frac{dy}{dx}$.

b Deduce from your answer to part **a** the least possible value of $\frac{dy}{dx}$. Justify your answer.

c Calculate the value(s) of $x$ where $\frac{dy}{dx}$ is:

i 7 ii 4 iii 31.

*continued on next page …*

8 The function $f(x) = x^3 - 13x + 12$ is shown below.

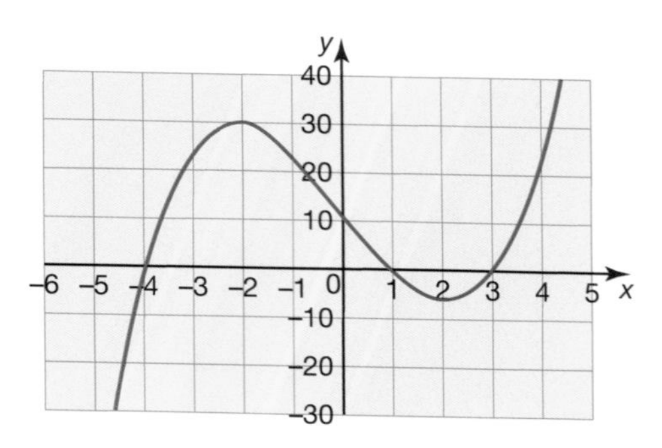


**a** Calculate the gradient function $f'(x)$.
**b** Calculate the gradient of the curve when $x = 3$.
**c** Give the gradient of the tangent to the curve at the point (3, 0).
**d** Calculate the equation of the tangent to the curve at the point (3, 0).
**e** Write down the gradient of the normal to the curve at the point (3, 0).
**f** Calculate the equation of the normal to the curve at the point (3, 0). Give your answer in the form $ax + by + c = 0$.

9 The function $f(x) = -x^2 - 2x + 8$ is shown below.

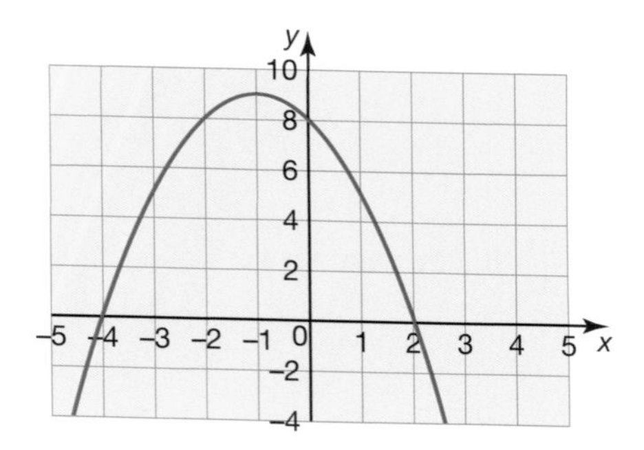


**a** Calculate the gradient function.
**b** Show that the points A(−2, 8) and B(1, 5) lie on the curve.
**c** Calculate the gradient of the curve at points A and B.
**d** Calculate the equation of the tangent to the curve at A.
**e** Calculate the equation of the normal to the curve at B.
**f** Calculate the coordinates of the point of intersection of the tangent at A and the normal to the curve at B.

Class 12Y

| Score | Frequency |
|---|---|
| 31–40 | 3 |
| 41–50 | 8 |
| 51–60 | 6 |
| 61–70 | 3 |
| 71–80 | 2 |
| 81–90 | 4 |
| 91–100 | 5 |

**b** Student's own response

2

| Number of apples | Frequency |
|---|---|
| 1–20 | 9 |
| 21–40 | 6 |
| 41–60 | 7 |
| 61–80 | 11 |
| 81–100 | 7 |
| 101–120 | 4 |
| 121–140 | 4 |
| 141–160 | 2 |

## Exercise 2.3.2

1

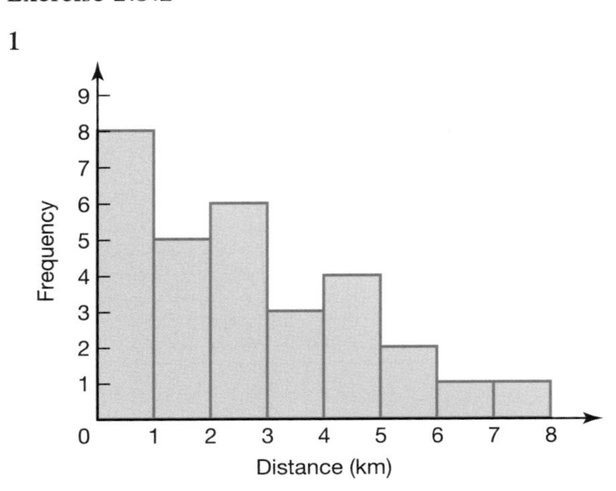


2

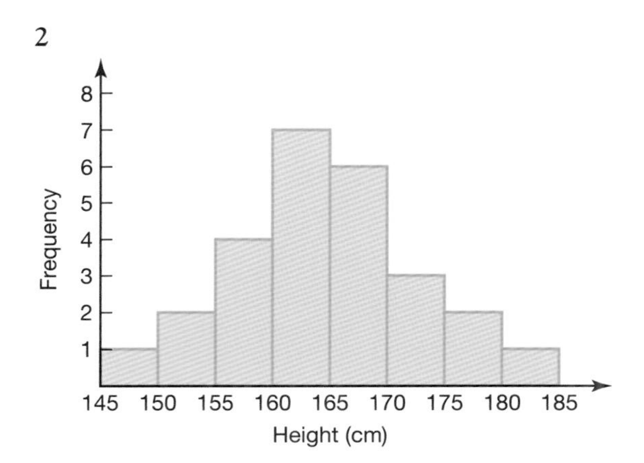


## Exercise 2.4.1

1 **a**

| Finishing time (h) | 0– | 0.5– | 1.0– | 1.5– | 2.0– | 2.5– | 3.0–3.5 |
|---|---|---|---|---|---|---|---|
| Frequency | 0 | 0 | 6 | 34 | 16 | 3 | 1 |
| Cumulative frequency | 0 | 0 | 6 | 40 | 56 | 59 | 60 |

**b**

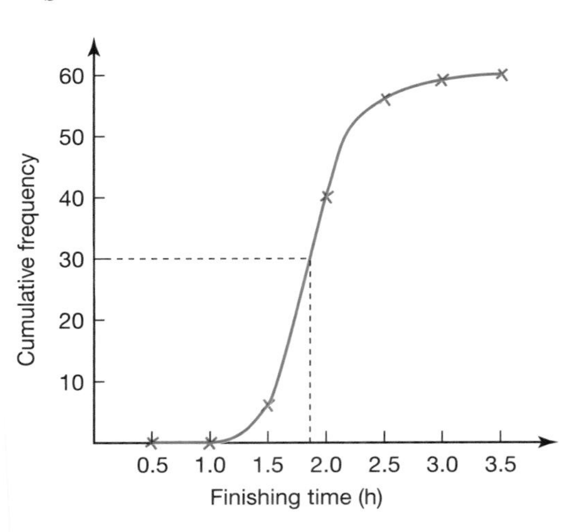


**c** Median ≈ 1.8 hours

**d** As many runners finished before as after the median.

2 a

| | Group A | | Group B | | Group C | |
|---|---|---|---|---|---|---|
| Score | Freq. | Cum. freq. | Freq. | Cum. freq. | Freq. | Cum. freq. |
| $0 \le x < 20$ | 1 | 1 | 0 | 0 | 1 | 1 |
| $20 \le x < 40$ | 5 | 6 | 0 | 0 | 2 | 3 |
| $40 \le x < 60$ | 6 | 12 | 4 | 4 | 2 | 5 |
| $60 \le x < 80$ | 3 | 15 | 4 | 8 | 4 | 9 |
| $80 \le x < 100$ | 3 | 18 | 4 | 12 | 8 | 17 |

b

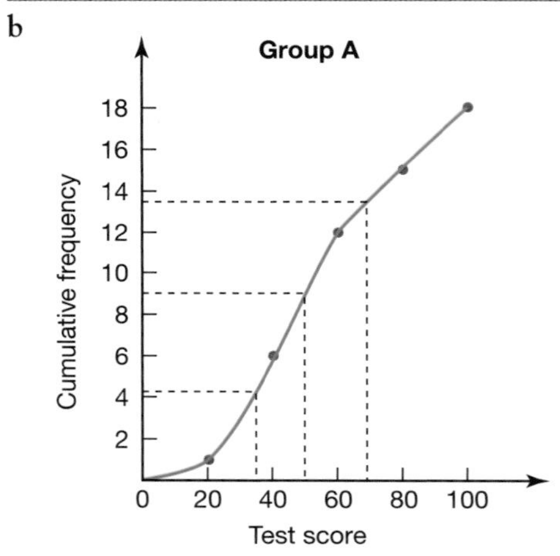


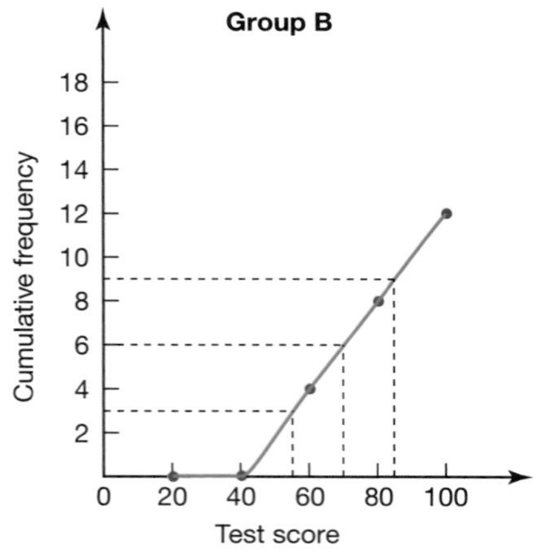


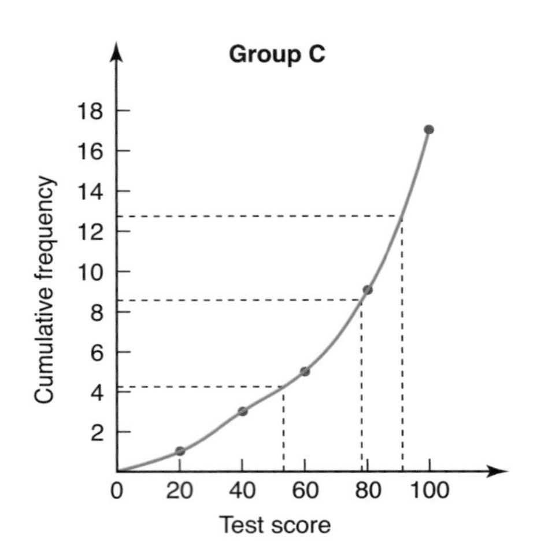


c Group A median ≈ 50
Group B median ≈ 70
Group C median ≈ 78

d As many students were above as below the median.

e Group A: Lower quartile ≈ 35,
Upper quartile ≈ 69
Group B: Lower quartile ≈ 55,
Upper quartile ≈ 85
Group C: Lower quartile ≈ 54,
Upper quartile ≈ 90

3 a

| | 2010 | | 2011 | | 2012 | |
|---|---|---|---|---|---|---|
| Height (cm) | Freq. | Cum. freq. | Freq. | Cum. freq. | Freq. | Cum. freq. |
| $150 \le h < 155$ | 6 | 6 | 2 | 2 | 2 | 2 |
| $155 \le h < 160$ | 8 | 14 | 9 | 11 | 6 | 8 |
| $160 \le h < 165$ | 11 | 25 | 10 | 21 | 9 | 17 |
| $165 \le h < 170$ | 4 | 29 | 4 | 25 | 8 | 25 |
| $170 \le h < 175$ | 1 | 30 | 3 | 28 | 2 | 27 |
| $175 \le h < 180$ | 0 | 30 | 2 | 30 | 2 | 29 |
| $180 \le h < 185$ | 0 | 30 | 0 | 30 | 1 | 30 |

b

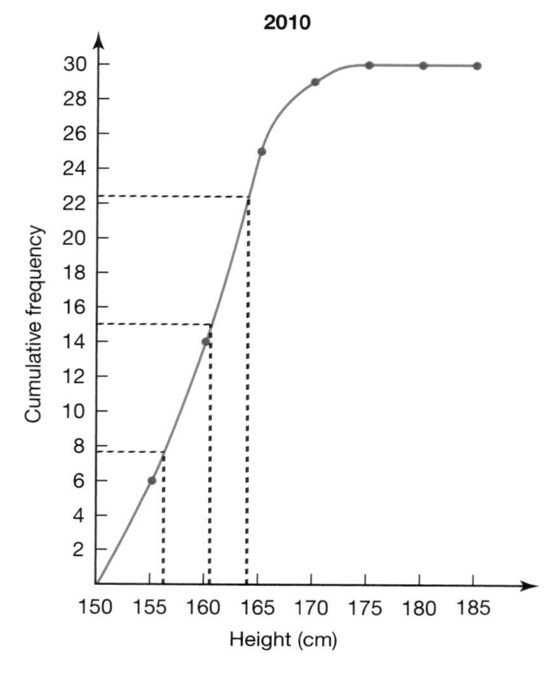


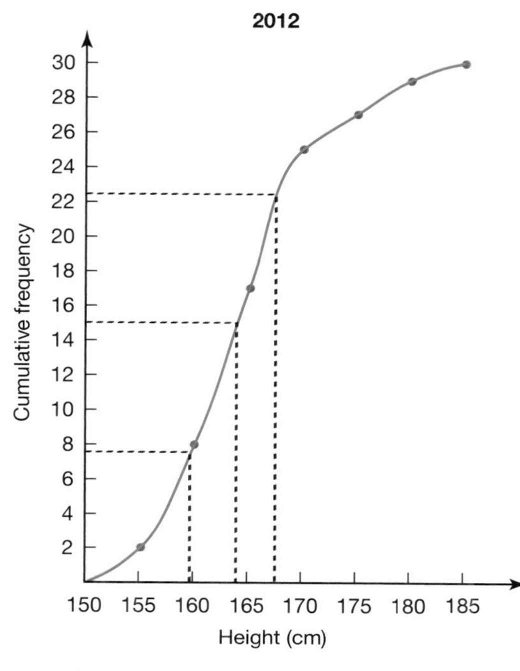


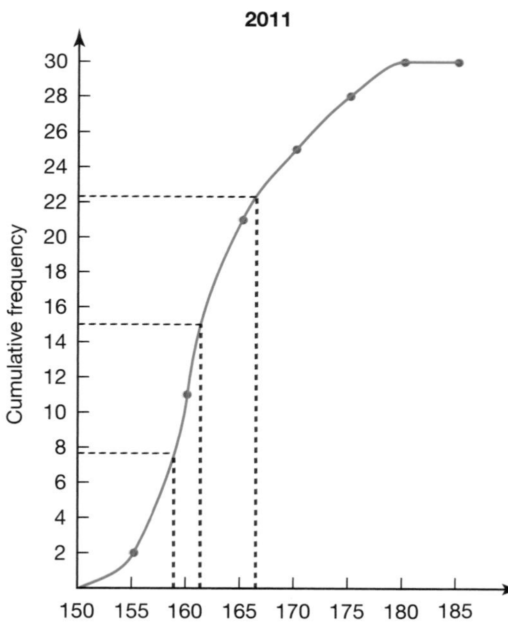


c Median (2010) ≈ 161 cm
Median (2011) ≈ 162 cm
Median (2012) ≈ 164 cm

d As many students are taller than the median as shorter than the median.

e 2010: Lower quartile ≈ 156,
Upper quartile ≈ 164
2011: Lower quartile ≈ 158,
Upper quartile ≈ 166
2012: Lower quartile ≈ 160,
Upper quartile ≈ 168

f Student's own comments

## Exercise 2.4.2

1 a

| | Goals scored | Goals let in |
|---|---|---|
| **i)** Mean | 1.15 | 2.00 |
| **ii)** Median | 1 | 2 |
| **iii)** $q_1$ | 0 | 1 |
| **iv)** $q_3$ | 1.5 | 3 |

b

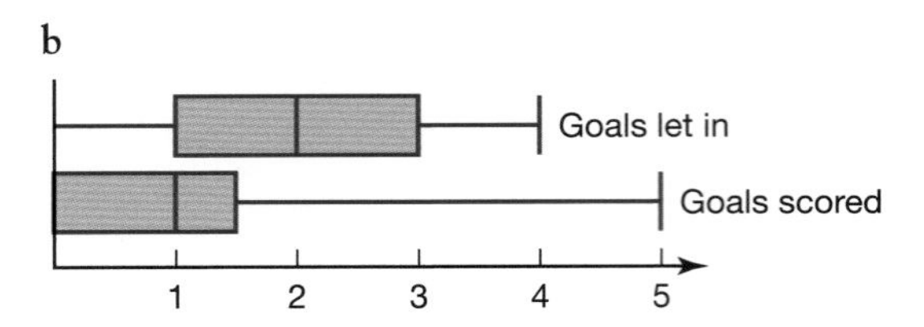


c Student's own report

2 a

| | Resort A | Resort B |
|---|---|---|
| i) Mean | 8.5 | 8.5 |
| ii) Median | 8 | 8 |
| iii) $q_1$ | 7 | 8 |
| iv) $q_3$ | 10 | 9 |

b

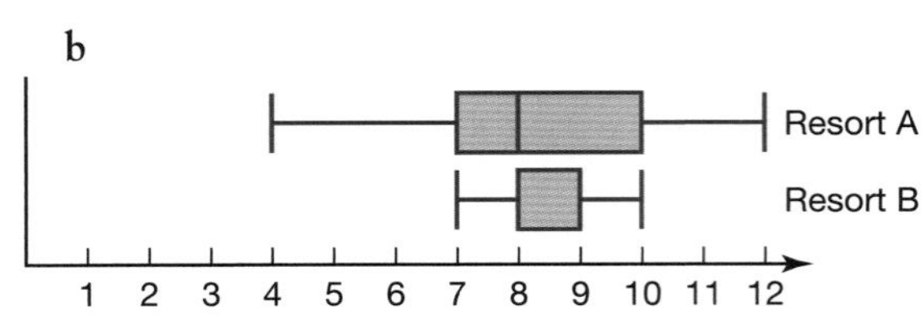


c Student's own response

3 Student's own explanation using the box and whisker diagrams below.

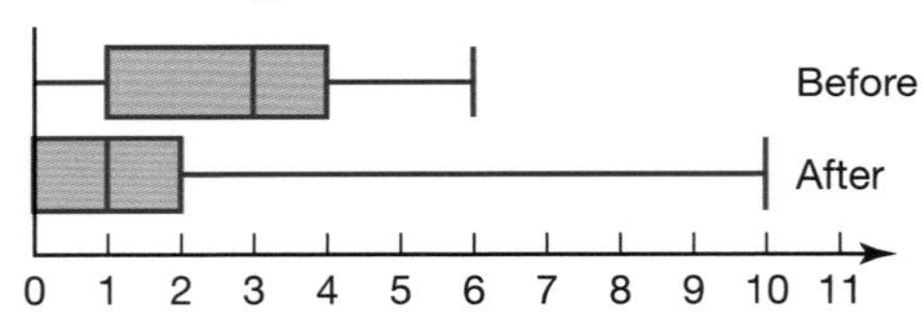


### Exercise 2.5.1

1 a Mean = 1.67 Median = 1
Mode = 1
b Mean = 6.2 Median = 6.5
Mode = 7
c Mean = 26.4 Median = 27
Mode = 28
d Mean = 13.95 s Median = 13.9 s
Mode = 13.8 s

2 91.1 kg

3 103 points

### Exercise 2.5.2

1 Mean = 3.35 Median = 3
Mode = 1 and 4

2 Mean = 7.03 Median = 7
Mode = 7

3 a Mean = 6.33 Median = 7
Mode = 8
b The mode as it gives the highest number of flowers per bush.

### Exercise 2.5.3

1 a 29.1
b 30–39

2 a 60.9
b 60–69

3 a 5 mins 50 secs
b 0–4
c Student's own comments

### Exercise 2.6.1

1 a

| Distance thrown (m) | 0– | 20– | 40– | 60– | 80–100 |
|---|---|---|---|---|---|
| Frequency | 4 | 9 | 15 | 10 | 2 |
| Cumulative frequency | 4 | 13 | 28 | 38 | 40 |

b

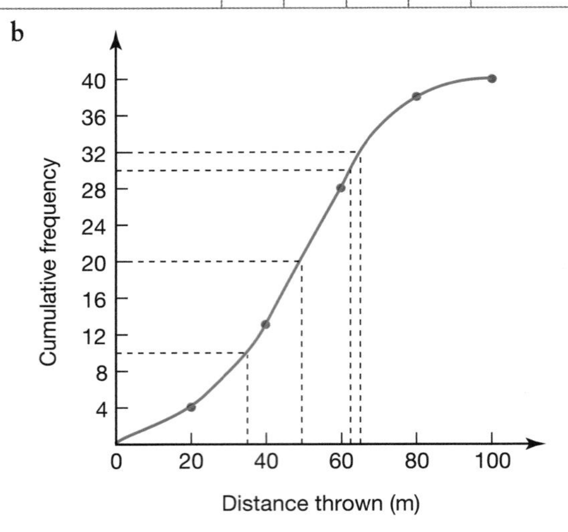

c Median ≈ 50 m
d Interquartile range ≈ 28 m
e Qualifying distance ≈ 66 m

2 a

| Type A | | |
|---|---|---|
| **Mass (g)** | **Frequency** | **Cum. freq.** |
| 75– | 4 | 4 |
| 100– | 7 | 11 |
| 125– | 15 | 26 |
| 150– | 32 | 58 |
| 175– | 14 | 72 |
| 200– | 6 | 78 |
| 225–250 | 2 | 80 |

| Type B | | |
|---|---|---|
| **Mass (g)** | **Frequency** | **Cum. freq.** |
| 75– | 0 | 0 |
| 100– | 16 | 16 |
| 125– | 43 | 59 |
| 150– | 10 | 69 |
| 175– | 7 | 76 |
| 200– | 4 | 80 |
| 225–250 | 0 | 80 |

b

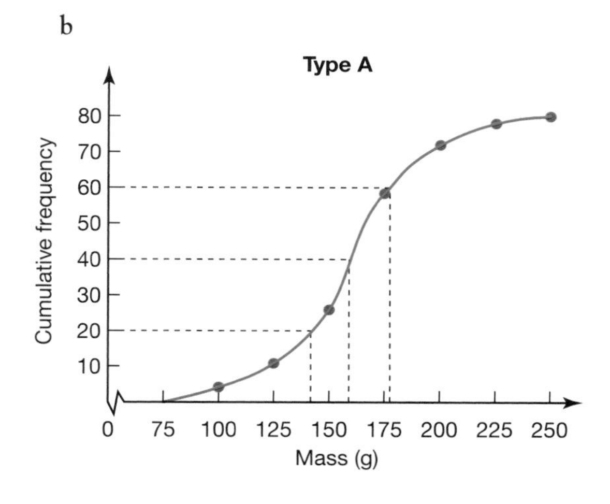


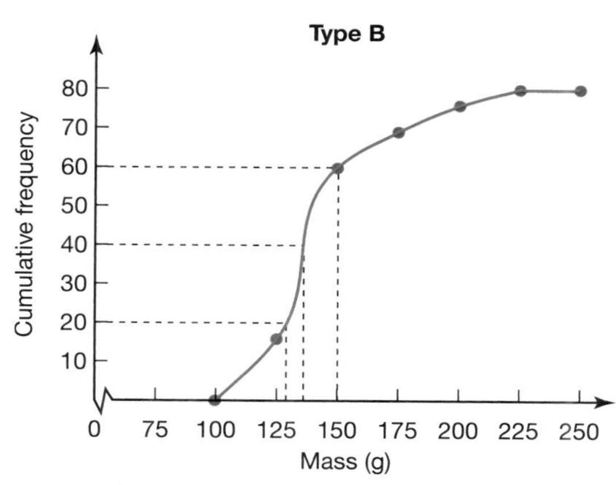


c Median type A ≈ 157 g
Median type B ≈ 137 g

d i) Lower quartile type A ≈ 140 g
Lower quartile type B ≈ 127 g
ii) Upper quartile type A ≈ 178 g
Upper quartile type B ≈ 150 g
iii) Interquartile type range type A ≈ 38 g
Interquartile type range type B ≈ 23 g

e Student's own report

3 a Student's own explanation
b Student's own explanation

**Exercise 2.6.2**

1 a i) 5.5 ii) 7 iii) 5
iv) 2.58
b i) 78.75 ii) 16 iii) 9
iv) 5.40
c i) 3.85 ii) 3.9 iii) 1.6
iv) 1.05

2 i) 2.31 ii) 6 iii) 2
iv) 1.43

3 i) 71.53 ii) 8 iii) 1
iv) 1.44

4 i) 2.72 ii) 6 iii) 3
iv) 1.69

5 a 6.5 b 0.18

**Student assessment 1**

1 a Discrete b Discrete c Continuous
d Discrete e Continuous f Continuous
g Continuous

2 a

| Mark (%) | Frequency | Cumulative frequency |
|---|---|---|
| 31–40 | 21 | 21 |
| 41–50 | 55 | 76 |
| 51–60 | 125 | 201 |
| 61–70 | 74 | 275 |
| 71–80 | 52 | 327 |
| 81–90 | 45 | 372 |
| 91–100 | 28 | 400 |

b

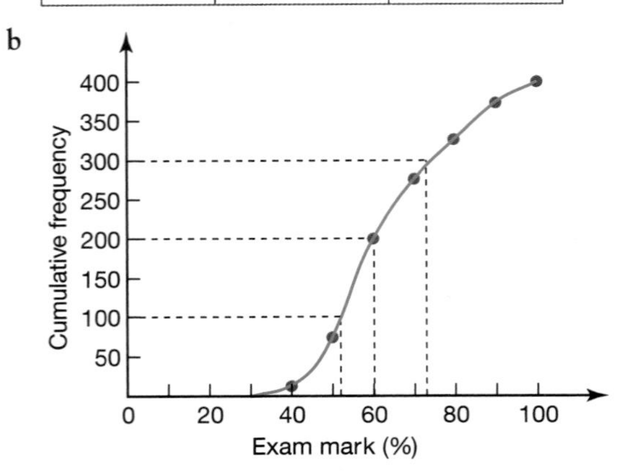


c i) Median ≈ 60%

ii) Lower quartile ≈ 52%

Upper quartile ≈ 74%

iii) IQR ≈ 22%

3 a

| Mark (%) | Frequency | Cumulative frequency |
|---|---|---|
| 1–10 | 10 | 10 |
| 11–20 | 30 | 40 |
| 21–30 | 40 | 80 |
| 31–40 | 50 | 130 |
| 41–50 | 70 | 200 |
| 51–60 | 100 | 300 |
| 61–70 | 240 | 540 |
| 71–80 | 160 | 700 |
| 81–90 | 70 | 770 |
| 91–100 | 30 | 800 |

b

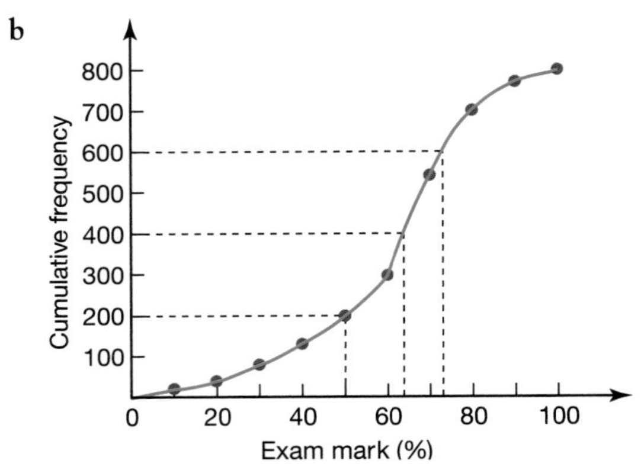


c A grade ≥ 72%

d Fail < 50%

e 200 students failed.

f 200 students achieved an A grade.

**4 a**

| Time (mins) | $10 \leq t < 15$ | $15 \leq t < 20$ | $20 \leq t < 25$ | $25 \leq t < 30$ | $30 \leq t < 35$ | $35 \leq t < 40$ | $40 \leq t < 45$ |
|---|---|---|---|---|---|---|---|
| Motorway frequency | 3 | 5 | 7 | 2 | 1 | 1 | 1 |
| Motorway cumulative frequency | 3 | 8 | 15 | 17 | 18 | 19 | 20 |
| Country lanes frequency | 0 | 0 | 9 | 10 | 1 | 0 | 0 |
| Country lanes cumulative frequency | 0 | 0 | 9 | 19 | 20 | 20 | 20 |

**b**

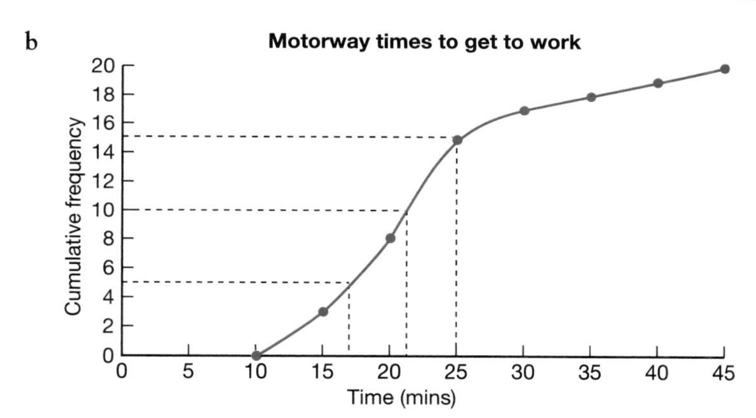


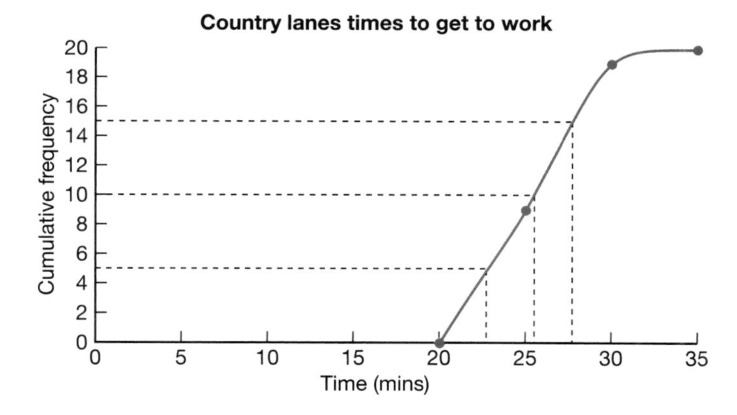


**c i)** Motorway median ≈ 21mins
**ii)** Motorway lower quartile ≈ 17 mins
Motorway upper quartile ≈ 25 mins
**iii)** Motorway IQR ≈ 8 mins

Country lanes median ≈ 26 mins
Country lanes lower quartile ≈ 23 mins
Country lanes upper quartile ≈ 28 mins
Country lanes IQR ≈ 5 mins

**d** Student's explanation
**e** Student's explanation

5 Box and whisker diagram A is likely to belong to mixed ability class + student's explanation. Box and whisker diagram B is likely to belong to the other class + student's explanation.

**Student assessment 2**

1 a Mean = 5.4 Median = 5
Mode = 5
b Mean = 75.4 Median = 72
Mode = 72
c Mean = 13.8 Median = 15
Mode = 18
d Mean = 6.1 Median = 6
Mode = 3

2 61 kg

3 a 2.83 b 3 c 3

4 a 2.4 b 2 c 3

5 a

| Group | Mid-interval value | Frequency |
|---|---|---|
| 0–19 | 9.5 | 0 |
| 20–39 | 29.5 | 4 |
| 40–59 | 49.5 | 9 |
| 60–79 | 69.5 | 3 |
| 80–99 | 89.5 | 6 |
| 100–119 | 109.5 | 4 |
| 120–139 | 129.5 | 4 |

b 75.5

6 a

| Group | Mid-interval value | Frequency |
|---|---|---|
| 10–19 | 14.5 | 3 |
| 20–29 | 24.5 | 7 |
| 30–39 | 34.5 | 9 |
| 40–49 | 44.5 | 2 |
| 50–59 | 54.5 | 4 |
| 60–69 | 64.5 | 3 |
| 70–79 | 74.5 | 2 |

b 39.2

7 3.8

8 a 2.05 b 5 c 1.26

**Examination questions**

1 a

| *x* | 0–20 | 20–40 | 40–60 | 60–80 | 80–100 |
|---|---|---|---|---|---|
| **Freq.** | 14 | 26 | 58 | 16 | 6 |

b 50
c $45\frac{2}{3}$ or 45.7

2 a i 30 ii 32 iii 28
b $0.25 \times 56 = 14$

3 a 55
b i $62.\dot{5}$ or 62.6 ii 8.86
c 36.0

**b** **i)** $\frac{1}{27}$ **ii)** $\frac{1}{3}$ **iii)** $\frac{1}{9}$

**iv)** $\frac{1}{3}$ **v)** $\frac{5}{9}$ **vi)** $\frac{1}{3}$

**2** **a**

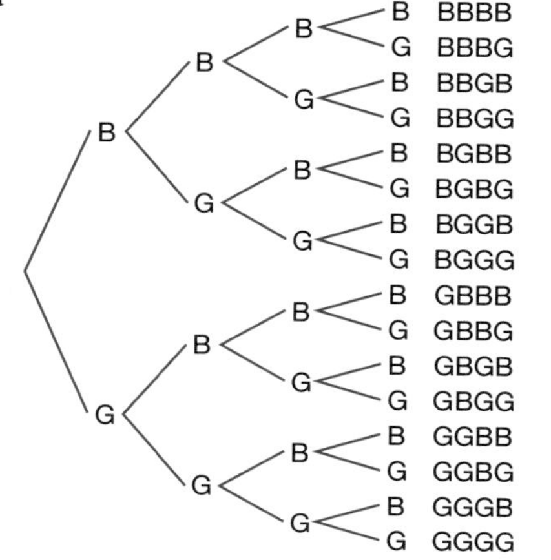


**b** **i)** $\frac{1}{16}$ **ii)** $\frac{3}{8}$ **iii)** $\frac{15}{16}$

**iv)** $\frac{5}{16}$

**3** **a**

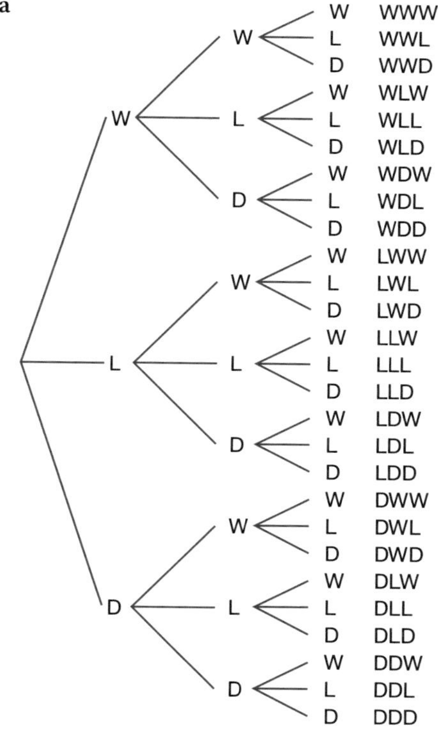


**b** **i)** $\frac{1}{27}$ **ii)** $\frac{10}{27}$ **iii)** $\frac{19}{27}$

**iv)** $\frac{8}{27}$

**c** Student's own answer

**4** **a**

G: G GG, W GW, B GB, O GO

W: G WG, W WW, B WB, O WO

B: G BG, W BW, B BB, O BO

O: G OG, W OW, B OB, O OO

**b** **i)** $\frac{1}{16}$ **ii)** $\frac{1}{8}$ **iii)** $\frac{1}{8}$

## Exercise 3.7.3

**1** **a**

| Roll 1 | Roll 2 | Outcomes | Probability |
|---|---|---|---|
| $\frac{1}{6}$ Six | $\frac{1}{6}$ Six | Six, Six | $\frac{1}{6}\times\frac{1}{6}=\frac{1}{36}$ |
| | $\frac{5}{6}$ Not six | Six, Not six | $\frac{1}{6}\times\frac{5}{6}=\frac{5}{36}$ |
| $\frac{5}{6}$ Not six | $\frac{1}{6}$ Six | Not six, Six | $\frac{5}{6}\times\frac{1}{6}=\frac{5}{36}$ |
| | $\frac{5}{6}$ Not six | Not six, Not six | $\frac{5}{6}\times\frac{5}{6}=\frac{25}{36}$ |

**b** **i)** $\frac{1}{6}$ **ii)** $\frac{11}{16}$ **iii)** $\frac{5}{36}$

**iv)** $\frac{125}{216}$ **v)** $\frac{91}{216}$

**c** They add up to 1, because either iv or v.

**2** **a** $\frac{4}{25}$ **b** $\frac{54}{125}$ **c** $\frac{98}{125}$

**3** **a**

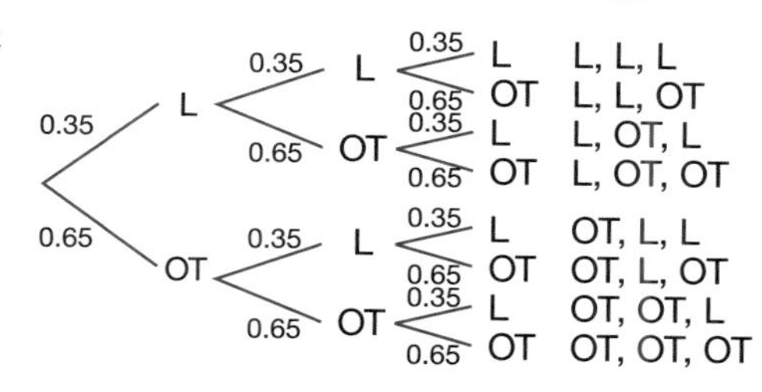


**b** **i)** 0.275 **ii)** 0.123 **iii)** 0.444

**iv)** 0.718

**4** 0.027

**5** **a** $0.75^2 = 0.56$

**b** $0.75^3 = 0.42$

**c** $0.75^{10} = 0.06$

### Exercise 3.7.4

1 a $\frac{25}{81}$ b $\frac{16}{81}$ c $\frac{20}{81}$
d $\frac{40}{81}$

2 a $\frac{5}{18}$ b $\frac{1}{6}$ c $\frac{5}{18}$
d $\frac{5}{9}$

3 a $\frac{2}{50}$ b $\frac{3}{10}$

4 a $\frac{1}{45}$ b $\frac{1}{3}$

5 a $\frac{920}{9312}$ b $\frac{8372}{9312}$

6 a

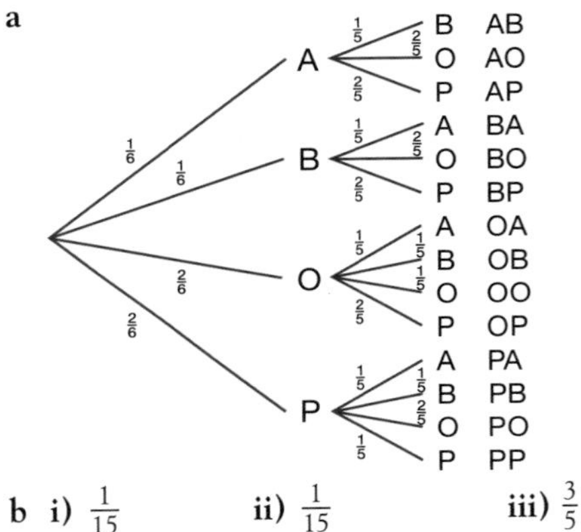


b i) $\frac{1}{15}$ ii) $\frac{1}{15}$ iii) $\frac{3}{5}$

### Exercise 3.7.5

1 a

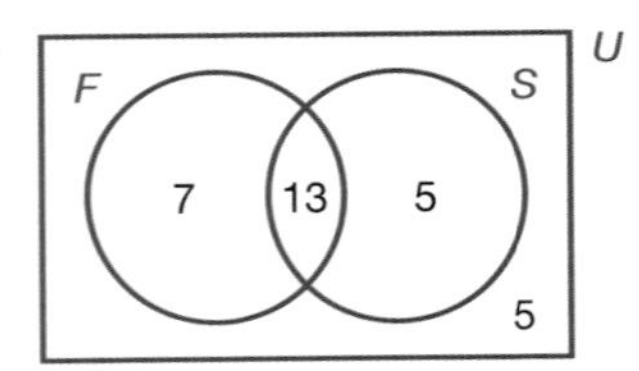


b $\frac{13}{30}$

2 a $\frac{5}{35} = \frac{1}{7}$ b $\frac{14}{35} = \frac{2}{5}$ c $\frac{13}{35}$

3 $\frac{45}{108}$

4 a

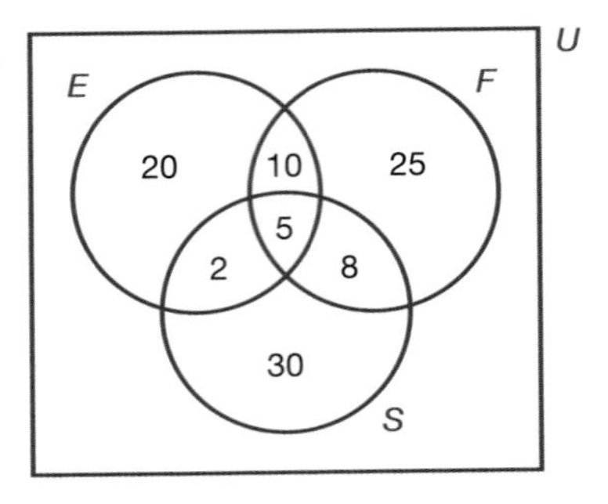


b i) $\frac{5}{100}$ ii) $\frac{20}{100}$ iii) $\frac{25}{100}$

### Exercise 3.7.6

1 a No (mutually exclusive)
b 0.8

2 a Yes
b i) $\frac{1}{6}$ ii) $\frac{2}{3}$ iii) $\frac{1}{2}$

3 $\frac{1}{144}$ if each month is equally likely, or $\frac{36}{5329}$ if taken as 30 days out of 365. (Leap years excluded – though this could be an extension.)

4 a Yes
b i) 0.1 ii) 0.9 iii) 0.3
iv) 0.6

5 a 0.675 b 0.2 c 0.875
d 0.77

6 0.93

7 0.66

8 e Only 23 people are needed before the probability of two sharing the same birthday is greater than 50%.

2 a 65.6
  b 32.4
  c $47.2 \leq C \geq 52.8$

3 a

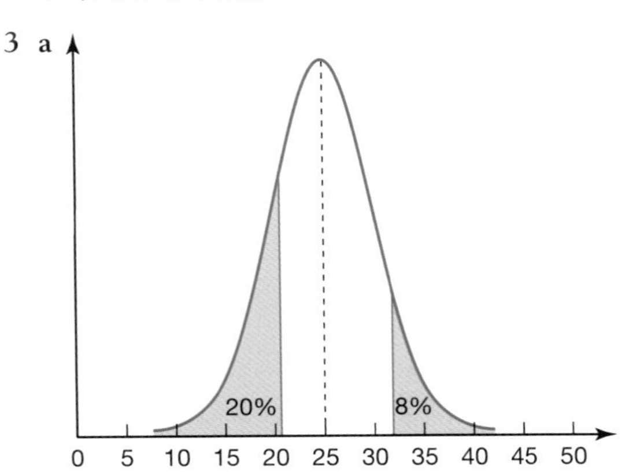


  b 20.8 cm
  c 32.0 cm

4 $191.4 \leq L \geq 268.6$ cm

5 a 14 260
  b 64.7 g

**Exercise 4.2.1**

1 Students' answers may differ from those given below.
  a Possible positive correlation (strength depending on topics tested)
  b No correlation
  c Positive correlation (likely to be quite strong)
  d Negative correlation (likely to be strong)
  e Depends on age range investigated. 0–16 years likely to be a positive correlation. Ages 16+ little correlation.
  f Strong positive correlation

2 a

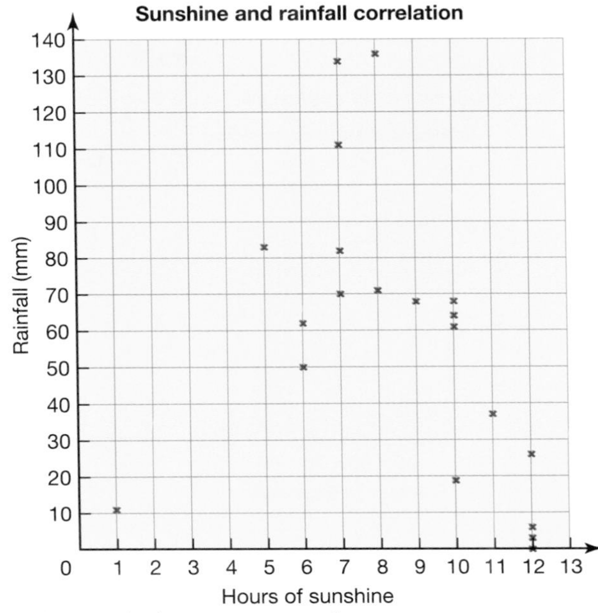


  b Graph shows a very weak negative correlation. Student's answers as to whether this is what they expected.

3 a

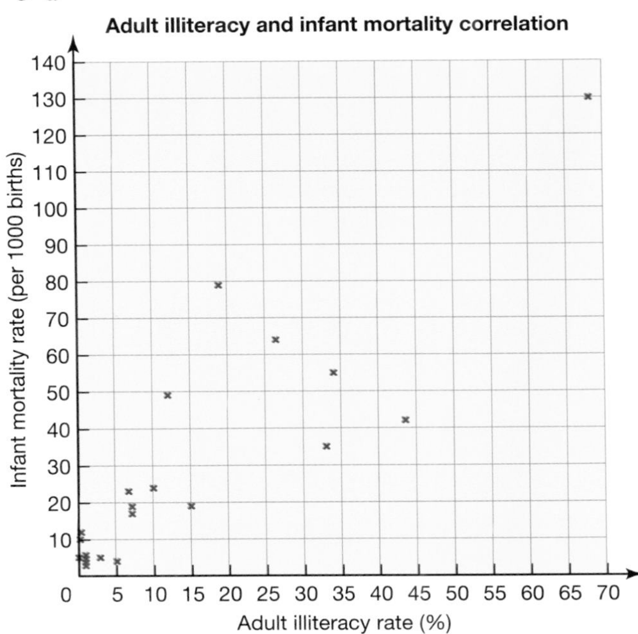

**b** Student's answer, however it is important to stress that although there is a correlation, it doesn't imply that one variable affects the other.

**c** Student's own explanation

**d**

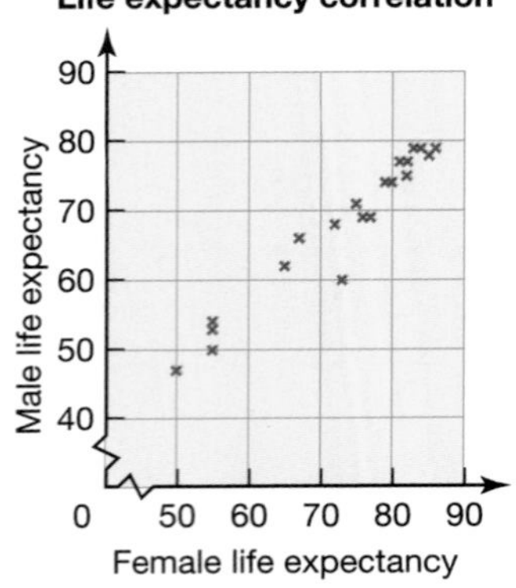


**4 a d**

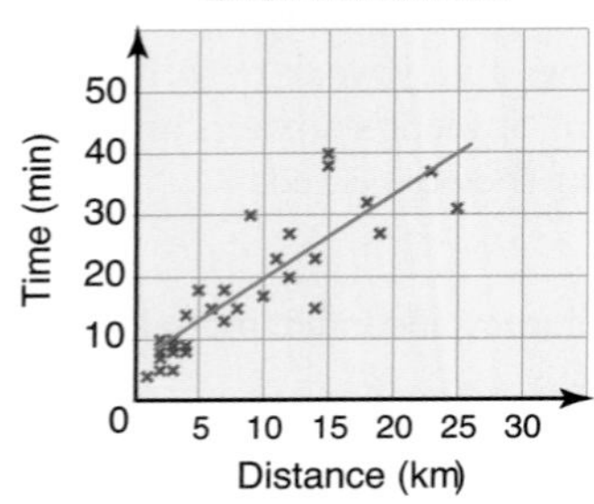


**b** (Strong) positive correlation

**c** Student's own explanation

**e** Approximately 10 km

**Exercise 4.2.2**

**1 a c**

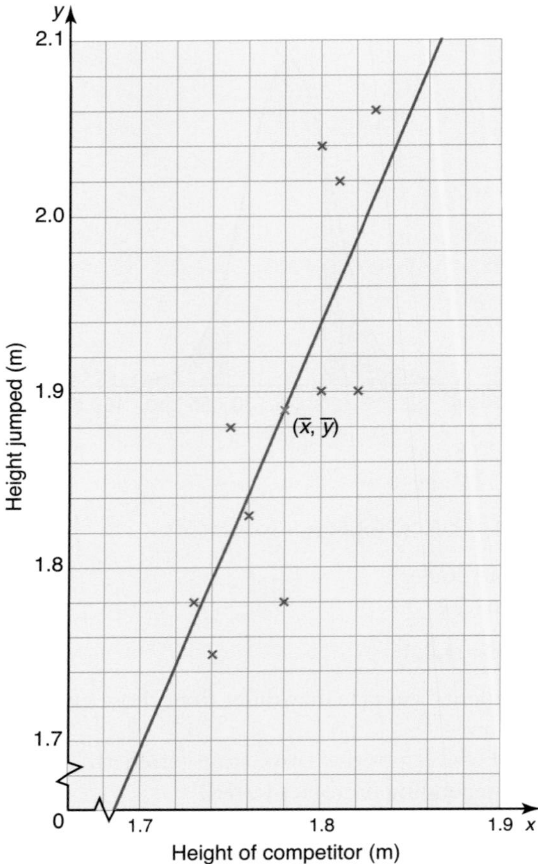


**b** $\bar{x} = 1.78\,\text{m}$ $\quad \bar{y} = 1.89\,\text{m}$

**d** $r = 0.79$

This implies a fairly good correlation, i.e. the taller the competitor, generally the greater the height jumped.

2 a c

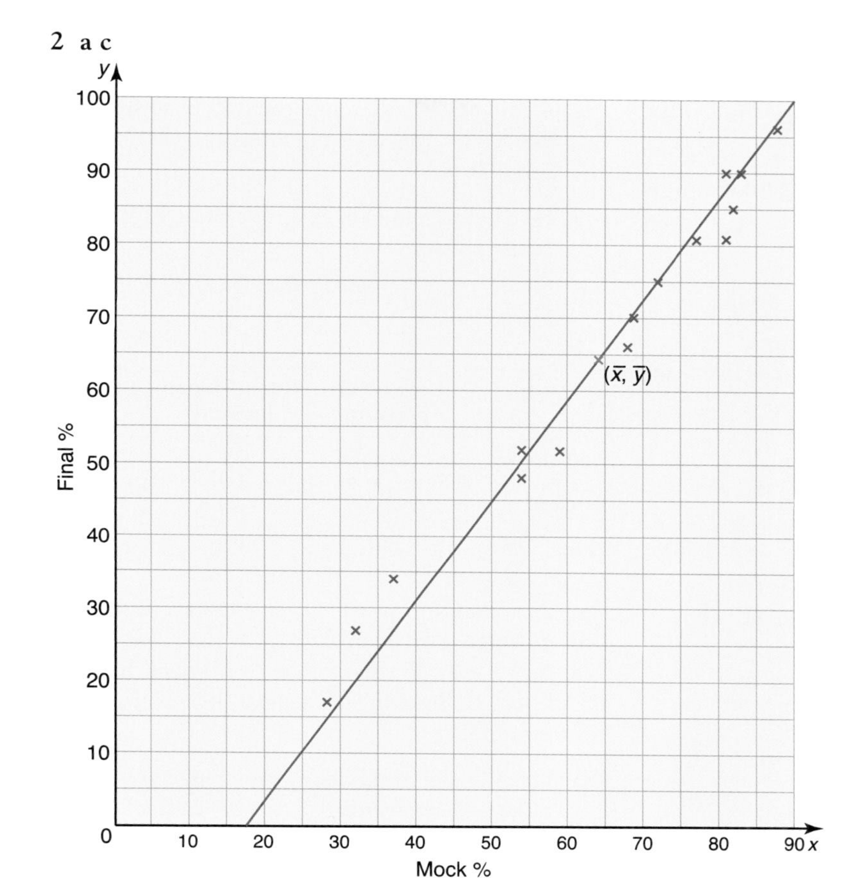


**b** $\bar{x} = 64.3\%$ $\bar{y} = 64.3\%$

**d** $r = 0.99$

This implies a very strong correlation between the mock % and the final % for English.

**3 a** $r = 0.89$

This implies a strong correlation between the mock % and final % for Mathematics.

**b** Although both show a strong correlation between mock % and final %, the results appear to suggest that it is stronger for English.

**4 a b** Student's solutions and justifications

**5 a–d** Student's own data, graph and analysis.

**e** $r$ is likely to be lower for 11-year-old students than for 17-year-old students as children of 11 are at different stages of development.

## Exercise 4.3.1

**1 a** $r = -0.959$

**b** $y = -1.468x + 92.742$

**c** $y = 66$ minutes

**d** A valid estimate as 18 hours is within the data range collected.

**2 a** $r = 0.970$ **b** $y = 0.866x - 24.433$

**c** IQ: 95 → IB: 58% IQ: 155 → IB: 110%

**d** The estimate for the IQ of 95 is valid as it falls in the range of the data collected. The estimation for the IQ of 155 is an extrapolation producing an IB result greater than 100% and is therefore invalid.

**3 a** $r = 0.973$ indicates a strong positive correlation between the salary and the number of years of experience.

**b** $y = 0.00116x - 28.552$

**c** $x = \$33\,200$ – this is a valid estimate as 10 years falls within the data range collected.

d 87 years experience – this value falls outside the data range collected. The result implies a firefighter approximately 100 years old, and is therefore invalid.

4 a $r = -0.946$ indicates a strong negative correlation, i.e. as height increases, temperature decreases.

b $y = -0.00189x + 7.4$

c Height = 41 000 m – although −70°C is slightly outside the data range, this is marginal and therefore the answer is likely to be valid.

### Exercise 4.4.1

1 a $H_0$ : A person's opinion regarding the wearing of safety helmets is independent of whether they are a cyclist or not.
$H_1$ : A person's opinion regarding the wearing of safety helmets is dependent on whether they are a cyclist or not.

b

| | Helmet compulsory | Helmet voluntary | Total |
|---|---|---|---|
| Cyclist | 125 | 175 | 300 |
| Non-cyclist | 125 | 175 | 300 |
| Total | 250 | 350 | 600 |

c $\chi^2 = 98.743$

d 1 degree of freedom

e The table gives a critical value of 3.841. $98.743 > 3.841$, therefore the null hypothesis is rejected. The opinions are dependent on whether they are cyclists or not.

2 a $H_0$ : Being given the drug and living for more than 3 months are independent events.
$H_1$ : Being given the drug does affect the chance of surviving for longer than 3 months.

b

| | Alive after 3 months | Not alive after 3 months | Total |
|---|---|---|---|
| Given drug | 78.81 | 61.19 | 140 |
| Given placebo | 73.19 | 56.81 | 130 |
| Total | 152 | 118 | 270 |

c $\chi^2 = 4.040$

d 1 degree of freedom

e $4.040 < 6.635$, therefore the null hypothesis is valid. The drug does not affect the chances of survival.

3 a $H_0$ : Being a smoker does not cause high blood pressure.
$H_1$ : Being a smoker does cause high blood pressure.

b

| | High blood pressure | Normal blood pressure | Total |
|---|---|---|---|
| Non-smoker | 96.63 | 33.37 | 130 |
| Smoker | 349.37 | 120.63 | 470 |
| Total | 446 | 154 | 600 |

c $\chi^2 = 8.215$

d 1 degree of freedom

e $8.215 > 6.635$ therefore the null hypothesis is rejected. Smoking does cause high blood pressure.

4 a $H_0$ : Gender and holiday preference are independent events.
$H_1$ : Holiday preference is affected by gender.

b

| | Beach | Walking | Cruise | Sail | Ski | Total |
|---|---|---|---|---|---|---|
| Male | 68.93 | 40.17 | 30.59 | 26.93 | 43.37 | 210 |
| Female | 82.07 | 47.83 | 36.41 | 32.07 | 51.63 | 250 |
| Total | 151 | 88 | 67 | 59 | 95 | 460 |

c $\chi^2 = 12.233$

d 4 degrees of freedom

e $12.233 > 7.779$ therefore the null hypothesis is rejected. Holiday preference is dependent upon gender.

5 a $H_0$ : Age and musical preference are independent.
$H_1$ : Musical preference is dependent on age.

b $\chi^2 = 49.077$

c 8 degrees of freedom

d $49.077 > 15.507$ therefore the null hypothesis is rejected. Musical preference is dependent upon age.

## Student assessment 1

1 Student's answers may differ from those given below.

 a Negative correlation (likely to be strong). Assume that motorcycles are not rare or vintage.

 b Factors such as social class, religion and income are likely to affect results therefore little correlation is likely.

2 a B has the greater mean as its peak is to the right of A's.

 b B has the greater standard deviation as the distribution is more spread out than A.

3 a

 b 30.9%

4 a

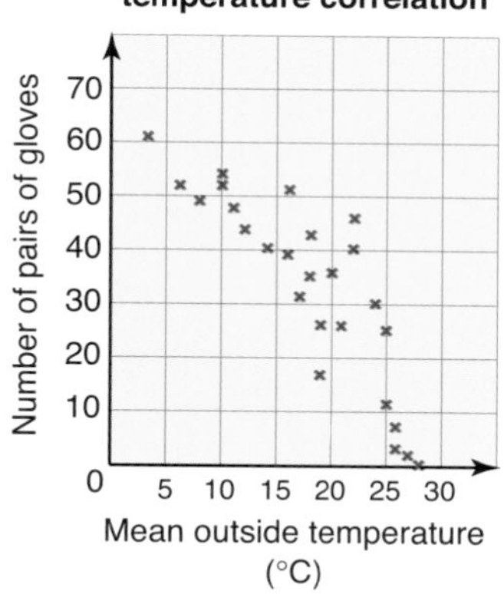


 b The graph indicates a negative correlation.

 c Student's own explanation

5 a $H_0$ : The drug has no effect on the dog's condition.

 $H_1$ : The drug improves the dog's condition.

 b

| | Improved | Did not improve | Total |
|---|---|---|---|
| Given drug | 168.68 | 91.32 | 260 |
| Not given drug | 97.32 | 52.68 | 150 |
| Total | 266 | 144 | 410 |

 c $\chi^2 = 2.061$

 d 1 degree of freedom

 e $2.061 < 3.841$ therefore the null hypothesis is supported, i.e. the drug does not significantly improve the dog's condition.

## Student assessment 2

1 a Likely to be fairly strongly positively correlated, although there will be exceptions, e.g. unemployed, people who borrow money, people who save a lot.

 b No correlation

2 a 46.5%

 b

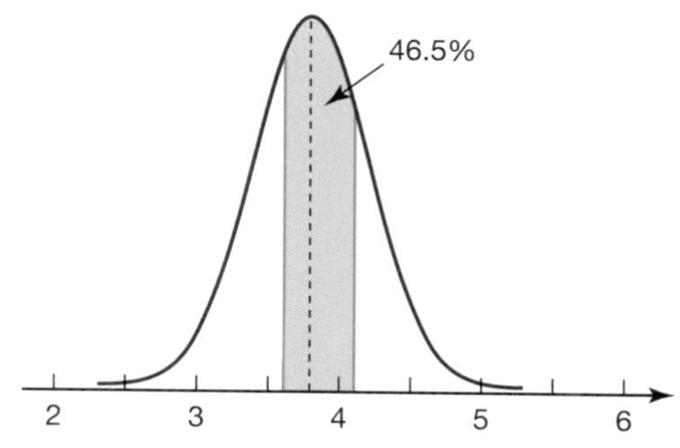


3 a 13.8%

 b

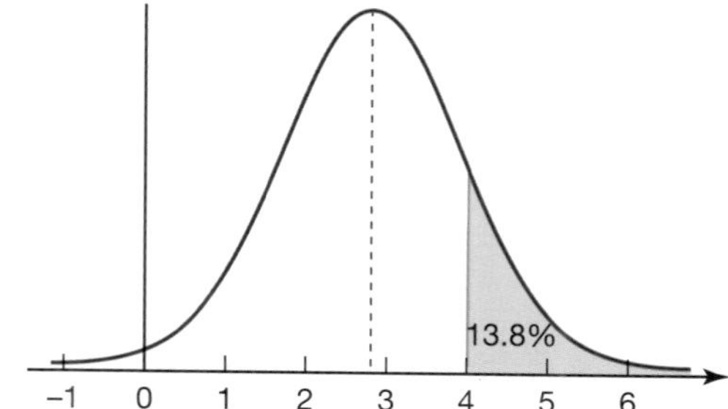


 c Approximately 99.7% of data that is normally distributed falls within three standard deviations of the mean. Three standard deviations below the mean in this example would give −0.5 m, i.e. a negative length, which is not possible. This is also shown on the sketch where the left 'tail' is negative. Therefore the data is not truly normally distributed.

**4 a** $y = 76241x + 3.76 \times 10^6$
**b** $r = 0.277$
**c** There doesn't appear to be a correlation between a footballer's salary and his popularity. This does not support the statement in the newspaper report.

**5 a** $H_0$ : The opinion on whether to ban fox hunting is independent of whether you live in the city or the country.
$H_1$ : The opinion on whether to ban fox hunting is dependent on whether you live in the city or the country.
**b** $\chi^2 = 1683.01$
**c** $1683.01 > 6.635$ therefore the null hypothesis is rejected, i.e. the opinion to ban fox hunting is dependent on whether you live in the country or the city.

**Examination questions**

**1 a** Chosen profession is independent of gender. *Or* There is no association between gender and chosen profession.
**b** 2
**c** 36
**d** $p$-value > 0.05 so accept $H_0$.

**2 a c e**

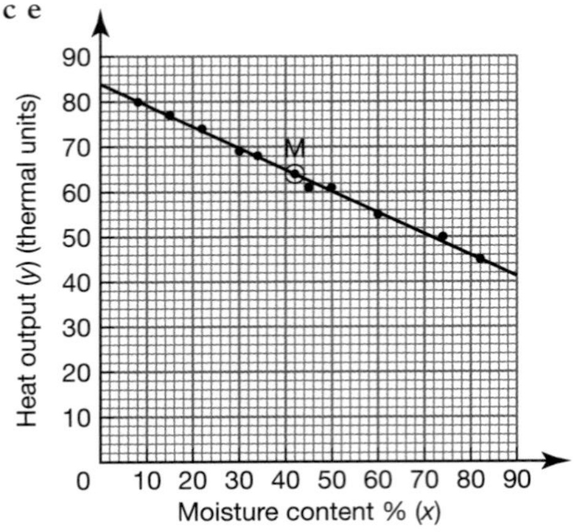

Heat output ($y$) (thermal units)
Moisture content % ($x$)

**b i** $\bar{x} = 42$
**ii** $\bar{y} = 64$
**d** −0.998

**f** 72.0 (or 71.95 or 72)
**g** Yes since 25% lies within the data set and $r$ is close to −1.

**3 Part A**

**a i** 50
**ii** 16.8
**iii** 30.5 cm
**iv** 12.3 cm
**b** 0.911 (or 0.912 or 0.910)
**c** $y = 0.669x - 2.95$
**d** 33.8 cm
**e i** 64.0 (or 63.95 or 63.9)
**ii** It is not valid. It lies too far outside the values that are given.

**Part B**

**a** 28
**b** $\frac{29 \times 45}{100}\left(\frac{28}{100} \times \frac{45}{100} \times 100\right) = 12.6$
**c i** The favourite car colour is independent of gender.
**ii** 2
**iii** 5.991 (or 5.99)
**iv** Accept the null hypothesis since $1.367 < 5.991$.

## Topic 5

**Revision exercise**

**1 a ii)** 5.66 units **iii)** (3, 4)
**b ii)** 4.24 units **iii)** (4.5, 2.5)
**c ii)** 5.66 units **iii)** (3, 6)
**d ii)** 8.94 units **iii)** (2, 4)
**e ii)** 6.32 units **iii)** (3, 4)
**f ii)** 6.71 units **iii)** (−1.5, 4)
**g ii)** 8.25 units **iii)** (−2, 1)
**h ii)** 8.94 units **iii)** (0, 0)
**i ii)** 7 units **iii)** (0.5, 5)
**j ii)** 6 units **iii)** (2, 3)
**k ii)** 8.25 units **iii)** (0, 4)
**l ii)** 10.82 units **ii)** (0, 1.5)

**2 a i)** 4.25 units **ii)** (2.5, 2.5)
**b i)** 5.66 units **ii)** (5, 4)
**c i)** 8.94 units **ii)** (4, 2)
**d i)** 8.94 units **ii)** (5, 0)
**e i)** 4.24 units **ii)** (−1.5, 4.5)
**f i)** 4.47 units **ii)** (−4, −3)
**g i)** 7.21 units **ii)** (0, 3)

**h** i) 7.21 units ii) (5, −1)
**i** i) 12.37 units ii) (0, 2.5)
**j** i) 8.49 units ii) (1, −1)
**k** i) 11 units ii) (0.5, −3)
**l** i) 8.25 units ii) (4, 2)

### Exercise 5.1.1

**1 a** i) 1 ii) −1
**b** i) 1 ii) −1
**c** i) 1 ii) −1
**d** i) 2 ii) $-\frac{1}{2}$
**e** i) 3 ii) $-\frac{1}{3}$
**f** i) 2 ii) $-\frac{1}{2}$
**g** i) 4 ii) $-\frac{1}{4}$
**h** i) $\frac{1}{2}$ ii) −2
**i** i) 0 ii) infinite
**j** i) infinite ii) 0
**k** i) $\frac{1}{4}$ ii) −4
**l** i) $\frac{3}{2}$ ii) $-\frac{2}{3}$

**2 a** i) −1 ii) 1
**b** i) −1 ii) 1
**c** i) −2 ii) $\frac{1}{2}$
**d** i) $-\frac{1}{2}$ ii) 2
**e** i) −1 ii) 1
**f** i) −2 ii) $\frac{1}{2}$
**g** i) $-\frac{3}{2}$ ii) $\frac{2}{3}$
**h** i) $\frac{2}{3}$ ii) $-\frac{3}{2}$
**i** i) $-\frac{1}{4}$ ii) 4
**j** i) −1 ii) 1
**k** i) 0 ii) infinite
**l** i) −4 ii) $\frac{1}{4}$

### Exercise 5.1.2

**1 a** $y = 7$ **b** $y = 2$
**c** $x = 7$ **d** $x = 3$
**e** $y = x$ **f** $y = \frac{1}{2}x$
**g** $y = -x$ **h** $y = -2x$

**2 a** $y = x + 1$ **b** $y = x + 3$
**c** $y = x - 2$ **d** $y = 2x + 2$
**e** $y = \frac{1}{2}x + 5$ **f** $y = \frac{1}{2}x - 1$

**3 a** $y = -x + 4$ **b** $y = -x - 2$
**c** $y = -2x - 2$ **d** $y = -\frac{1}{2}x + 3$
**e** $y = -\frac{3}{2}x + 2$ **f** $y = -4x + 1$

**4 a** 2 a 1 b 1 c 1 d 2 e $\frac{1}{2}$ f $\frac{1}{2}$
3 a −1 b −1 c −2 d $-\frac{1}{2}$ e $-\frac{3}{2}$ f −4
**b** The gradient is equal to the coefficient of $x$.
**c** The constant being added/subtracted indicates where the line intersects the $y$-axis.

**5 b** Only the intercept $c$ is different.

**6** The lines are parallel.

### Exercise 5.1.3

**1 a** $m = 2$ $c = 1$ **b** $m = 3$ $c = 5$
**c** $m = 1$ $c = -2$ **d** $m = \frac{1}{2}$ $c = 4$
**e** $m = -3$ $c = 6$ **f** $m = -\frac{2}{3}$ $c = 1$
**g** $m = -1$ $c = 0$ **h** $m = -1$ $c = -2$
**i** $m = -2$ $c = 2$

**2 a** $m = 3$ $c = 1$ **b** $m = -\frac{1}{2}$ $c = 2$
**c** $m = -2$ $c = -3$ **d** $m = -2$ $c = -4$
**e** $m = \frac{1}{4}$ $c = 6$ **f** $m = 3$ $c = 2$
**g** $m = 1$ $c = -2$ **h** $m = -8$ $c = 6$
**i** $m = 3$ $c = 1$

**3 a** $m = 2$ $c = -3$ **b** $m = \frac{1}{2}$ $c = 4$
**c** $m = 2$ $c = -4$ **d** $m = -8$ $c = 12$
**e** $m = 2$ $c = 0$ **f** $m = -3$ $c = 3$
**g** $m = 2$ $c = 1$ **h** $m = -\frac{1}{2}$ $c = 2$
**i** $m = 2$ $c = -\frac{1}{2}$

**4 a** $m = 2$ $c = -4$ **b** $m = 1$ $c = 6$
**c** $m = -3$ $c = -1$ **d** $m = -1$ $c = 4$
**e** $m = 10$ $c = -2$ **f** $m = -3$ $c = \frac{3}{2}$
**g** $m = -9$ $c = 2$ **h** $m = 6$ $c = -14$
**i** $m = 2$ $c = -\frac{3}{2}$

**5 a** $m = 2$ $c = -2$ **b** $m = 2$ $c = 3$
**c** $m = 1$ $c = 0$ **d** $m = \frac{3}{2}$ $c = 6$
**e** $m = -1$ $c = \frac{2}{3}$ **f** $m = -4$ $c = 2$
**g** $m = 3$ $c = -12$ **h** $m = 0$ $c = 0$
**i** $m = -3$ $c = 0$

6 a $m = 1$ $c = 0$ b $m = -\frac{1}{2}$ $c = -2$
c $m = -3$ $c = 0$ d $m = 1$ $c = 0$
e $m = -2$ $c = -\frac{2}{3}$ f $m = \frac{2}{3}$ $c = -4$
g $m = -\frac{2}{5}$ $c = 0$ h $m = -\frac{1}{3}$ $c = -\frac{7}{6}$
i $m = 3$ $c = 0$ j $m = -\frac{2}{3}$ $c = -\frac{8}{3}$

### Exercise 5.1.4

1 a i) $y = 2x - 1$ ii) $2x - y - 1 = 0$
b i) $y = 3x + 1$ ii) $3x - y + 1 = 0$
c i) $y = 2x + 3$ ii) $2x - y + 3 = 0$
d i) $y = x - 4$ ii) $x - y - 4 = 0$
e i) $y = 4x + 2$ ii) $4x - y + 2 = 0$
f i) $y = -x + 4$ ii) $x + y - 4 = 0$
g i) $y = -2x + 2$ ii) $2x + y - 2 = 0$
h i) $y = -3x - 1$ ii) $3x + y + 1 = 0$
i i) $y = \frac{1}{2}x$ ii) $x - 2y = 0$

2 a i) $y = \frac{1}{7}x + \frac{26}{7}$ ii) $x - 7y + 26 = 0$
b i) $y = \frac{6}{7}x + \frac{4}{7}$ ii) $6x - 7y + 4 = 0$
c i) $y = \frac{3}{2}x + \frac{15}{2}$ ii) $3x - 2y + 15 = 0$
d i) $y = 9x - 13$ ii) $9x - y - 13 = 0$
e i) $y = -\frac{1}{2}x + \frac{5}{2}$ ii) $x + 2y - 5 = 0$
f i) $y = -\frac{3}{13}x + \frac{70}{13}$ ii) $3x + 13y - 70 = 0$
g i) $y = 2$ ii) $y - 2 = 0$
h i) $y = -3x$ ii) $3x + y = 0$
i i) $x = 6$ ii) $x - 6 = 0$

### Exercise 5.1.5

1 a

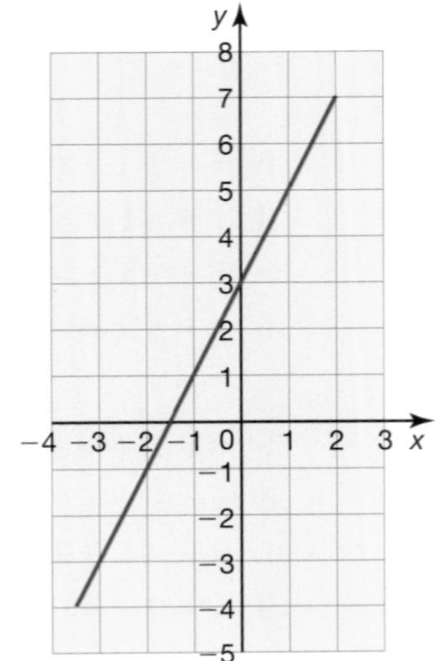

b

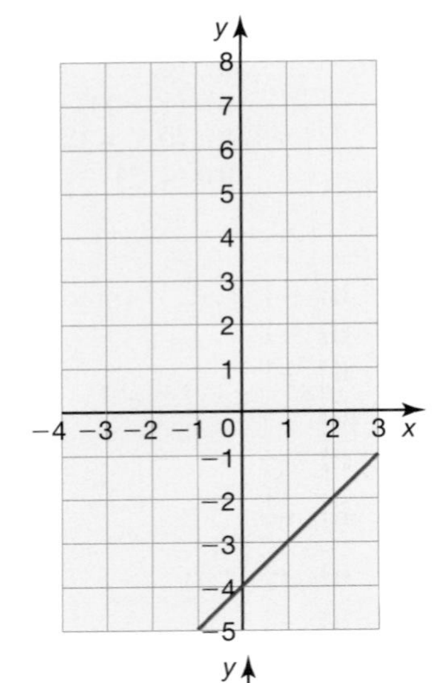

c

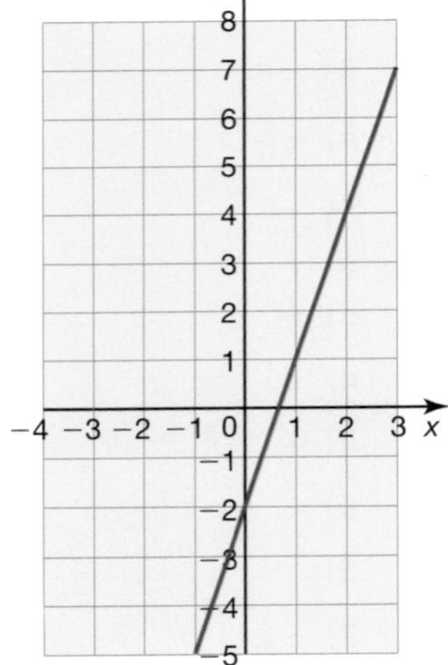

d

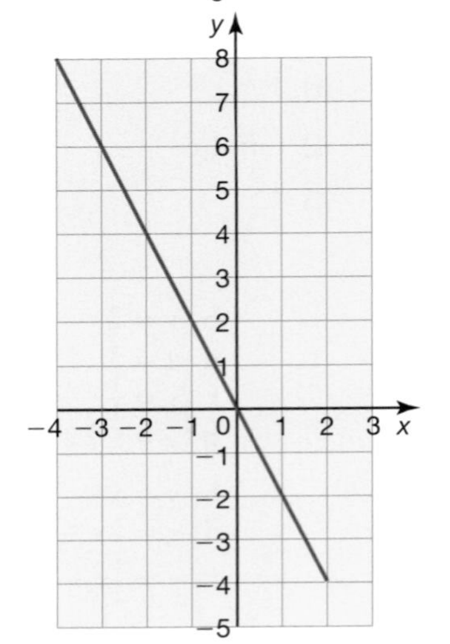

e

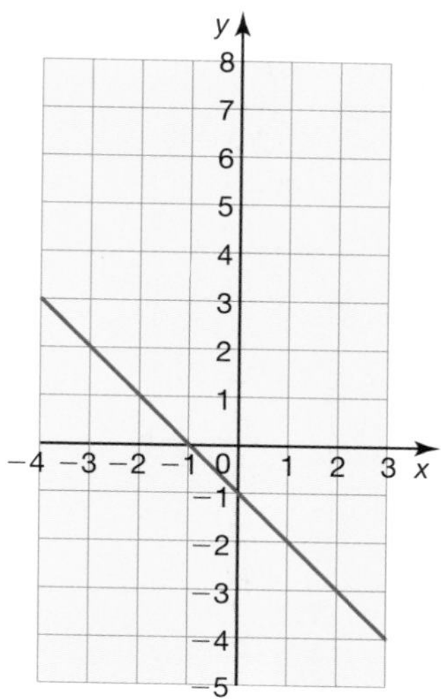
y
8
7
6
5
4
3
2
1
−4 −3 −2 −1 0 1 2 3 x
−1
−2
−3
−4
−5

f

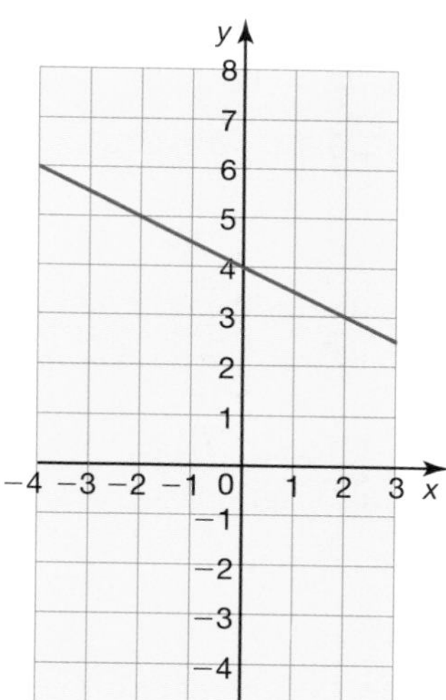
y
8
7
6
5
4
3
2
1
−4 −3 −2 −1 0 1 2 3 x
−1
−2
−3
−4
−5

g

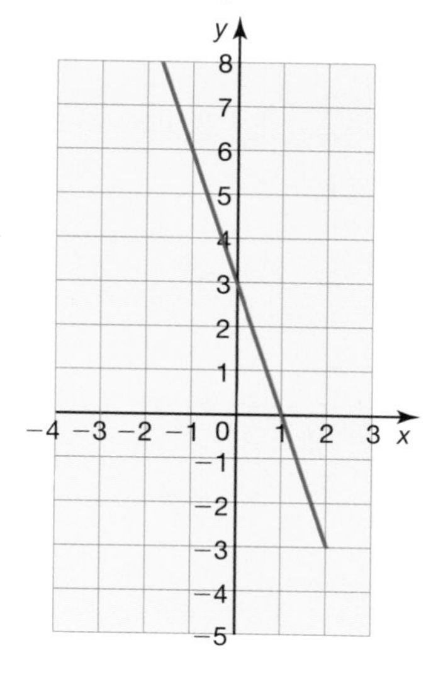
y
8
7
6
5
4
3
2
1
−4 −3 −2 −1 0 1 2 3 x
−1
−2
−3
−4
−5

h

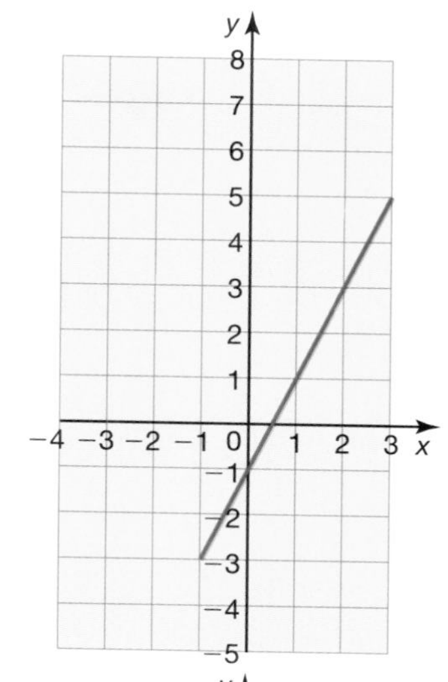
y
8
7
6
5
4
3
2
1
−4 −3 −2 −1 0 1 2 3 x
−1
−2
−3
−4
−5

i

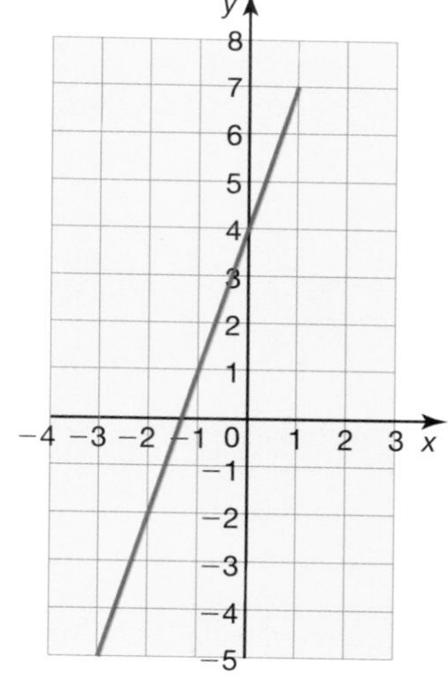
y
8
7
6
5
4
3
2
1
−4 −3 −2 −1 0 1 2 3 x
−1
−2
−3
−4
−5

2 a

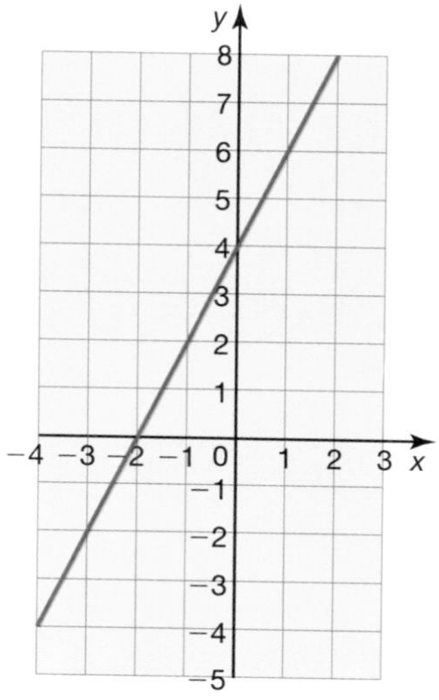
y
8
7
6
5
4
3
2
1
−4 −3 −2 −1 0 1 2 3 x
−1
−2
−3
−4
−5

b

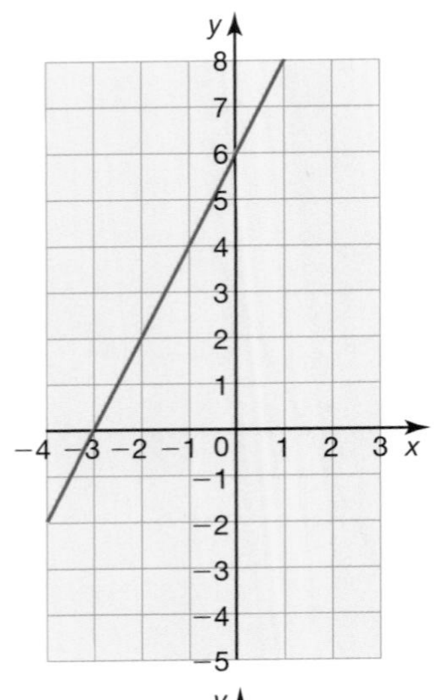
y
8 7 6 5 4 3 2 1
−1 −2 −3 −4 −5
−4 −3 −2 −1 0 1 2 3 x

c

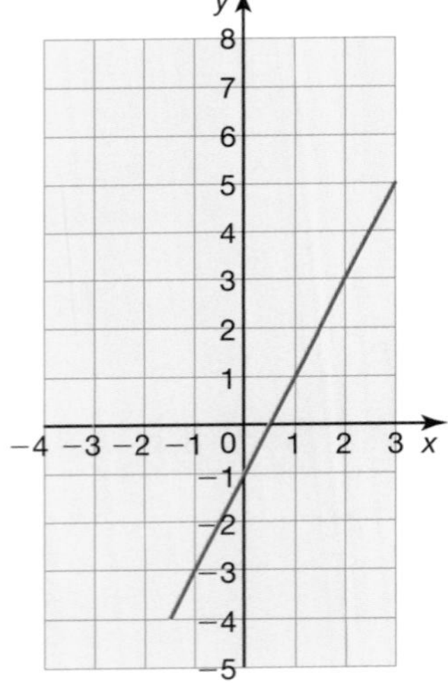
y
8 7 6 5 4 3 2 1
−1 −2 −3 −4 −5
−4 −3 −2 −1 0 1 2 3 x

d

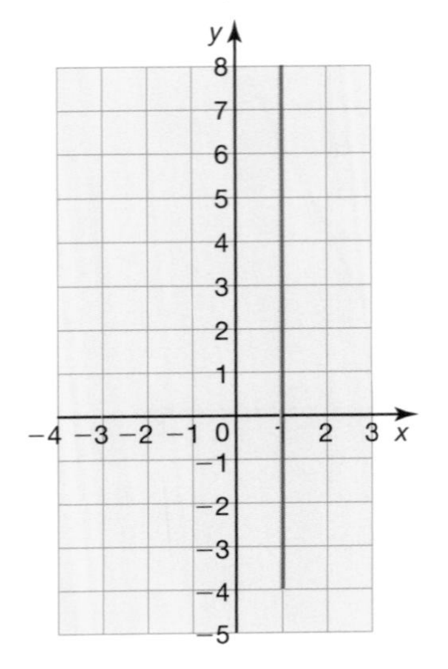
y
8 7 6 5 4 3 2 1
−1 −2 −3 −4 −5
−4 −3 −2 −1 0 2 3 x

e

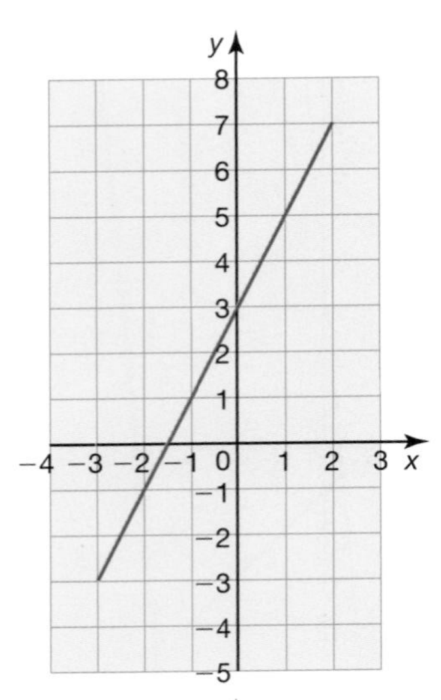
y
8 7 6 5 4 3 2 1
−1 −2 −3 −4 −5
−4 −3 −2 −1 0 1 2 3 x

f

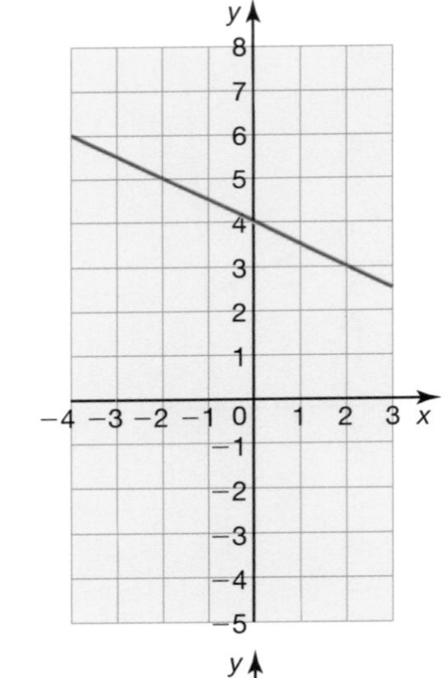
y
8 7 6 5 4 3 2 1
−1 −2 −3 −4 −5
−4 −3 −2 −1 0 1 2 3 x

g

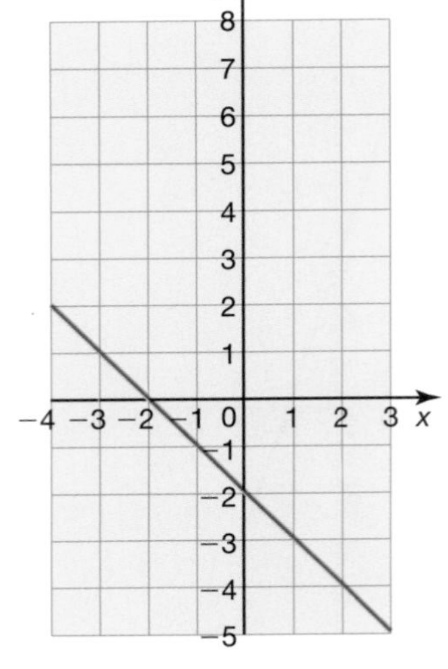
y
8 7 6 5 4 3 2 1
−1 −2 −3 −4 −5
−4 −3 −2 −1 0 1 2 3 x

**h**

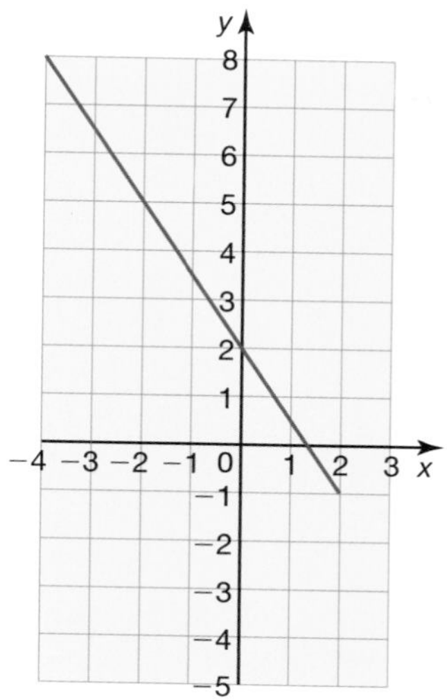
y
8
7
6
5
4
3
2
1
−4 −3 −2 −1 0 1 2 3 x
−1
−2
−3
−4
−5

**i**

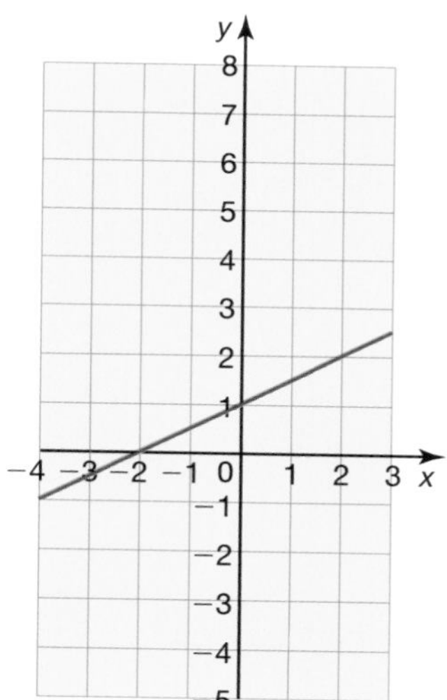
y
8
7
6
5
4
3
2
1
−4 −3 −2 −1 0 1 2 3 x
−1
−2
−3
−4
−5

**3 a**

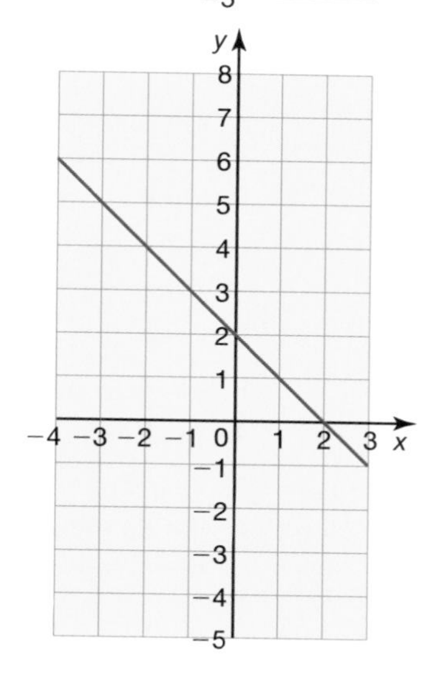
y
8
7
6
5
4
3
2
1
−4 −3 −2 −1 0 1 2 3 x
−1
−2
−3
−4
−5

**b**

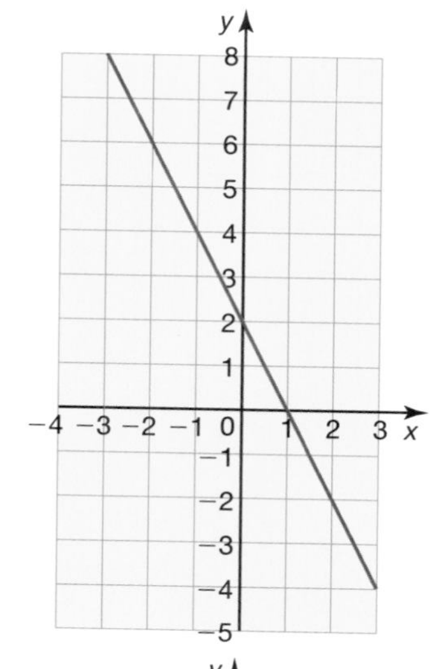
y
8
7
6
5
4
3
2
1
−4 −3 −2 −1 0 1 2 3 x
−1
−2
−3
−4
−5

**c**

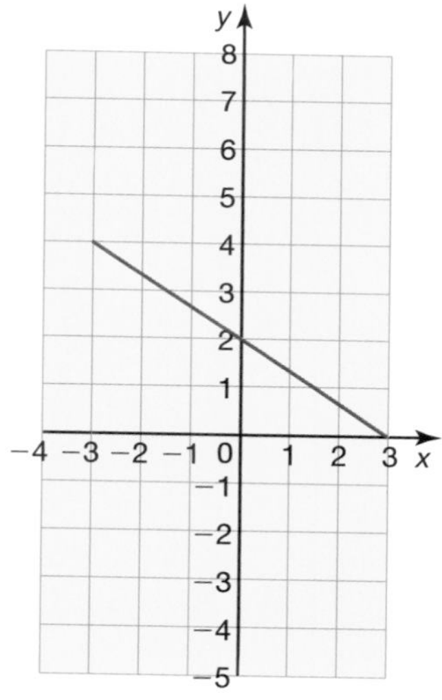
y
8
7
6
5
4
3
2
1
−4 −3 −2 −1 0 1 2 3 x
−1
−2
−3
−4
−5

**d**

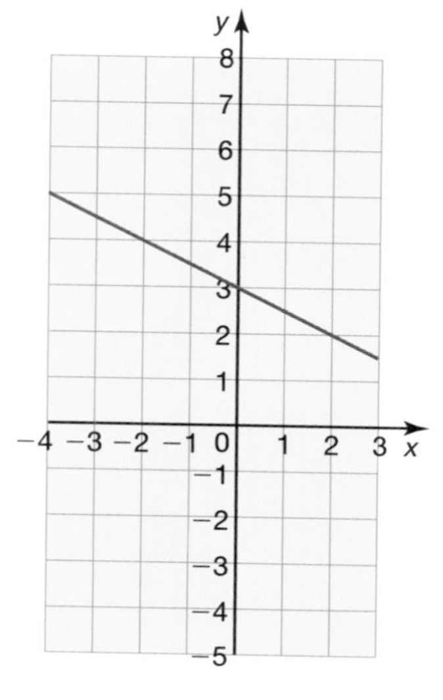
y
8
7
6
5
4
3
2
1
−4 −3 −2 −1 0 1 2 3 x
−1
−2
−3
−4
−5

e

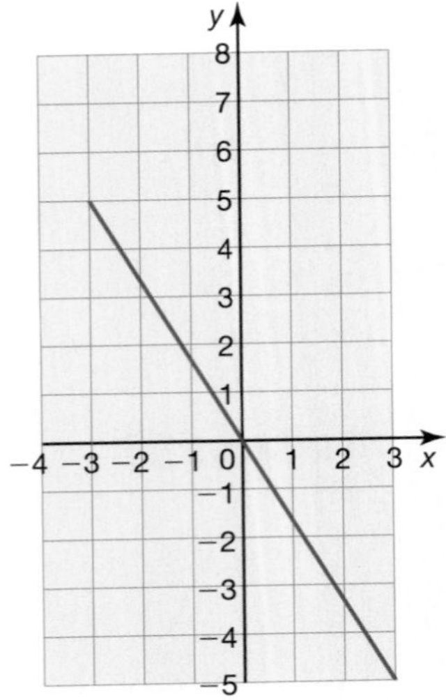


f

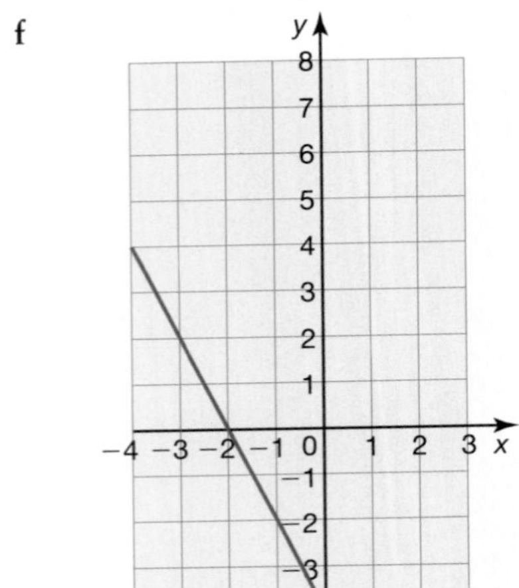


g

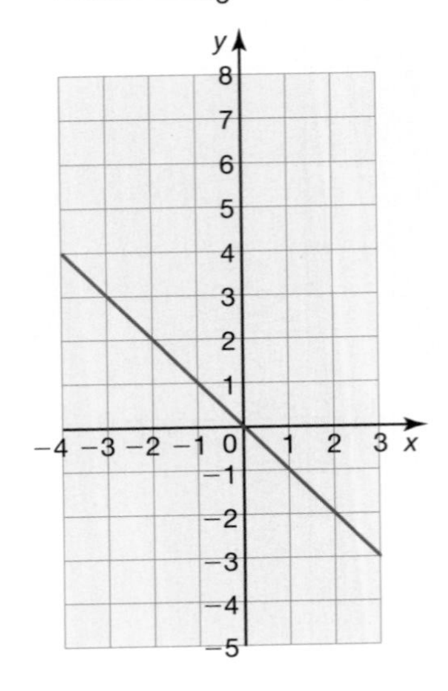


h

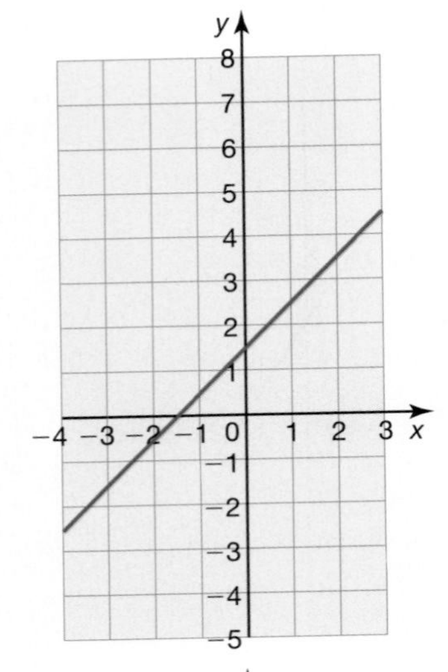


i

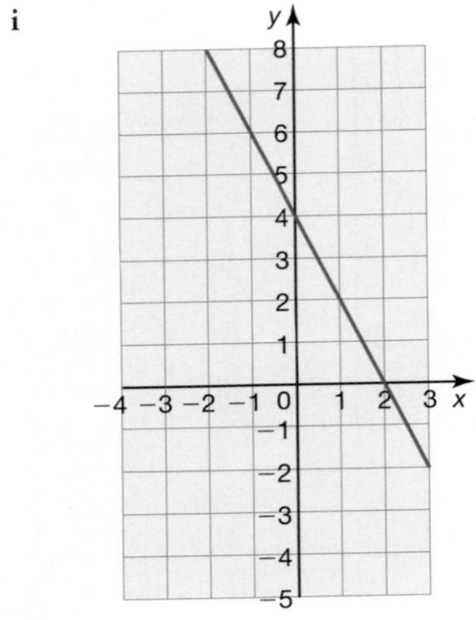


**Exercise 5.1.6**

**1** **a** $x = 4$ $y = 2$ **b** $x = 6$ $y = 5$
**c** $x = 6$ $y = -1$ **d** $x = 5$ $y = 2$
**e** $x = 5$ $y = 2$ **f** $x = 4$ $y = 9$

**2** **a** $x = 3$ $y = 2$ **b** $x = 7$ $y = 4$
**c** $x = 1$ $y = 1$ **d** $x = 1$ $y = 5$
**e** $x = 1$ $y = 10$ **f** $x = 8$ $y = 2$

**3** **a** $x = 5$ $y = 4$ **b** $x = 4$ $y = 3$
**c** $x = 10$ $y = 5$ **d** $x = 6$ $y = 4$
**e** $x = 4$ $y = 4$ **f** $x = 10$ $y = -2$

**4** **a** $x = 5$ $y = 4$ **b** $x = 4$ $y = 2$
**c** $x = 5$ $y = 3$ **d** $x = 5$ $y = -2$
**e** $x = 1$ $y = 5$ **f** $x = -3$ $y = -3$

**5** **a** $x = -5$ $y = -2$ **b** $x = -3$ $y = -4$
**c** $x = 4$ $y = 3\frac{2}{3}$ **d** $x = 2$ $y = 7$
**e** $x = 1$ $y = 1$ **f** $x = 2$ $y = 9$

**6** **a** $x = 2$ $y = 3$ **b** $x = 5$ $y = 10$
**c** $x = 4$ $y = 6$ **d** $x = 4$ $y = 4$

**7** **a** $x = 1$ $y = -1$ **b** $x = 11\frac{2}{3}$ $y = 8$
**c** $x = 4$ $y = 0$ **d** $x = 3$ $y = 4$
**e** $x = 2$ $y = 8$ **f** $x = 1$ $y = 1$

### Exercise 5.1.7

**1** **a** $x = 2$ $y = 3$ **b** $x = 1$ $y = 4$
**c** $x = 5$ $y = 2$ **d** $x = 3$ $y = 3$
**e** $x = 4$ $y = 2$ **f** $x = 6$ $y = 1$

**2** **a** $x = 1$ $y = 4$ **b** $x = 5$ $y = 2$
**c** $x = 3$ $y = 3$ **d** $x = 6$ $y = 1$
**e** $x = 2$ $y = 3$ **f** $x = 2$ $y = 3$

**3** **a** $x = 0$ $y = 3$ **b** $x = 5$ $y = 2$
**c** $x = 1$ $y = 7$ **d** $x = 6$ $y = 4$
**e** $x = 2$ $y = 5$ **f** $x = 3$ $y = 0$

**4** **a** $x = 1$ $y = 0.5$ **b** $x = 2.5$ $y = 4$
**c** $x = \frac{1}{5}$ $y = 4$ **d** $x = \frac{3}{4}$ $y = \frac{1}{2}$
**e** $x = 5$ $y = \frac{1}{3}$ **f** $x = \frac{1}{2}$ $y = 1$

### Exercise 5.2.1

**1** **a** 19.2 cm **b** 15.1 cm **c** 43.8 cm
**d** 31.8 cm **e** 6.2 cm **f** 2.1 cm

**2** **a** 81.1° **b** 63.4° **c** 38.7°

**3** **a** 43.6° **b** 19.5 cm **c** 42.5°

**4** **a** 20.8 km **b** 215°

**5** **a** 228 km **b** 102 km **c** 103 km
**d** 147 km **e** 415 km **f** 217°

**6** **a** 6.7 m **b** 19.6 m **c** 15.3 m

**7** **a** 48.2° **b** 41.8° **c** 8 cm
**d** 8.9 cm **e** 76.0 cm$^2$

**8** **a** 342 m **b** 940 m

**9** 6.9 km

**10** **a** 225 m **b** 48.4°

### Exercise 5.3.1

**1** **a** sin 120° **b** sin 100° **c** sin 65°
**d** sin 40° **e** sin 52° **f** sin 13°

**2** **a** sin 145° **b** sin 130° **c** sin 150°
**d** sin 132° **e** sin 76° **f** sin 53°

**3** **a** 19°, 161° **b** 82°, 98° **c** 5°, 175°
**d** 72°, 108° **e** 13°, 167° **f** 28°, 152°

**4** **a** 70°, 110° **b** 9°, 171° **c** 53°, 127°
**d** 34°, 146° **e** 16°, 164° **f** 19°, 161°

### Exercise 5.3.2

**1** **a** −cos 160° **b** −cos 95° **c** −cos 148°
**d** −cos 85° **e** −cos 33° **f** −cos 74°

**2** **a** −cos 82° **b** −cos 36° **c** −cos 20°
**d** −cos 37° **e** −cos 9° **f** −cos 57°

**3** **a** cos 80° **b** −cos 90° **c** cos 70°
**d** cos 135° **e** cos 58° **f** cos 155°

**4** **a** −cos 55° **b** −cos 73° **c** cos 60°
**d** cos 82° **e** cos 88° **f** cos 70°

### Exercise 5.3.3

**1** **a** 8.9 cm **b** 8.9 cm **c** 6.0 mm
**d** 8.6 cm

**2** **a** 33.2° **b** 127.3° **c** 77.0°
**d** 44.0°

**3** **a** 25.3°, 154.7°
**b**

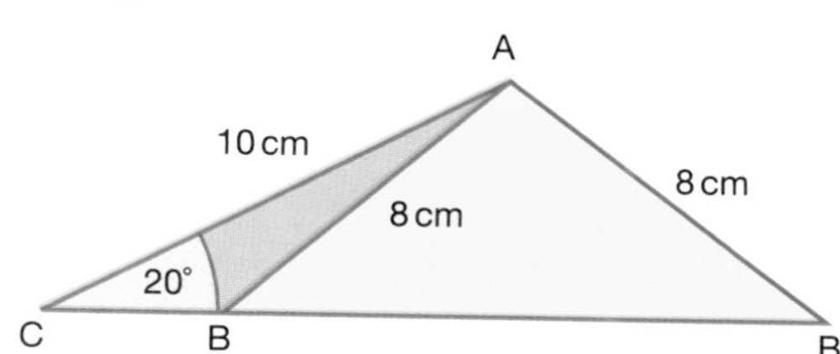


**4** **a** 74.6°, 105.4°
**b**

P
6 cm
4 cm
4 cm
40°
Q
R
R

**Exercise 5.3.4**

1 a 4.7 m　b 12.1 cm　c 9.1 cm
　d 3.1 cm　e 10.7 cm

2 a 125.1°　b 108.2°　c 33.6°
　d 37.0°　e 122.9°

3 a 42.9 m　b 116.9°　c 24.6°
　d 33.4°　e 35.0 m

4 370 m

5 a 669 m　b 546 m　c 473 m

6 73.9 m

**Exercise 5.3.5**

1 a 70.0 $cm^2$　b 70.9 $mm^2$　c 121.6 $cm^2$
　d 17.0 $cm^2$

2 a 24.6°　b 13.0 cm　c 23.1 cm
　d 63.2°

3 16 800 $m^2$

4 a 3.9 $m^2$　b 222 $m^3$

**Exercise 5.4.1**

1 a 5.7 cm　b 6.9 cm　c 54.7°
　d 2.8 cm

2 a 5.8 cm　b 6.2 cm　c 18.9°

3 a 6.4 cm　b 13.6 cm　c 61.9°

4 a 75.3°　b 56.3°

5 a i) 7.2 cm　ii) 21.1°
　b i) 33.7°　ii) 68.9°

6 a i) 8.5 cm　ii) 28.3°
　b i) 20.6°　ii) 61.7°

7 a 6.5 cm　b 11.3 cm　c 70.7 cm

8 a 11.7 cm　b 7.6 cm

9 a 25.0 cm　b 20.5 cm　c 26.0 cm
　d 12.5 cm　e 22.0 cm

10 a TU = TQ = 10 cm, QU = 8.5 cm
　b $Q = U = 64.9°$, $T = 50.2°$
　c 38.4 $cm^2$

**Exercise 5.4.2**

1 a RW　b TQ　c SQ
　d WU　e QV　f SV

2 a JM　b KN　c HM
　d HO　e JO　f MO

3 a ∠TPS　b ∠UPQ　c ∠VSW
　d ∠RTV　e ∠SUR　f ∠VPW

4 a 5.83 cm　b 31.0°

5 a 10.2 cm　b 29.2°　c 51.3°

6 a 6.71 cm　b 61.4°

7 a 7.81 cm　b 11.3 cm　c 12.5°

8 a 14.1 cm　b 8.49 cm　c 7.5 cm
　d 69.2°

9 a 17.0 cm　b 5.66 cm　c 7.00 cm
　d 51.1°

**Exercise 5.5.1**

1 a Volume = 27.6 $cm^3$, surface area = 60.8 $cm^2$
　b Volume = 277.1 $cm^3$, surface area = 235.3 $cm^2$
　c Volume = 42 $cm^3$, surface area = 95.4 $cm^2$

2 a 16 cm　b 4096 $cm^3$　c 3217 $cm^3$
　d 21.5%

3 a 42 $cm^2$　b 840 $cm^3$　c 564 $cm^2$

4 6.3 cm

5 2.90 $m^3$

6 a 24 $cm^2$　b 2 cm

7 a 216 $cm^2$　b 15.2 cm　c 25.0 $cm^3$

8 a 94.2 $cm^2$　b 14 cm　c 12.6 $cm^3$

9 4.4 cm

**Exercise 5.5.2**

1 a Volume = 905 $cm^3$, surface area = 452 $cm^2$
　b Volume = 3591 $cm^3$, surface area = 1134 $cm^2$
　c Volume = 2309.6 $cm^3$ (1 d.p.), surface area = 845 $cm^2$
　d Volume = 1.4 $cm^3$ (1 d.p.), surface area = 6.16 $cm^2$

2 a 5.6 cm b 0.4 m

3 a 1.15 cm b 3.09 mm

4 6.30 cm

5 86.7 $cm^3$

6 11.9 cm

7 a 4190 $cm^3$ b 8000 $cm^3$ c 47.6%

8 10.0 cm

9 A = 4.1 cm, B = 3.6 cm, C = 3.1 cm

10 3 : 2

11 1 : 4

12 a 157 $cm^2$ b 15.0 cm c 707 $cm^2$

13 a 804.2 $cm^2$ b 5.9 cm

**Exercise 5.5.3**

1 a 40 $cm^3$ b 133 $cm^3$ c 64 $cm^3$
d 70 $cm^3$

2 Volume = 147.1 $cm^3$, surface area = 189.2 $cm^2$

3 Volume = 44.0 $cm^3$, surface area = 73.3 $cm^2$

4 a 8 cm b 384 $cm^3$ c 378 $cm^3$

5 1120.9 $cm^2$

6 7 cm

7 4 cm

8 a 3.6 cm b 21.7 $cm^3$ c 88.7 $cm^3$

9 6.93 $cm^2$

10 a 693 $cm^2$ b 137 $cm^2$ c 23.6 cm

**Exercise 5.5.4**

1 a Arc length = 6.3 cm, sector area = 25.1 $cm^2$
b Arc length = 2.1 cm, sector area = 15.7 $cm^2$
c Arc length = 11.5 cm, sector area = 34.6 $cm^2$
d Arc length = 23.6 cm, sector area = 58.9 $cm^2$

2 a 4.19 cm b 114 $cm^2$ c 62.8 $cm^3$

3 a 56.5 $cm^3$ b 264 $cm^3$ c 1.34 $cm^3$
d 166 $cm^3$

4 6.91 cm

5 a 15.9 cm b 8.41 cm c 2230 $cm^3$

6 a Volume = 559.2 $cm^3$, surface area = 414.7 $cm^2$
b Volume = 3117.0 $cm^3$, surface area = 1649.3 $cm^2$

7 3.88 cm

8 a 33.0 cm b 5.25 cm c 7.31 cm
d 211 $cm^3$ e 148 $cm^2$

9 1131 $cm^2$

10 a 314 $cm^2$ b 4.5

11 a 2304 $cm^3$ b 603.2 $cm^3$ c 1700.8 $cm^3$

12 a 81.6 $cm^3$ b 275 $cm^3$ c 8 : 27

13 a 81.8 $cm^3$ b 101 $cm^2$

14 a 771 $m^3$ b 487 $m^2$

15 3166.7 $cm^3$

16 a 654.5 $cm^3$ b 12.5 cm c 2945.2 $cm^3$

**Student assessment 1**

1 a i) 13 units ii) (0, 1.5)
b i) 10 units ii) (4, 6)
c i) 5 units ii) (0, 4.5)
d i) 26 units ii) (−5, 2)

2 a b c d

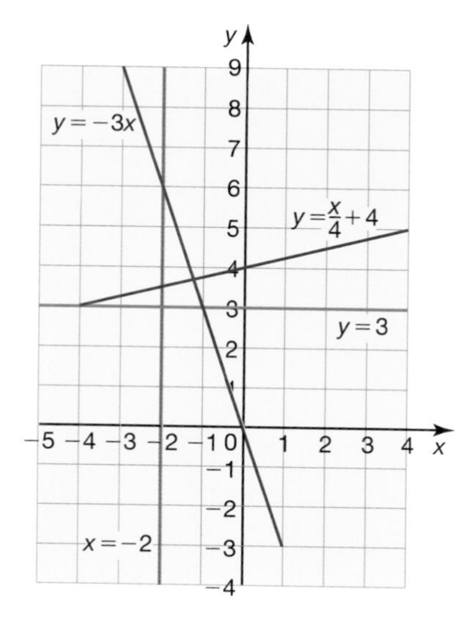

**3 a** i) $m = 1$ $c = 1$

ii)

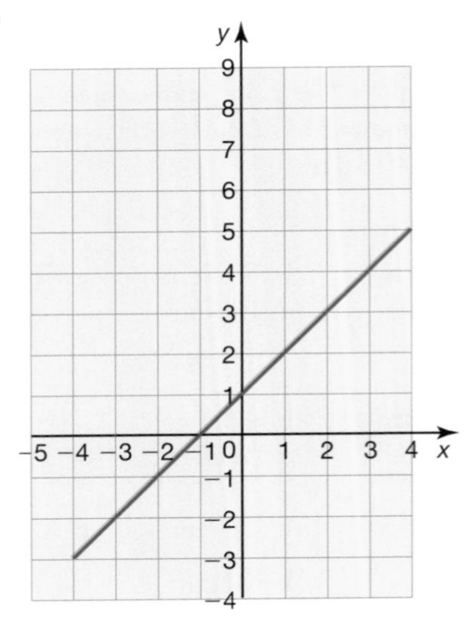


**b** i) $m = -3$ $c = 3$

ii)

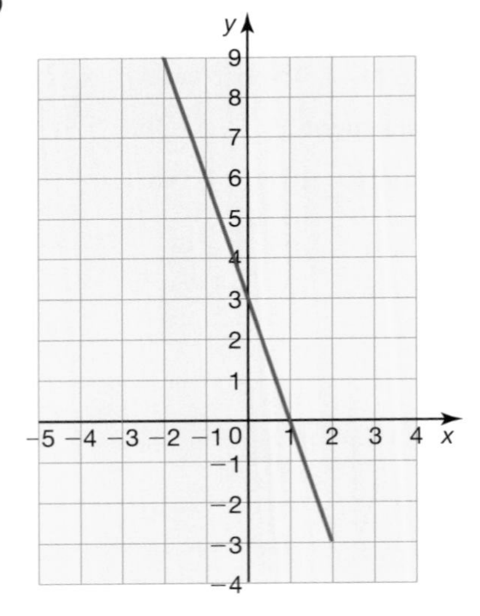


**c** i) $m = 2$ $c = 4$

ii)

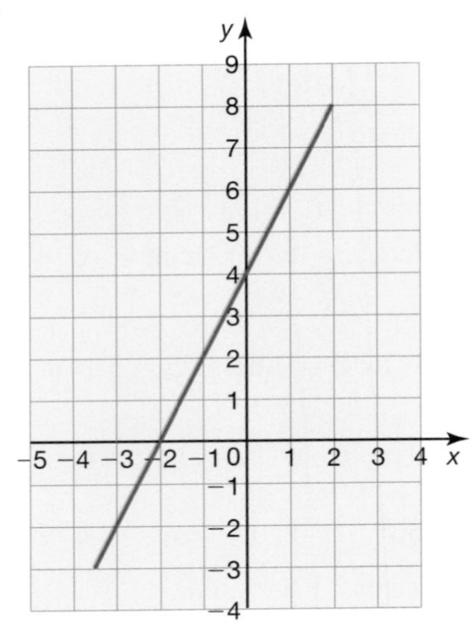


**d** i) $m = \frac{5}{2}$ $c = 4$

ii)

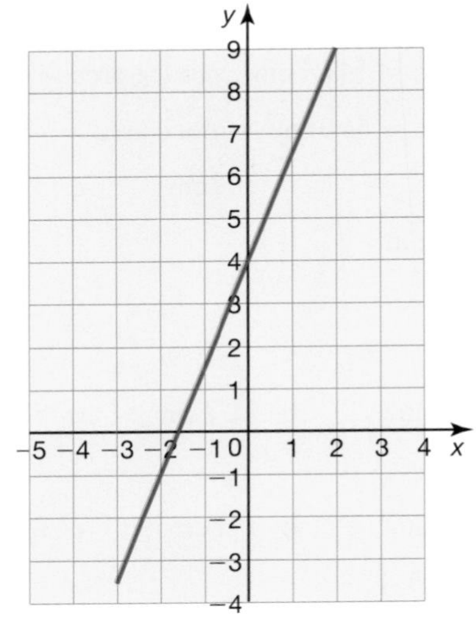


**4 a** $y = 3x - 4$ **b** $y = -2x + 7$
**c** $y = 2x - 5$ **d** $y = -4x + 3$

5 a $x = 2$ $y = 2$
b $x = 1$ $y = -1$
c $x = -2$ $y = 4$
d $x = -2$ $y = 0$

**Student assessment 2**

1 a 4.0 cm b 43.9 cm c 20.8 cm
d 3.9 cm

2 a 37° b 56° c 31°
d 34°

3 a 5.0 cm b 6.6 cm c 9.3 cm
d 28.5 cm

**Student assessment 3**

1 a 160.8 km b 177.5 km

2 a $\tan\theta = \frac{5}{x}$ b $\tan\theta = \frac{7.5}{(x+16)}$
c $\frac{5}{x} = \frac{7.5}{(x+16)}$ d 32 m e 8.88°

3 a 285 m b 117° c 297°

4 a 1.96 km b 3.42 km c 3.57 km

5 a 4003 m b 2.35°

6 Student's graph

7 a sin 130° b sin 30° c −cos 135°
d −cos 60°

8 134°

**Student assessment 4**

1 a 11.7 cm b 12.3 cm c 29°

2 a 10.8 cm b 11.9 cm c 30°
d 49°

3 Student's graph

4 a −cos 52° b cos 100°

5 a 9.8 cm b 30° c 19.6 cm

6 a 678.4 m b 11.6° c 718.1 m

**Student assessment 5**

1 a 18.0 m b 27° c 28.8 m
d 277.1 m²

2 a 12.7 cm b 67° c 93.4 cm²
d 14.7 cm

3 a 38°, 322° b 106°, 254°

4 a

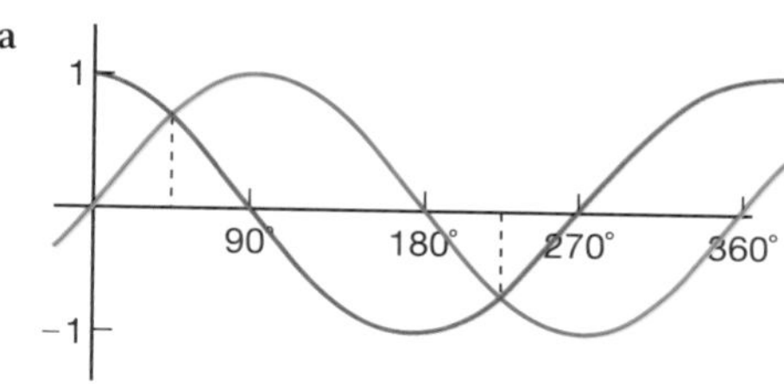


b $\theta = 45°, 225°$

5 a 5.8 cm b 6.7 cm c 7.8 cm
d 47° e 19 cm² f 37°

6 a 31.2 cm b 25.6 cm c 32.5 cm
d 15.6 cm e 27.5 cm

**Student assessment 6**

1 a 530.9 cm² b 1150.3 cm³

2 a 1210.7 cm² b 2592.0 cm³

3 a 22.9 cm b 229.2 cm² c 985.1 cm²
d 1833.5 cm³

4 a 904.8 cm³ b 12.0 cm c 13.4 cm
d 958.2 cm²

5 a 10.0 cm b 82.1 cm³ c 71.8 cm³
d 30.8 cm³ e 41.1 cm³

**Examination questions**

1

| Condition | Line |
|---|---|
| $m > 0$ and $c > 0$ | $L_5$ |
| $m < 0$ and $c > 0$ | $L_4$ |
| $m < 0$ and $c < 0$ | $L_1$ |
| $m > 0$ and $c < 0$ | $L_3$ |

2 a 60°
b 97.4 cm²
c $97.4 \times 120 = 11\,700$ cm³

3 a 12 m
b 12.5 m
c 16.3°

4 **Part A**

a 5 cm
b 9.43 cm
c 58.0°

**Part B**

a 720 m
b 72 300 m²
c 88.3°

## Topic 6

### Exercise 6.1.1

1 a, b and c

2 a domain: $-1 \le x \le 3$ range: $-3 \le f(x) \le 5$
b domain: $-4 \le x \le 0$ range: $-10 \le f(x) \le 2$
c domain: $-3 \le x \le 3$ range: $2 \le f(x) \le 11$
d domain: $y > 3$ range: $0 < g(y) < \frac{1}{3}$
e domain: $t \in \mathbb{R}$ range: $h(t) \in \mathbb{R}$
f domain: $y \in \mathbb{R}$ range: $f(y) = 4$
g domain: $n \in \mathbb{R}$ range: $f(n) \le 2$

### Exercise 6.2.1

1 a 2.375 b 0.5 c 0.125
d −4

2 a −1 b −6 c −3.5
d −16

3 a −4 b −0.25 c 5
d 2.75

4 a 2.75 b 0.25 c −3.5
d 0.5

5 a $f(x) = 50 + 100x$
b €300

6 a

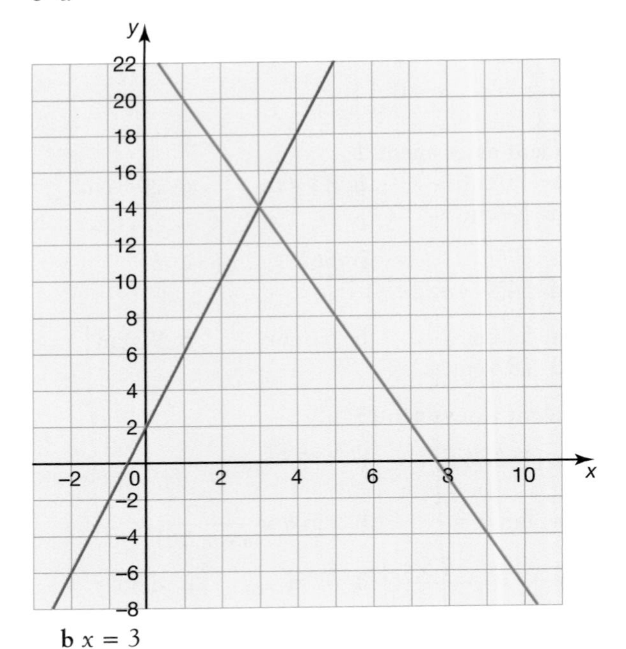


b $x = 3$

### Exercise 6.3.1

1 a

| $x$ | −4 | −3 | −2 | −1 | 0 | 1 | 2 | 3 |
|---|---|---|---|---|---|---|---|---|
| $y$ | 10 | 4 | 0 | −2 | −2 | 0 | 4 | 10 |

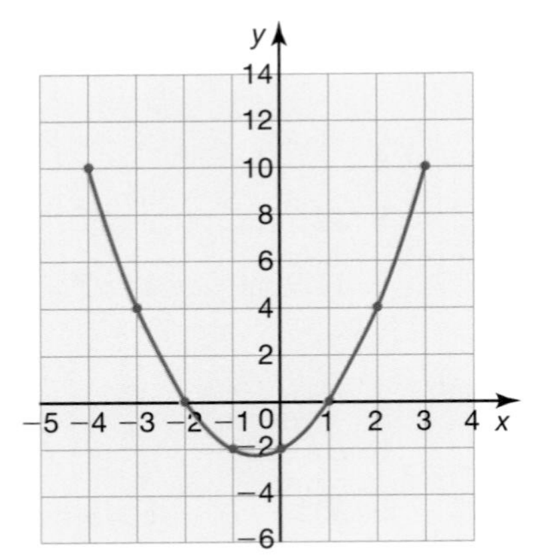


b $x = -\frac{1}{2}$, (−0.5, − 2.25), minimum

**2 a**

| $x$ | −3 | −2 | −1 | 0 | 1 | 2 | 3 | 4 | 5 |
|---|---|---|---|---|---|---|---|---|---|
| $y$ | −12 | −5 | 0 | 3 | 4 | 3 | 0 | −5 | −12 |

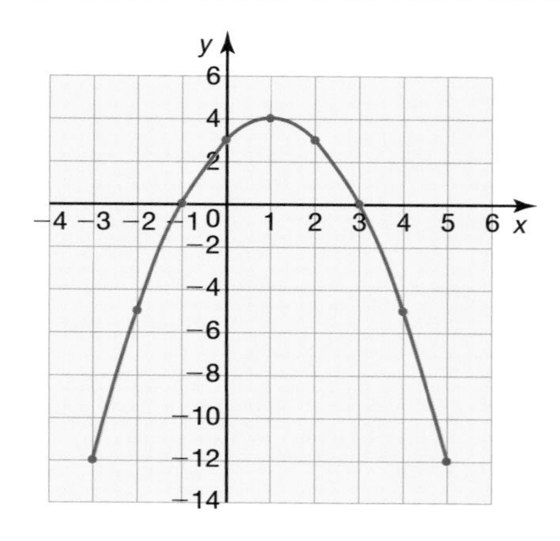

**b** $x = 1$, $(1, 4)$, maximum

**3 a**

| $x$ | −1 | 0 | 1 | 2 | 3 | 4 | 5 |
|---|---|---|---|---|---|---|---|
| $y$ | 9 | 4 | 1 | 0 | 1 | 4 | 9 |

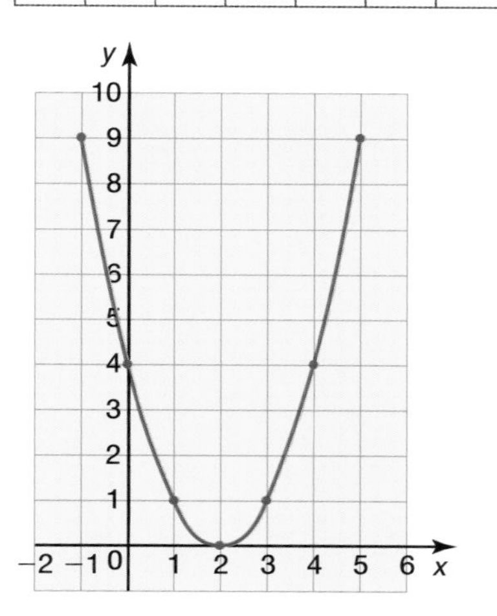

**b** $x = 2$, $(2, 0)$, minimum

**4 a**

| $x$ | −4 | −3 | −2 | −1 | 0 | 1 | 2 |
|---|---|---|---|---|---|---|---|
| $y$ | −9 | −4 | −1 | 0 | −1 | −4 | −9 |

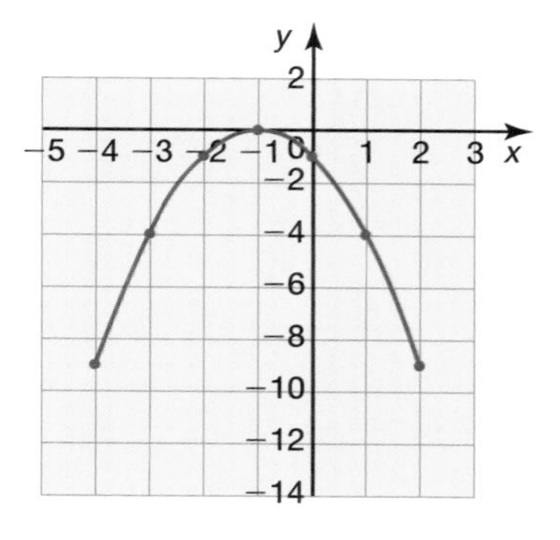

**b** $x = -1$, $(-1, 0)$, maximum

**5 a**

| $x$ | −4 | −3 | −2 | −1 | 0 | 1 | 2 | 3 | 4 | 5 | 6 |
|---|---|---|---|---|---|---|---|---|---|---|---|
| $y$ | 9 | 0 | −7 | −12 | −15 | −16 | −15 | −12 | −7 | 0 | 9 |

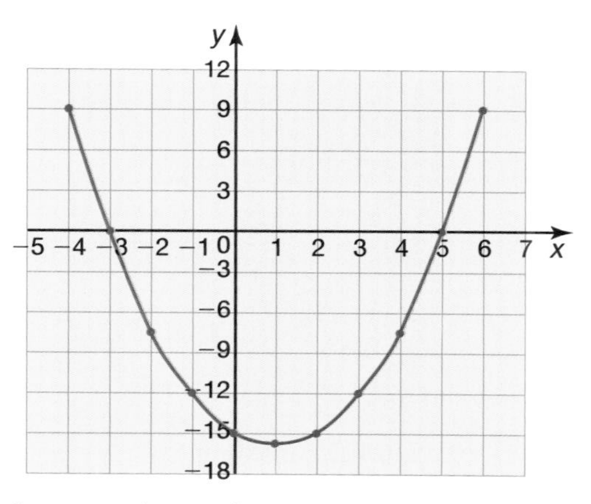

**b** $x = 1$, $(1, -16)$, minimum

**6 a**

| $x$ | −2 | −1 | 0 | 1 | 2 | 3 |
|---|---|---|---|---|---|---|
| $y$ | 9 | 1 | −3 | −3 | 1 | 9 |

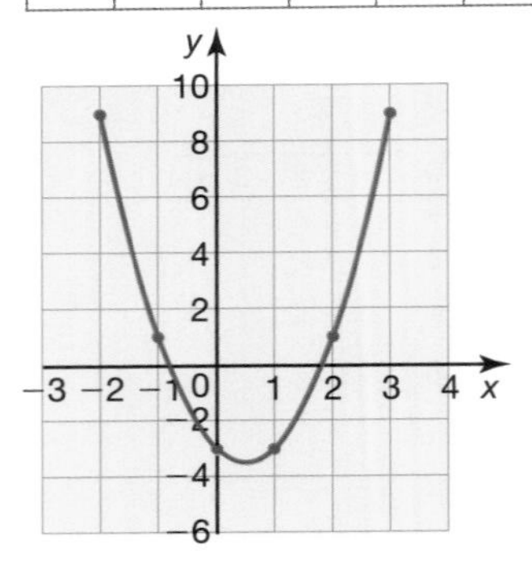


**b** $x = \frac{1}{2}$, (0.5, −3.5), minimum

**7 a**

| $x$ | −3 | −3 | −1 | 0 | 1 | 2 | 3 |
|---|---|---|---|---|---|---|---|
| $y$ | −15 | −4 | 3 | 6 | 5 | 0 | −9 |

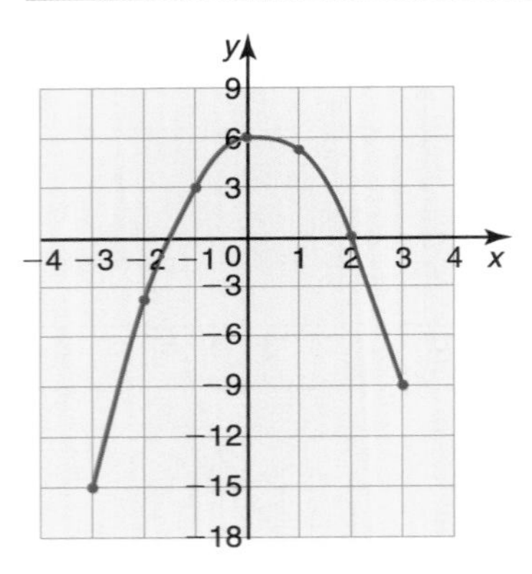


**b** $x = \frac{1}{4}$, (0.25, 6.125), maximum

**8 a**

| $x$ | −2 | −1 | 0 | 1 | 2 | 3 |
|---|---|---|---|---|---|---|
| $y$ | 12 | 0 | −6 | −6 | 0 | 12 |

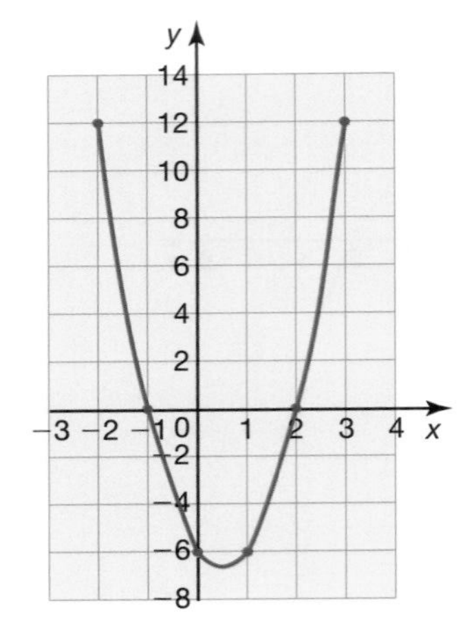


**b** $x = \frac{1}{2}$, (0.5, −6.75), minimum

**9 a**

| $x$ | −1 | 0 | 1 | 2 | 3 |
|---|---|---|---|---|---|
| $y$ | 7 | −4 | −7 | −2 | 11 |

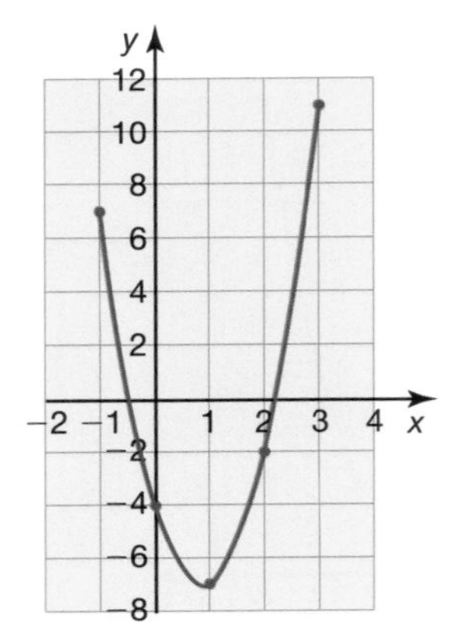


**b** $x = \frac{7}{8}$, (0.875, −7.0625), minimum

10 a

| $x$ | −2 | −1 | 0 | 1 | 2 | 3 |
|---|---|---|---|---|---|---|
| $y$ | −25 | −9 | −1 | −1 | −9 | −25 |

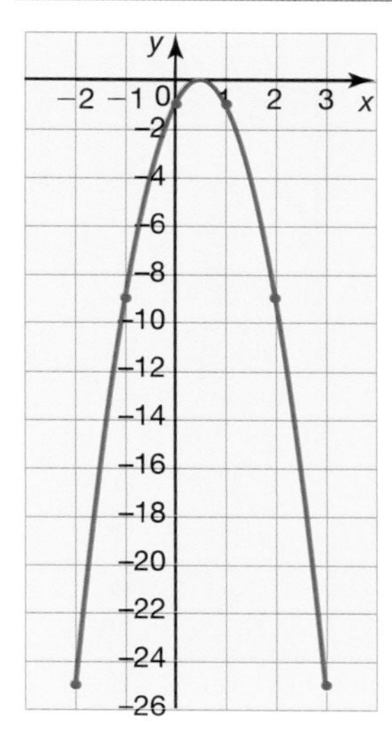


b $x = \frac{1}{2}$, (0.5, 0), maximum

**Exercise 6.3.2**

1 −2 and 3

2 −1 and 1

3 3

4 No solution (graph does not cross $x$-axis)

5 2

6 0.5 and 3

7 1

8 $-\frac{1}{3}$ and 2

**Exercise 6.3.3**

1 −1.6 and 2.6

2 No solution

3 2 and 4

4 −1 and 0

5 0.3 and 3.7

6 0 and 3.5

7 −0.2 and 2.2

8 $-\frac{1}{3}$ and 2

**Exercise 6.3.4**

1 a $(x + 4)(x + 3)$ b $(x + 6)(x + 2)$
c $(x + 12)(x + 1)$ d $(x - 3)(x - 4)$
e $(x - 6)(x - 2)$ f $(x - 12)(x - 1)$

2 a $(x + 5)(x + 1)$ b $(x + 4)(x + 2)$
c $(x + 3)^2$ d $(x + 5)^2$
e $(x + 11)^2$ f $(x - 6)(x - 7)$

3 a $(x + 12)(x + 2)$ b $(x + 8)(x + 3)$
c $(x - 6)(x - 4)$ d $(x + 12)(x + 3)$
e $(x + 18)(x + 2)$ f $(x - 6)^2$

4 a $(x + 5)(x - 3)$ b $(x - 5)(x + 3)$
c $(x + 4)(x - 3)$ d $(x - 4)(x + 3)$
e $(x + 6)(x - 2)$ f $(x - 12)(x - 3)$

5 a $(x - 4)(x + 2)$ b $(x - 5)(x + 4)$
c $(x + 6)(x - 5)$ d $(x + 6)(x - 7)$
e $(x + 7)(x - 9)$ f $(x + 9)(x - 6)$

6 a $2(x + 1)^2$ b $(2x + 3)(x + 2)$
c $(2x - 3)(x + 2)$ d $(2x - 3)(x - 2)$
e $(3x + 2)(x + 2)$ f $(3x - 1)(x + 4)$
g $(2x + 3)^2$ h $(3x - 1)^2$
i $(3x + 1)(2x - 1)$

**Exercise 6.3.5**

1 a −4 and −3 b −2 and −6
c −5 and 2 d −2 and 5
e −3 and −2 f −3
g −2 and 4 h −4 and 5
i −6 and 5 j −6 and 7

2 a −3 and 3 b −5 and 5
c −12 and 12 d −2.5 and 2.5
e −2 and 2 f $-\frac{1}{3}$ and $\frac{1}{3}$
g −4 and −2 h 2 and 4
i −4 and 6 j −6 and 8

3 a −9 and 4 b −1
c 0 and 8 d 0 and 7
e −1.5 and −1 f −1 and 2.5
g −12 and 0 h −9 and −3
i −6 and 6 j −10 and 10

4 −4 and 3

5 2

6 4

7 $x = 6$, height = 3 cm, base length = 12 cm

8 $x = 10$, height = 20 cm, base length = 2 cm

9 $x = 6$, base = 6 cm, height = 5 cm

10 11 m × 6 m

**Exercise 6.3.6**

1 a −3.14 and 4.14 b −5.87 and 1.87
c −6.14 and 1.14 d No solution
e −6.89 and 1.89 f 3.38 and 5.62

2 a −5.30 and −1.70 b −5.92 and 5.92
c 2.5 d No solution
e −4.77 and 3.77 f −2.83 and 2.83

3 a −0.73 and 2.73 b −1.87 and 5.87
c −1.79 and 2.79 d −3.83 and 1.83
e 0.38 and 2.62 f 0.39 and 7.61

4 a −0.85 and 2.35 b −1.40 and 0.90
c 0.14 and 1.46 d −2 and −0.5
e −0.39 and 1.72 f −1.54 and 1.39

**Exercise 6.4.1**

1 a i

| x | −3 | −2 | −1 | 0 | 1 | 2 | 3 |
|---|---|---|---|---|---|---|---|
| f(x) | 0.04 | 0.11 | 0.33 | 1 | 3 | 9 | 27 |

ii

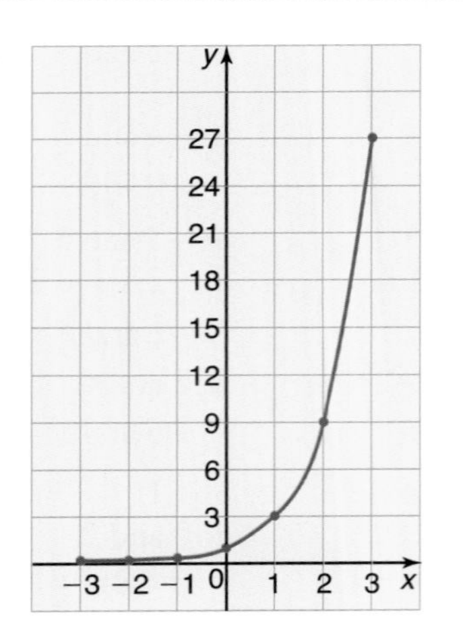


b i

| x | −3 | −2 | −1 | 0 | 1 | 2 | 3 |
|---|---|---|---|---|---|---|---|
| f(x) | 1 | 1 | 1 | 1 | 1 | 1 | 1 |

ii

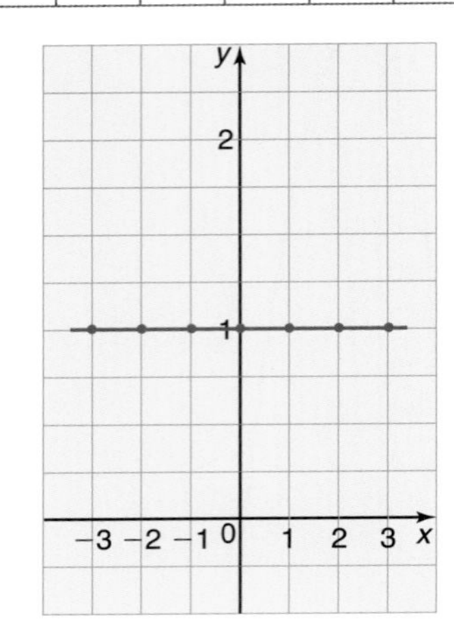


c i

| x | −3 | −2 | −1 | 0 | 1 | 2 | 3 |
|---|---|---|---|---|---|---|---|
| f(x) | 3.125 | 3.25 | 3.5 | 4 | 5 | 7 | 11 |

ii

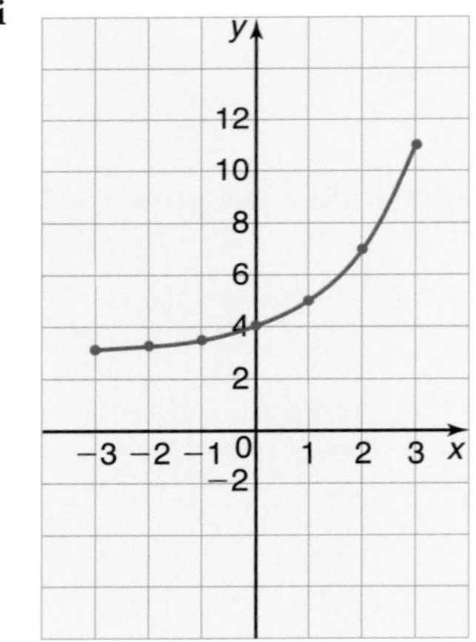

d i

| $x$ | −3 | −2 | −1 | 0 | 1 | 2 | 3 |
|---|---|---|---|---|---|---|---|
| $f(x)$ | −2.875 | −1.75 | −0.5 | 1 | 3 | 6 | 11 |

ii

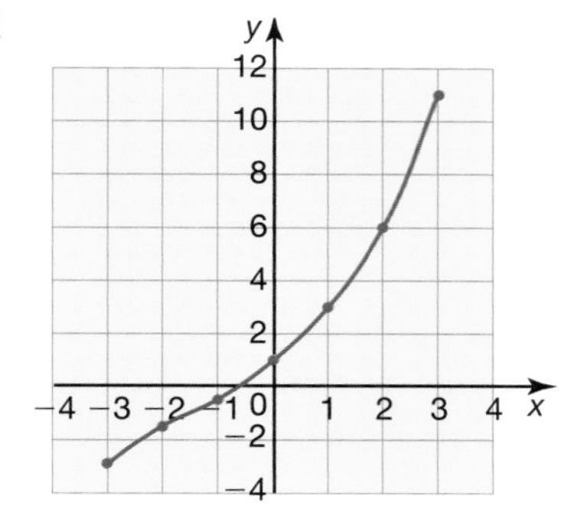

e i

| $x$ | −3 | −2 | −1 | 0 | 1 | 2 | 3 |
|---|---|---|---|---|---|---|---|
| $f(x)$ | 3.125 | 2.25 | 1.5 | 1 | 1 | 2 | 5 |

ii

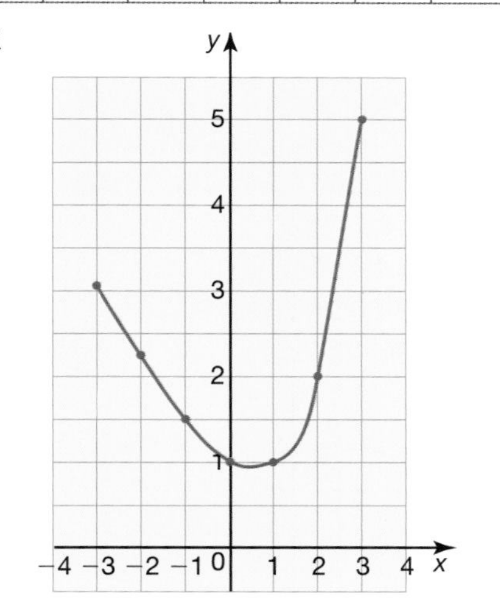

f i

| $x$ | −3 | −2 | −1 | 0 | 1 | 2 | 3 |
|---|---|---|---|---|---|---|---|
| $f(x)$ | −8.96 | −3.89 | −0.67 | 1 | 2 | 5 | 18 |

ii

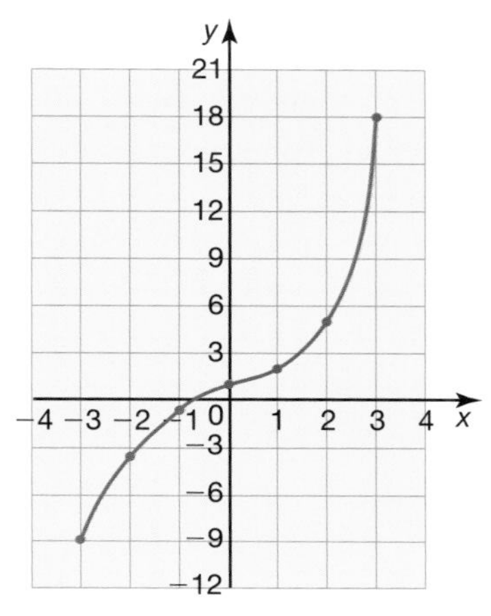

2 a 7 cm  b 0 cm  c 5 hours
e Approximately $5\frac{1}{2}$ hours

3 a $x = 2.5$  b $x = -\frac{1}{2}$

4 $x = 4.3$

**Exercise 6.4.2**

1 Translation $\begin{pmatrix} 0 \\ c \end{pmatrix}$

2 Stretch parallel to $y$-axis of scale factor $k$.

3 Changes the slope of the curve.

4 Reflection in the $y$-axis.

**Exercise 6.4.3**

1 16 777 216

2 240 000 years

3 4900 million

4 0.098 g

5 8

6 a 358 000 km²
b 7 years

7 1.0

8 9%

**Exercise 6.5.1**

1 The graph is stretched by a scale factor '$a$' parallel to the $y$-axis.

2 The graph is translated $\begin{pmatrix} -b \\ 0 \end{pmatrix}$.

**Exercise 6.5.2**

1 a i) Vertical asymptote: $x = -1$
Horizontal asymptote: $y = 0$
ii) Graph intercepts $y$-axis at $(0, 1)$
iii)

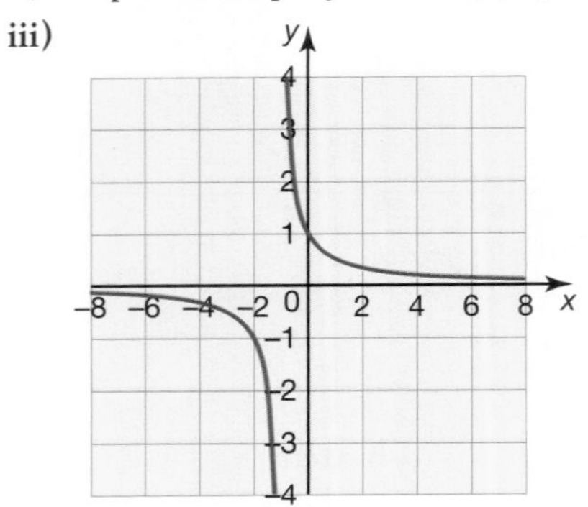


b i) Vertical asymptote: $x = -3$
Horizontal asymptote: $y = 0$
ii) Graph intercepts $y$-axis at $\left(0, \frac{1}{3}\right)$
iii)

c i) Vertical asymptote: $x = 4$
Horizontal asymptote: $y = 0$
ii) Graph intercepts $y$-axis at $\left(0, -\frac{1}{2}\right)$
iii)

d i) Vertical asymptote: $x = 3$
Horizontal asymptote: $y = 0$
ii) Graph intercepts $y$-axis at $\left(0, \frac{1}{3}\right)$
iii)

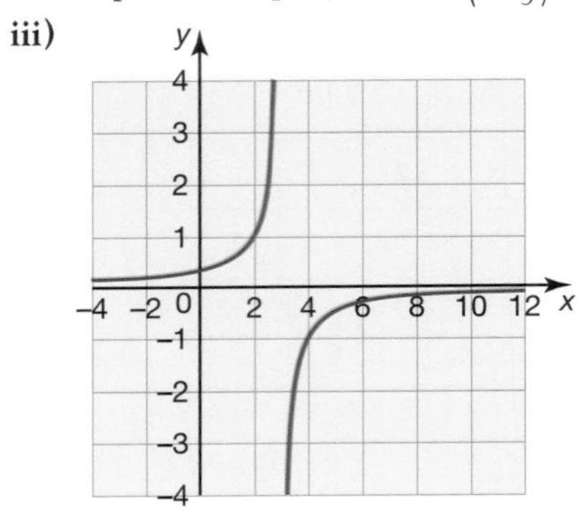


2 a i) Vertical asymptote: $x = 0$
Horizontal asymptote: $y = 2$
ii) Graph intercepts $x$-axis at $\left(-\frac{1}{2}, 0\right)$
iii)

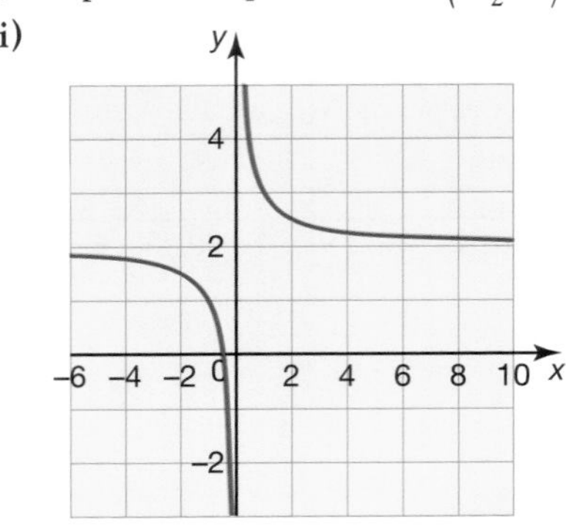

**b** **i)** Vertical asymptote: $x = 0$
Horizontal asymptote: $y = -3$
**ii)** Graph intercepts $x$-axis at $\left(\frac{1}{3}, 0\right)$
**iii)**

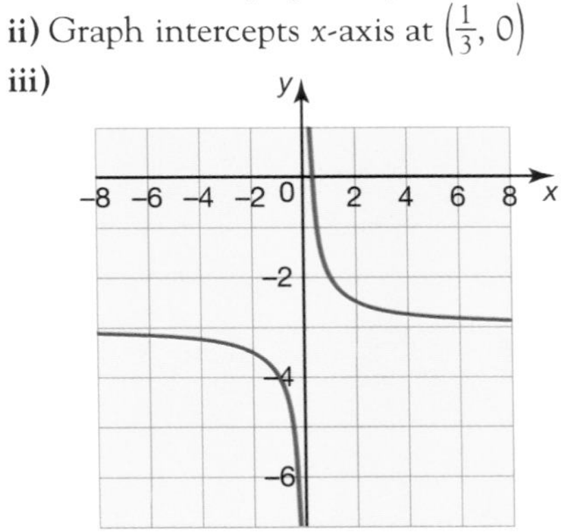

**c** **i)** Vertical asymptote: $x = 1$
Horizontal asymptote: $y = 4$
**ii)** Graph intercepts $x$-axis at $\left(\frac{3}{4}, 0\right)$ and $y$-axis at $(0, 3)$
**iii)**

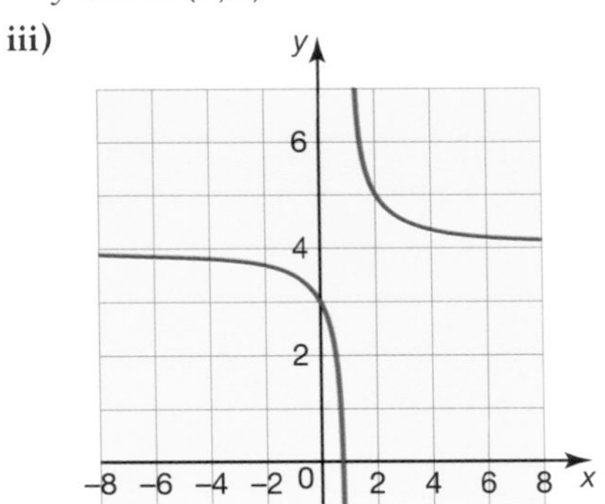

**d** **i)** Vertical asymptote: $x = -4$
Horizontal asymptote: $y = -1$
**ii)** Graph intercepts $x$-axis at $(-3, 0)$ and $y$-axis at $\left(0, -\frac{3}{4}\right)$
**iii)**

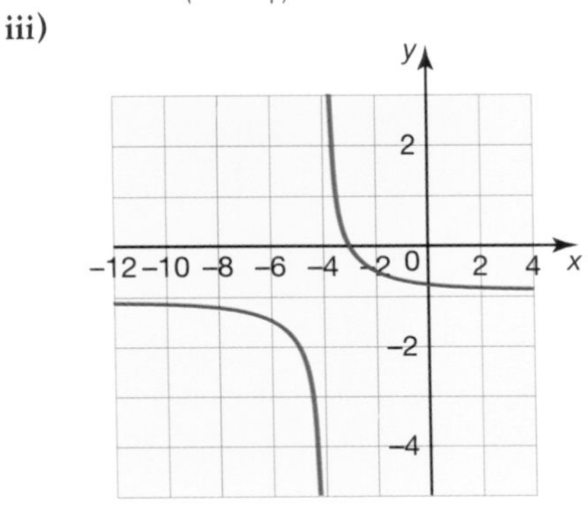

**3** **a** **i)** Vertical asymptote: $x = -\frac{1}{2}$
Horizontal asymptote: $y = 0$
**ii)** Graph intercepts $y$-axis at $(0, 1)$
**iii)**

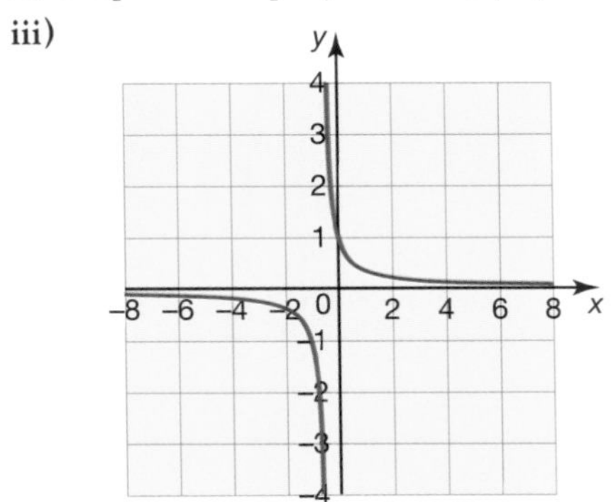

**b** **i)** Vertical asymptote: $x = \frac{1}{2}$
Horizontal asymptote: $y = 1$
**ii)** Graph intercepts axes at $(0,0)$
**iii)**

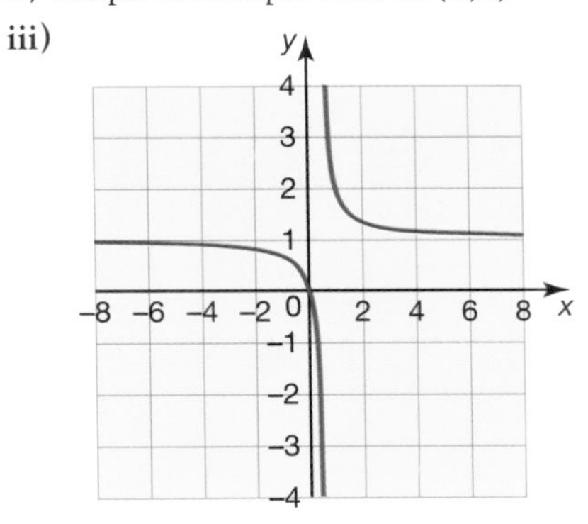

**c** **i)** Vertical asymptote: $x = \frac{1}{3}$
Horizontal asymptote: $y = 0$
**ii)** Graph intercepts $y$-axis at $(0, -2)$
**iii)**

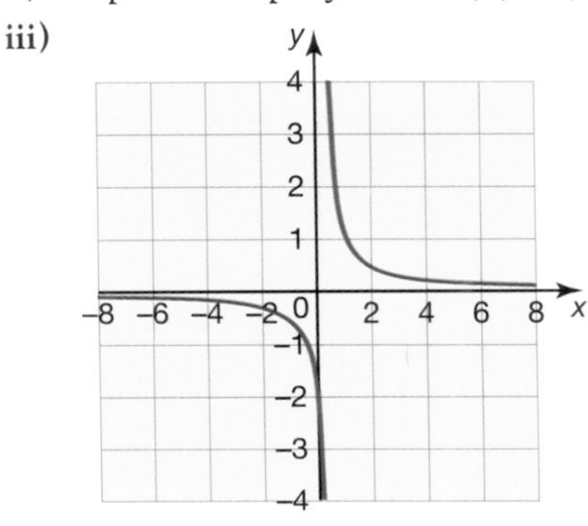

**d** i) Vertical asymptote: $x = \frac{1}{4}$
Horizontal asymptote: $y = 2$
ii) Graph intercepts $x$-axis at $\left(-\frac{3}{8}, 0\right)$ and the $y$-axis at $(0, 3)$
iii)

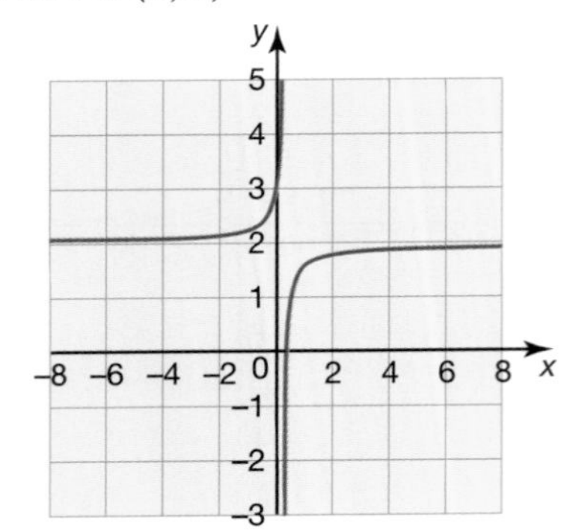

### Exercise 6.5.3

**1 a** i) Vertical asymptotes: $x = 1$ and $2$
Horizontal asymptote: $y = 0$
ii) Graph intercepts $y$-axis at $\left(0, \frac{1}{2}\right)$
iii)

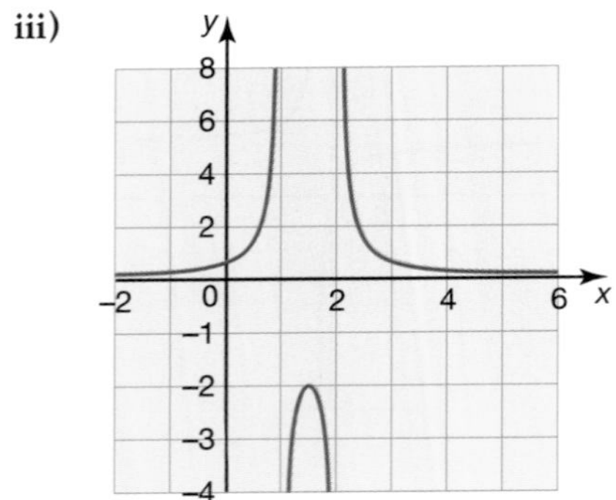

**b** i) Vertical asymptotes: $x = -3$ and $4$
Horizontal asymptote: $y = 0$
ii) Graph intercepts $y$-axis at $\left(0, -\frac{1}{12}\right)$
iii)

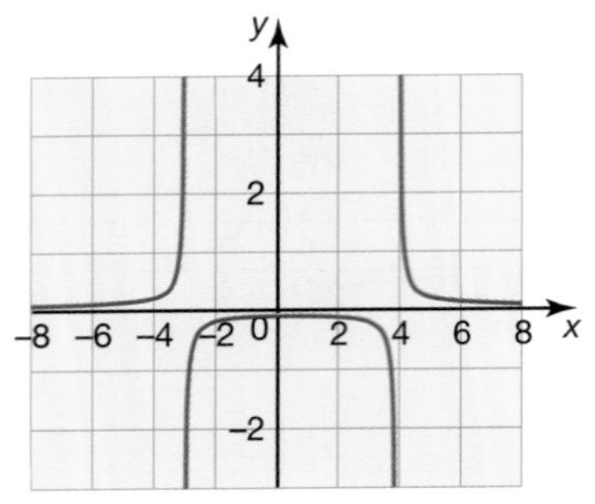

**c** i) Vertical asymptotes: $x = 0$ and $5$
Horizontal asymptote: $y = 1$
ii) Graph intercepts $x$-axis at $\left(\frac{5 \pm \sqrt{21}}{2}, 0\right)$
iii)

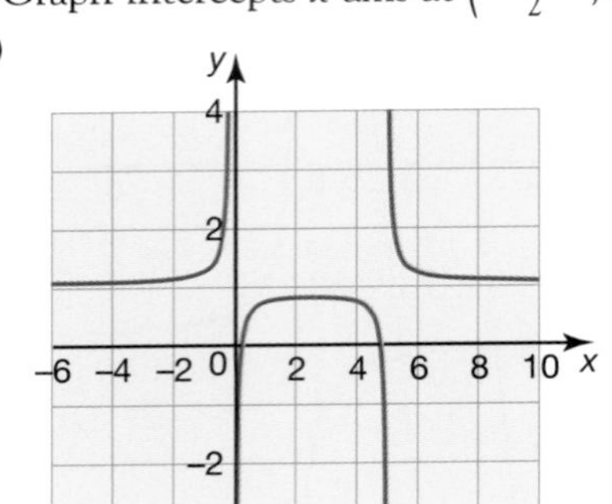

**d** i) Vertical asymptote: $x = -2$
Horizontal asymptote: $y = -3$
ii) Graph intercepts the $x$-axis at $\left(-2 \pm \sqrt{\frac{1}{3}}0\right)$ and the $y$-axis at $\left(0, -2\frac{3}{4}\right)$
iii)

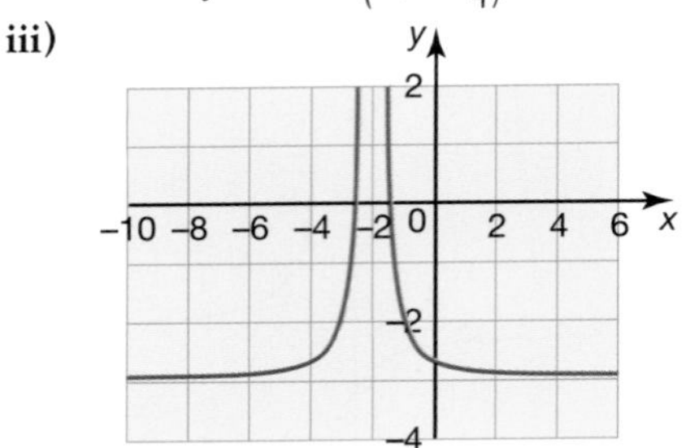

**2 a** i) Vertical asymptotes: $x = -4$ and $1$
Horizontal asymptote: $y = 0$
ii) Graph intercepts the $y$-axis at $\left(0, -\frac{1}{4}\right)$
iii)

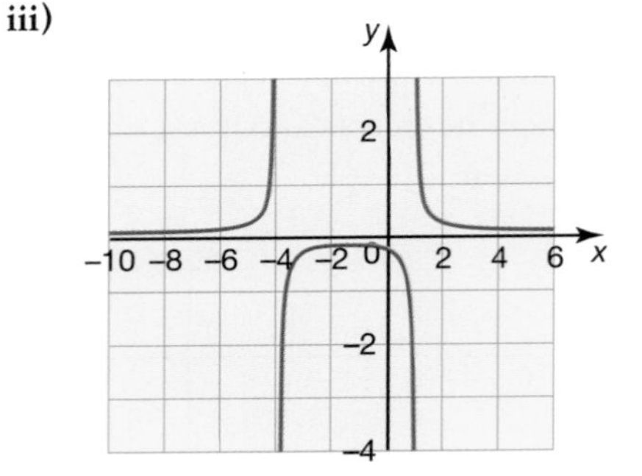

b i) Vertical asymptotes: $x = -5$ and 2

Horizontal asymptote: $y = 0$

ii) Graph intercepts the $y$-axis at $\left(0, -\frac{1}{10}\right)$

iii)

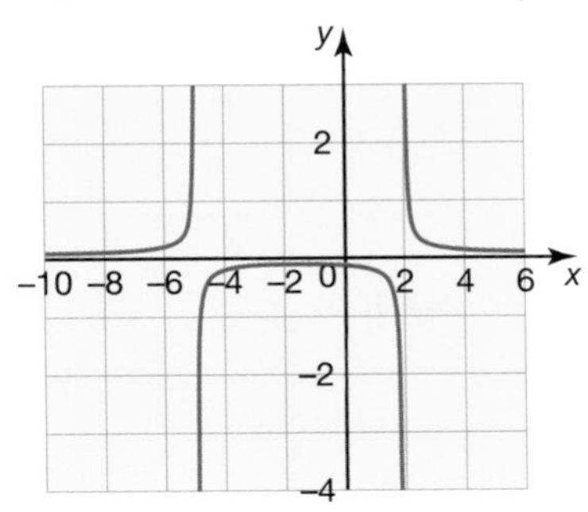

c i) Vertical asymptote: $x = -1$

Horizontal asymptote: $y = 3$

ii) Graph intercepts the $y$-axis at (0, 4)

iii)

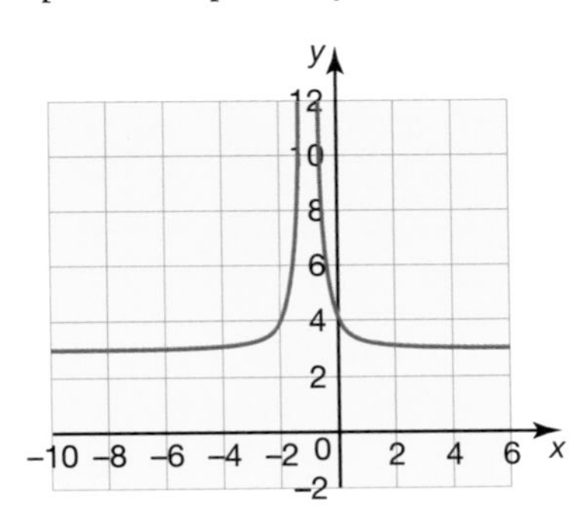

d i) Vertical asymptotes: $x = -\frac{1}{2}$ and 4

Horizontal asymptote: $y = -1$

ii) Graph intercepts the $x$-axis at $\left(\frac{7 \pm \sqrt{89}}{4}, 0\right)$ and the $y$-axis at $\left(0, -\frac{5}{4}\right)$

iii)

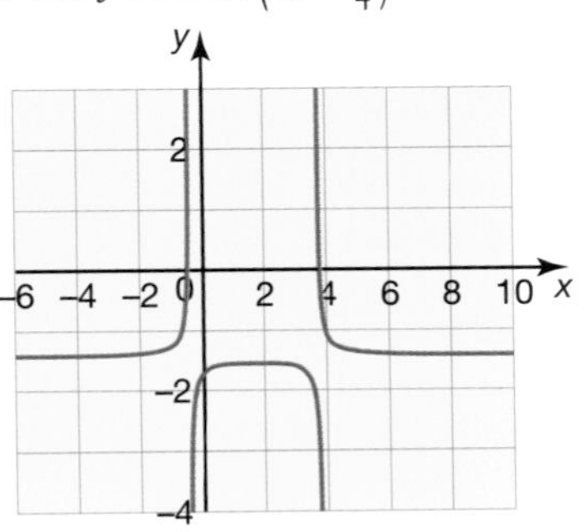

**Exercise 6.5.4**

1 a i)

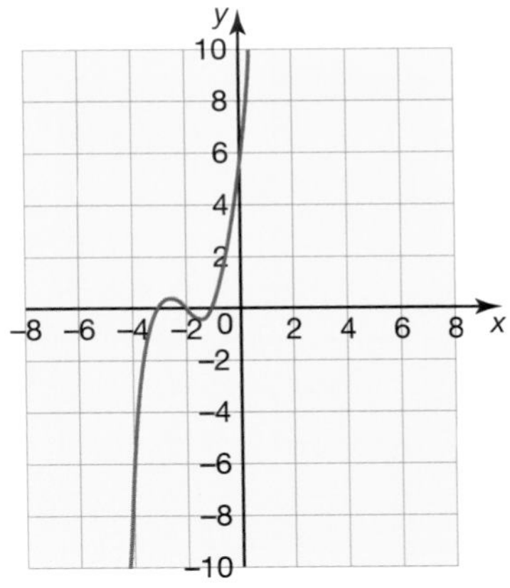

ii) Graph intercepts the $y$-axis at (0, 6)

iii) $(x + 3)(x + 2)(x + 1)$

b i)

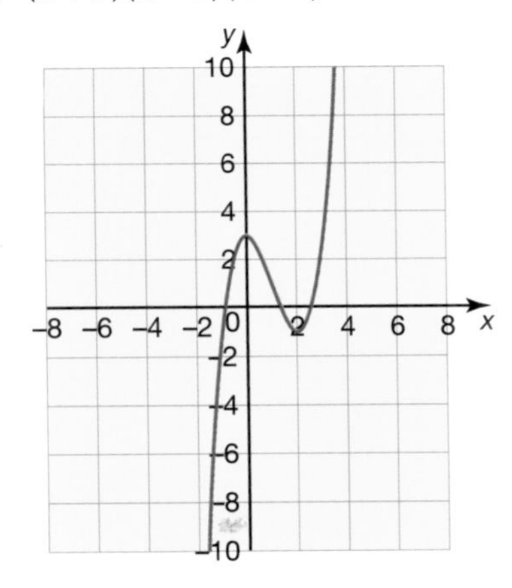

ii) Graph intercepts the $y$-axis at (0, 3)

iii) $(x + 0.88)(x - 1.35)(x - 2.53)$

c i)

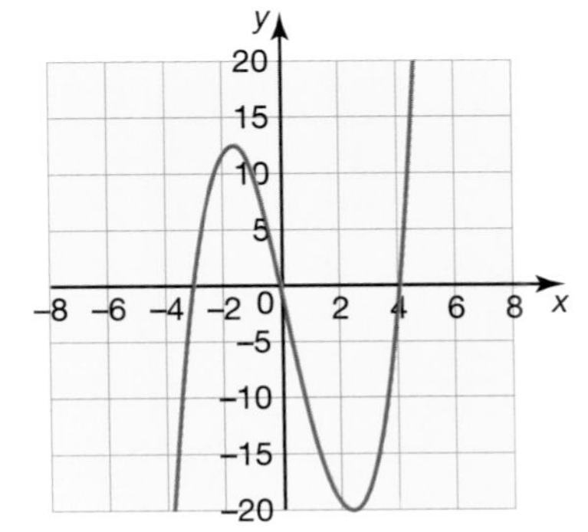

**ii)** Graph intercepts the $y$-axis at (0, 0)
**iii)** $x(x + 3)(x - 4)$

**d i)**

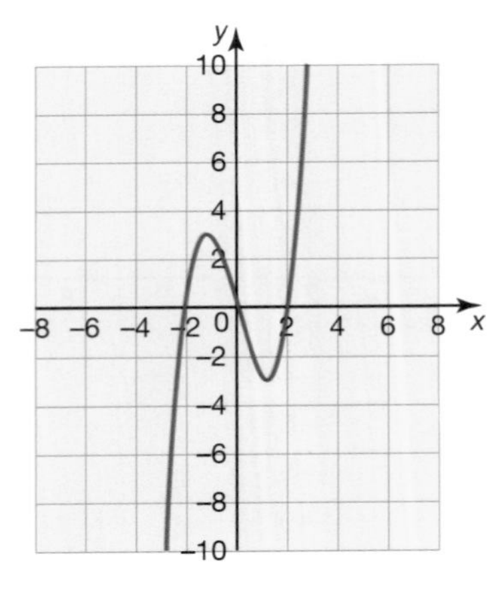


**ii)** Graph intercepts the $y$-axis at (0, 0)
**iii)** $x(x + 2)(x - 2)$

**2 a i)**

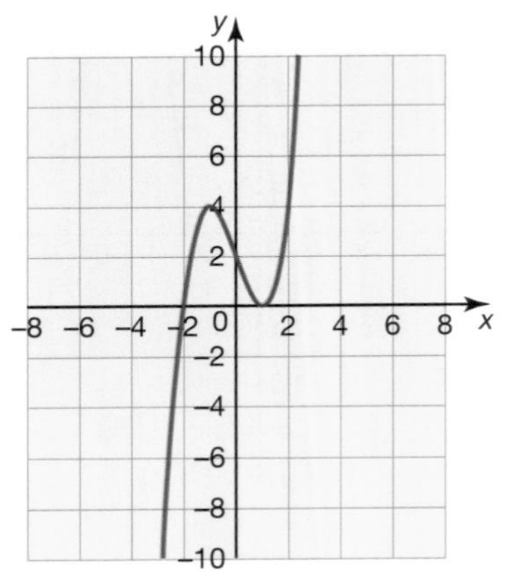


**ii)** Graph intercepts the $y$-axis at (0, 2)
**iii)** $(x + 2)(x - 1)(x - 1)$

**b i)**

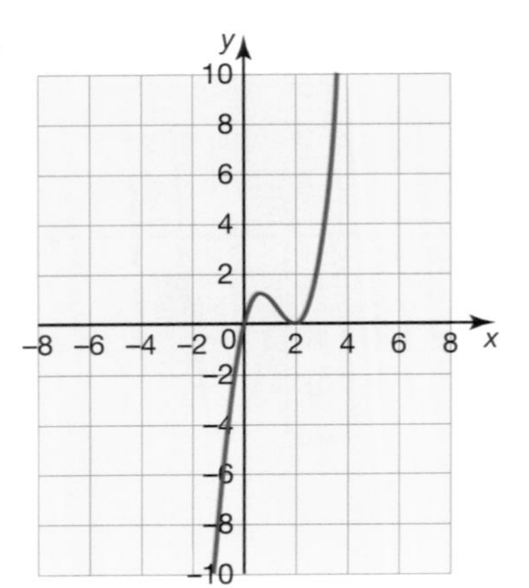


**ii)** Graph intercepts the $y$-axis at (0, 0)
**iii)** $x(x - 2)(x - 2)$

**c i)**

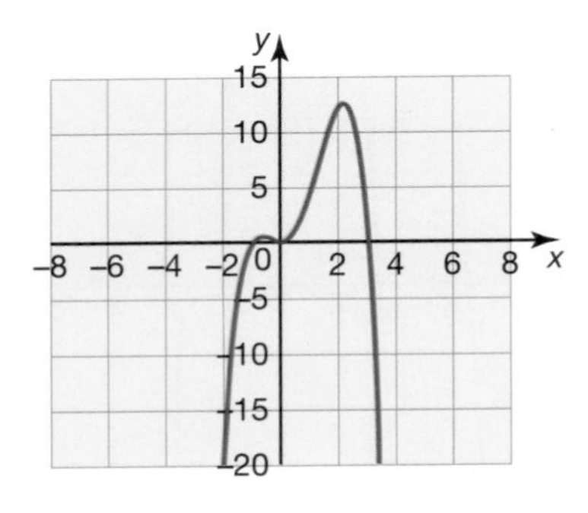


**ii)** Graph intercepts the $y$-axis at (0, 0)
**iii)** $-x^2(x + 1)(x - 3)$

**d i)**

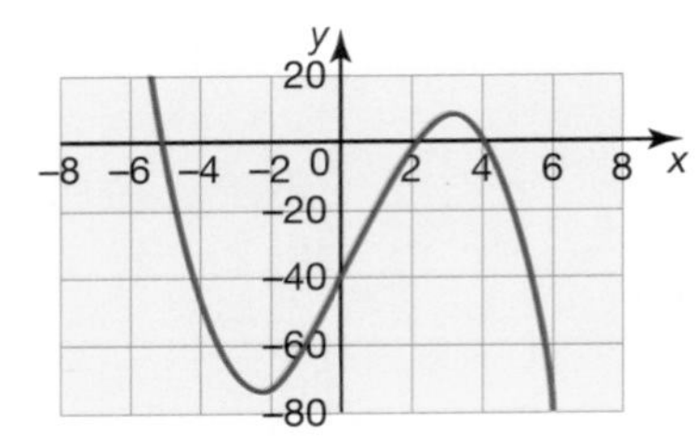


**ii)** Graph intercepts the $y$-axis at (0, −40)
**iii)** $-(x + 5)(x - 2)(x - 4)$

**Exercise 6.6.1**

**1 a** $x = 4$ and $-5$
**b** $x = -3$ and $\frac{1}{2}$
**c** $x = -\frac{1}{3}$ and $\frac{1}{2}$
**d** $x = -\frac{5}{2}$ and 7
**e** $x = 1$ and $\frac{1}{4}$
**f** $x = -2$, 1 and 4

**2 a** $x = 0.2$ and 4.9
**b** $x = -1.6$ and 1.0
**c** $x = -0.5$, 0.7 and 5.9
**d** $x = -0.8$ and 1.3
**e** $x = 0.4$ and 2.2
**f** $x = 17.6°$ and $162.4°$; $0° \le x \le 360°$

**Student assessment 1**

1 **b** is not a function as each input value should only produce one output value.

2 **a** domain: $-3 \leq x \leq 1$ range: $-11 \leq f(x) \leq 1$
**b** domain: $x \in R$ range: $g(x) \geq 0$

3 $3 \leq f(q) \leq 27$

4 **a** 5 **b** $-1$ **c** $-\frac{11}{2}$

5 **a** $(-3, 4)$ **b** $(3, 5)$ **c** $\left(0, \frac{7}{2}\right)$

6 **a**

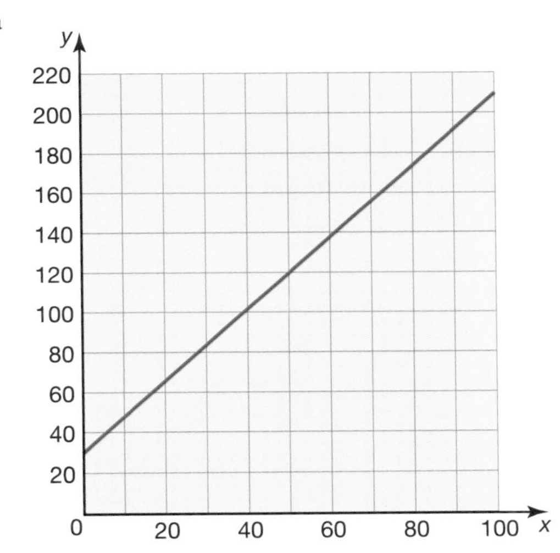


**b** **i)** ≈ 50 °F **ii)** ≈ 104 °F **iii)** ≈ 176 °F

7 **a** $R(n) = 12n$
**b**

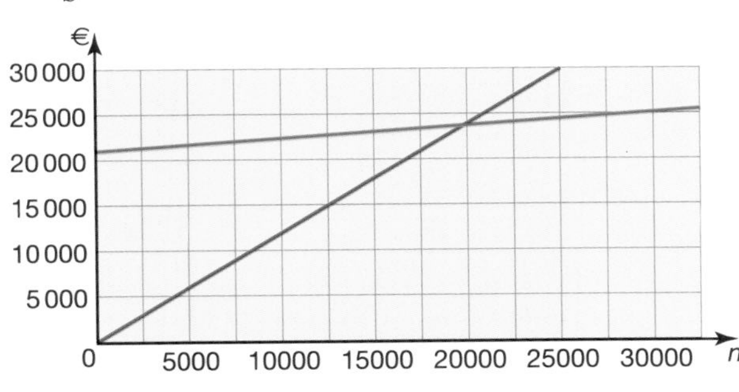


**c** 20 000 copies
**d** ≈ 2.4 million euros (Exact €2 415 000)

**Student assessment 2**

1

| $x$ | −3 | −2 | −1 | 0 | 1 | 2 | 3 |
|---|---|---|---|---|---|---|---|
| $f(x) = x^2 + 3x - 9$ | −9 | −11 | −11 | −9 | −5 | 1 | 9 |

**b**

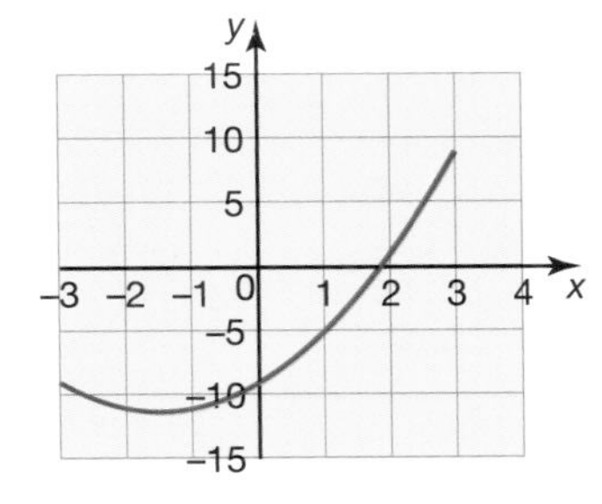


**c** Range: $-11.25 \leq f(x) \leq 9$

2 **a** $f(x) = (x - 3)(x - 6)$
**b** $h(y) = (3y - 2)(y + 1)$
**c** $f(x) = (x - 5)(x + 2)$
**d** $h(x) = (2x - 1)(x + 4)$

3 **a** $x = -4$ and $-2$ **b** $x = 1$ and $5$
**c** $x = -5$ **d** $x = 1$ and $\frac{4}{3}$

4 **a** $x = 0.191$ and $1.31$
**b** $x = 2.63$ and $-0.228$

5 €8244.13

6 $211.8x$

7 22 years

8 17 m

**Student assessment 3**

1 a

| $x$ | −3 | −2 | −1 | 0 | 1 | 2 | 3 |
|---|---|---|---|---|---|---|---|
| $f(x)$ | −1.875 | −1.75 | −1.5 | −1 | 0 | 2 | 6 |

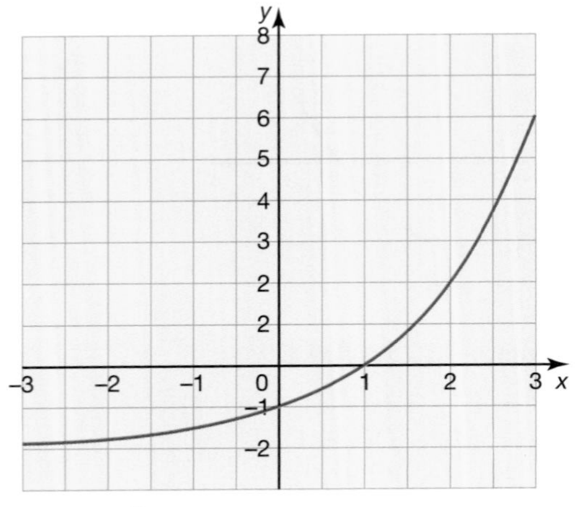


c $y = -2$

2 a

| $t$ | 0 | 1 | 2 | 3 | 4 | 5 | 6 | 7 | 8 |
|---|---|---|---|---|---|---|---|---|---|
| $h$ | 8.1 | 4.1 | 2.1 | 1.1 | 0.6 | 0.35 | 0.225 | 0.163 | 0.131 |

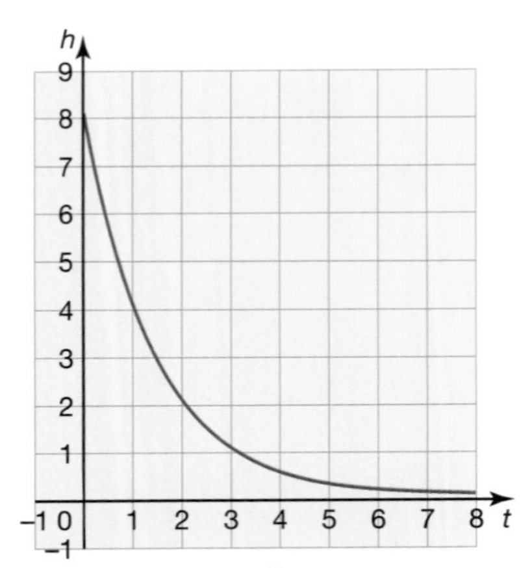


c When $t = 0$, $h = 8.1$ m

d $y = 0.1$

e There is always 10 cm (0.1 m) of grain left in the container.

3 a Vertical asymptote $x = 3$
Horizontal asymptote $y = 1$

b $\left(0, \frac{1}{3}\right)$ and $(1, 0)$

c

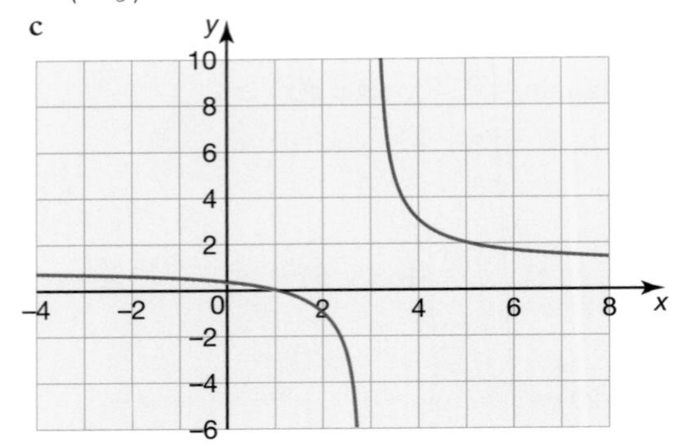


4 a Vertical asymptote $x = 2$
Horizontal asymptote $y = -1$

b $\left(0, -\frac{5}{4}\right)$ and $\left(\frac{5}{2}, 0\right)$

c

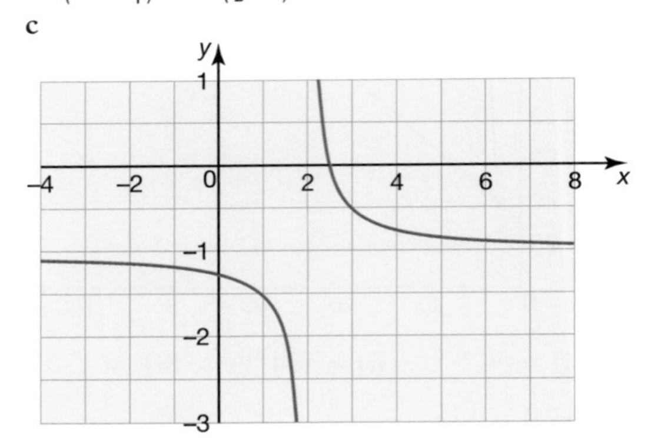


5 a Vertical asymptotes $x = -1$ and 5
Horizontal asymptote $y = 0$

b $\left(0, -\frac{1}{5}\right)$

c

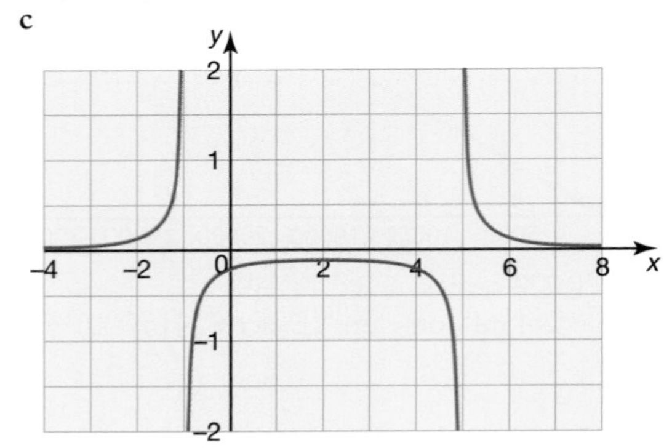

6 a Vertical asymptotes $x = 2$ and 3
Horizontal asymptote $y = 2$
b $\left(0, 2\frac{1}{6}\right)$
c

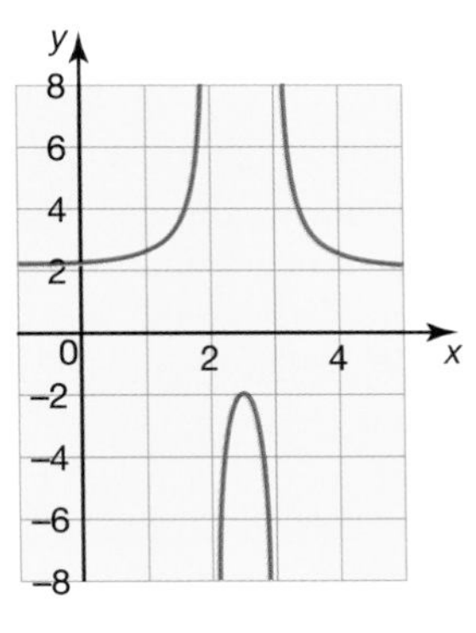


**Examination questions**

1 a $\frac{2}{9} \leq f(x) \leq 486$
b 4

2 a –5
b $b = 1, c = 30$

3 a i

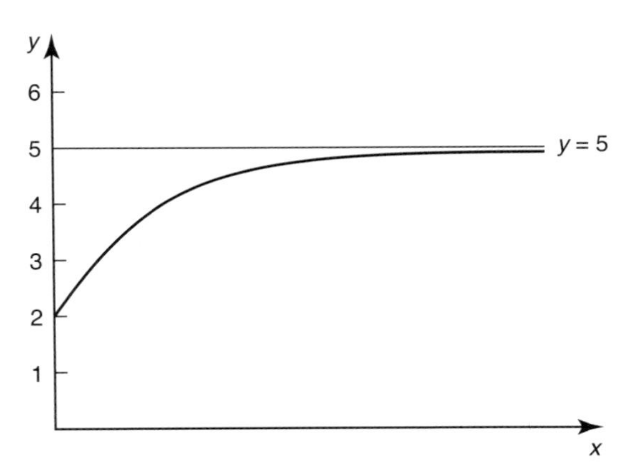


ii (0, 2)
b see graph
c zero

4 a 0.25
b 2
c $y = 1.25$

## Topic 7

**Exercise 7.1.1**

1 a 4 b 6 c −2
d $2x$

2 a 4 b 8 c −8
d $4x$

3 a 1 b 2 c 3
d $x$

**Exercise 7.1.2**

1 a $\frac{dy}{dx} = 3x^2$ b $\frac{dy}{dx} = 6x$
c $\frac{dy}{dx} = 2x + 2$ d $\frac{dy}{dx} = 2x$
e $\frac{dy}{dx} = 3$ f $\frac{dy}{dx} = 4x - 1$

2

| Function $f(x)$ | Gradient function $f'(x)$ |
|---|---|
| $x^2$ | $2x$ |
| $2x^2$ | $4x$ |
| $\frac{1}{2}x^2$ | $x$ |
| $x^2 + x$ | $2x + 1$ |
| $x^3$ | $3x^2$ |
| $3x^2$ | $6x$ |
| $x^2 + 2x$ | $2x + 2$ |
| $x^2 - 2$ | $2x$ |
| $3x - 3$ | 3 |
| $2x^2 - x + 1$ | $4x - 1$ |

3 If then $f(x) = ax^n$ then $\frac{dy}{dx} = anx^{n-1}$.

**Exercise 7.1.3**

1 a $4x^3$ b $5x^4$ c $6x$
d $15x^2$ e $18x^2$ f $56x^6$

2 a $x^2$ b $x^3$ c $\frac{1}{2}x$
d $2x^3$ e $\frac{6}{5}x^2$ f $\frac{2}{3}x^2$

### Exercise 7.2.1

1 a $15x^2$ b $14x$ c $24x^5$
d $\frac{1}{2}x$ e $4x^5$ f $\frac{15}{4}x^4$
g $0$ h $6$ i $0$

2 a $6x + 4$ b $15x^2 - 4x$
c $30x^2 - x$ d $18x^2 - 6x + 1$
e $48x^3 - 4x$ f $x^2 - x + 1$
g $-12x^3 + 8x$ h $-30x^4 + 12x^3 - 1$
i $-\frac{9}{2}x^5 + 2x^2$

3 a $2x + 1$ b $4$ c $6x + 1$
d $\frac{1}{2}x + \frac{1}{2}$ e $6x + 3$ f $6x^2 - 8x$
g $2x + 10$ h $4x + 7$ i $3x^2 - 4x - 3$

### Exercise 7.2.2

1 a $-x^{-2}$ b $-3x^{-4}$ c $-4x^{-3}$
d $2x^{-3}$ e $x^{-4}$ f $2x^{-6}$

2 a $x^{-1}$ b $2x^{-1}$ c $3x^{-2}$
d $\frac{2}{3}x^{-3}$ e $\frac{3}{7}x^{-2}$ f $\frac{2}{9}x^{-3}$

3 a $-3x^{-2} + 2$ b $4x - x^{-2}$
c $-3x^{-2} + 2x^{-3} + 2$ d $-3x^{-4} + 3x^2$
e $-8x^{-5} + 3x^{-4}$ f $x^{-3} - x^{-4}$

### Exercise 7.2.3

1 a $6t + 1$ b $6t^2 + 2t$ c $15t^2 - 2t$
d $-2t^{-2}$ e $-t^{-3}$ f $4t^3 + 2t^{-3}$

2 a $-3x^{-2}$ b $-2t^{-2} - 1$ c $-2r^{-3} + r^{-2}$
d $-2l^{-5} + 2$ e $\frac{1}{2} + n^{-4}$ f $-\frac{4}{5}t^{-3} - 3t^2$

3 a $2x + 4$ b $1 - 2t$ c $3t^2$
d $2 - 6r$ e $x - x^{-2}$ f $3t^{-4}$

4 a $2t$ b $4t$
c $-2t^{-3}$ d $3t^2 - 2 + 4t^{-3}$
e $2t^2 - \frac{4}{3}t + 1$ f $-9t^{-4} + \frac{45}{2}t^{-6} - 4t$

### Exercise 7.3.1

1 a $6$ b $-3$ c $0$
d $0$ e $-\frac{1}{2}$ and $-5$ f $6$

2 a $-\frac{1}{4}$ b $-2$
c $-3\frac{3}{16}$ d $-3$
e $3\frac{31}{32}$ f $-36$ and $-68$

3 a i) $4\frac{1}{2}$ ii) $31\frac{1}{2}$ iii) $72$
iv) $0$
b $\frac{dN}{dt} = 10t - \frac{3}{2}t^2$
c i) $8\frac{1}{2}$ ii) $16\frac{1}{2}$ iii) $6$
iv) $-50$
e The graph increases during the first 6–7 days, hence number of new infections increases. When $t = 10$ the graph is at zero, hence number of new infections is zero.
f The rate is initially increasing, i.e. the gradient of the graph is increasing. After approximately 4 days, the rate of increase starts to decrease, i.e. the gradient of curve is less steep. After 10 seconds the gradient is negative, hence the rate of increase is negative too.

4 a i) 243 m ii) 2000 m
b $\frac{dh}{dt} = 60t - 3t^2$
c i) $108\,\text{m}\,\text{h}^{-1}$ ii) $225\,\text{m}\,\text{h}^{-1}$ iii) $0\,\text{m}\,\text{h}^{-1}$
e The steepness of the graph indicates the rate at which the balloon is ascending. After 20 hours the graph has peaked therefore the rate of ascent is $0\,\text{m}\,\text{h}^{-1}$.
f The steepest part of the graph occurs when $t = 10$ hours. Therefore this represents when the balloon is climbing at its fastest.

### Exercise 7.3.2

1 a (3, 6) b (2.5, 19.75)
c (2, 10) d (0, −1)
e (3, 15) f (−5, −14)

2 a $\left(1, 4\frac{5}{6}\right)$ and $\left(-2, -8\frac{2}{3}\right)$
b $\left(-1, -4\frac{1}{3}\right)$ and $(-3, -9)$
c $\left(2, -5\frac{1}{3}\right)$
d $(1, 4)$ and $\left(-\frac{1}{3}, -1\frac{13}{27}\right)$

3 a $\frac{ds}{dt} = 4 + 10t$ b $\frac{1}{2}$ seconds
c 3 seconds d 57 m

4 a 20°C  b $\frac{dT}{dt} = 24t - 3t^2$
c i) 21 °C/min
ii) 48 °C/min
iii) 0 °C/min
d $t = 2$ or 6 minutes  e 236 °C

**Exercise 7.3.3**

1 a $f'(x) = 2x - 3$  b 1
c 1  d $y = x - 1$
e $-1$  f $y = -x + 3$

2 a 4  b $y = 4x - 10$
c $-\frac{1}{4}$  d $x + 4y + 6 = 0$

3 a $-4$  b $y = -4x - 2$
c $y = \frac{1}{4}x - 2$

4 a $-2x + 2$  b $(-6, 10)$
c $(-4, 10)$

5 a $-4$  b $y = -4x + 17$
c $y = 5$
d $N_1$: $y = \frac{1}{4}x$, $N_2$: $x = 2$
e $(2, \frac{1}{2})$

6 a $f'(x) = -x - 1$  b $-3$
c $(2, -8)$

**Exercise 7.4.1**

1 a i) $f'(x) = 2x$  ii) $x > 0$
b i) $f'(x) = 2x - 3$  ii) $x > \frac{3}{2}$
c i) $f'(x) = -2x + 10$  ii) $x < 5$
d i) $f'(x) = 3x^2 - 24x + 48$
ii) $x < 4$ and $x > 4$
e i) $f'(x) = -3x^2 + 25$
ii) $-\frac{5}{\sqrt{3}} < x < \frac{5}{\sqrt{3}}$
f i) $f'(x) = x^3 - x$
ii) $-1 < x < 0$ and $x > 1$

2 a $x < 0$  b $x < \frac{3}{2}$  c $x < 5$
d never  e $x < -\frac{5}{\sqrt{3}}$ and $x > \frac{5}{\sqrt{3}}$
f $x < -1$ and $0 < x < 1$

3 a $f'(x) = x^2 + \frac{1}{3}$
$x^2 \geqslant 0$ for all values of $x$, $\rightarrow x^2 + \frac{1}{3} > 0$ for all values of $x$
Therefore $f'(x)$ is an increasing function for all values of $x$.

b $f'(x) = x^2 - 1 - x^4 = x^2(1 - x^2) - 1$
When $x < -1$ or $x > 1 \rightarrow (1 - x^2) < 0$
therefore $f'(x) < 0$
When $-1 < x < 1 \rightarrow 0 < x^2(1 - x^2) < 1$
therefore $f'(x) < 0$
Hence $f'(x)$ is a decreasing function for all values of $x$.

4 $k < -\frac{1}{3}$

**Exercise 7.5.1**

1 a i) $f'(x) = 2x - 3$  ii) $(3, 4)$
b i) $f'(x) = 2x + 12$  ii) $(-6, -1)$
c i) $f'(x) = -2x + 8$  ii) $(4, 3)$
d i) $f'(x) = -6$  ii) no stationary points

2 a i) $f'(x) = 3x^2 - 24x + 48$
ii) $(4, 6)$
b i) $f'(x) = 3x^2 - 12$
ii) $(-2, 16)$ and $(2, -16)$
c i) $f'(x) = 3x^2 - 6x - 45$
ii) $(-3, 89)$ and $(5, -167)$
d i) $f'(x) = x^2 + 3x - 4$
ii) $\left(-4, 13\frac{2}{3}\right)$ and $\left(1, -7\frac{1}{6}\right)$

3 a i) $f'(x) = -4 - 2x$
ii) $(-2, 5)$
iii) $(-2, 5)$ is a maximum
iv) $(0, 1)$
v)

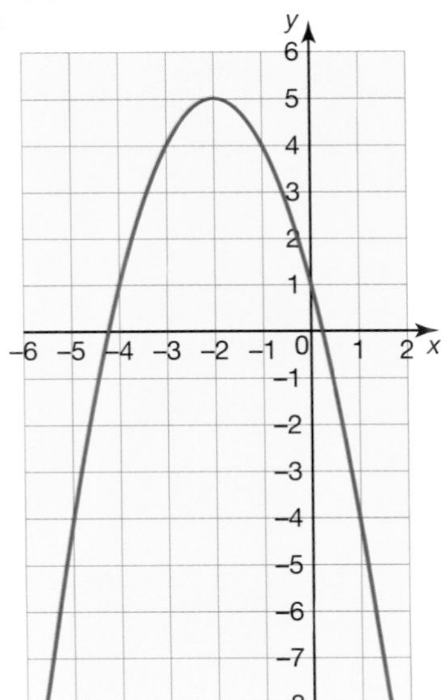

**b i)** $f'(x) = x^2 - 8x + 12$

**ii)** $\left(2, 7\frac{2}{3}\right)$ and $(6, -3)$

**iii)** $\left(2, 7\frac{2}{3}\right)$ is a maximum, $(6, -3)$ is a minimum

**iv)** $(0, -3)$

**v)**

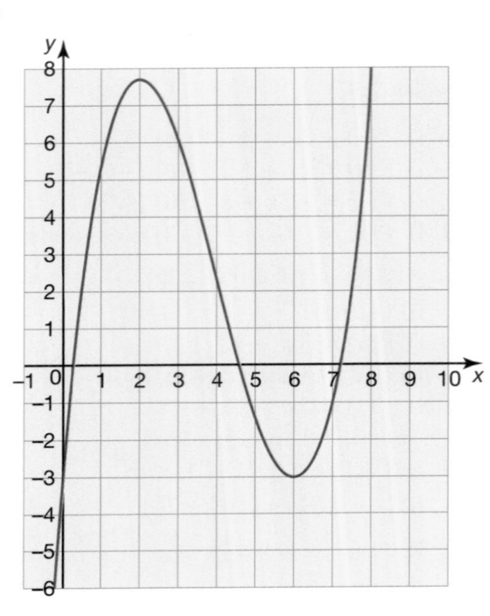


**c i)** $f'(x) = -2x^2 + 6x - 4$

**ii)** $\left(1, -1\frac{2}{3}\right)$ and $\left(2, -1\frac{1}{3}\right)$

**iii)** $\left(1, -1\frac{2}{3}\right)$ is a minimum point, $\left(2, -1\frac{1}{3}\right)$ is a maximum point

**iv)** $(0, 0)$

**v)**

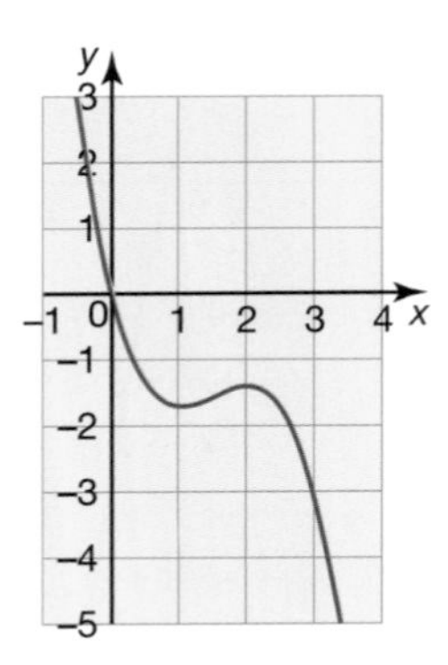


**d i)** $f'(x) = 3x^2 - 9x - 30$

**ii)** $(-2, 38)$ and $\left(5, -133\frac{1}{2}\right)$

**iii)** $(-2, 38)$ is a maximum point, $\left(5, -133\frac{1}{2}\right)$ is a minimum point

**iv)** $(0, 4)$

**v)**

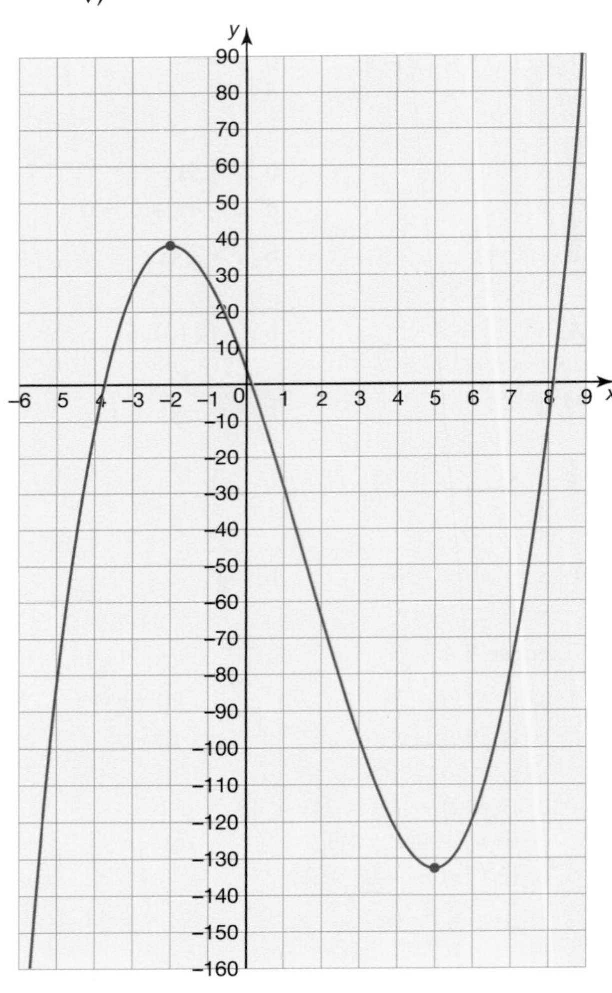

**4 a i)** $f'(x) = 3x^2 - 18x + 27$ **ii)** $(3, -3)$

**iii)** $(3, -3)$ is a point of inflexion

**iv)** $(0, -30)$

**v)**

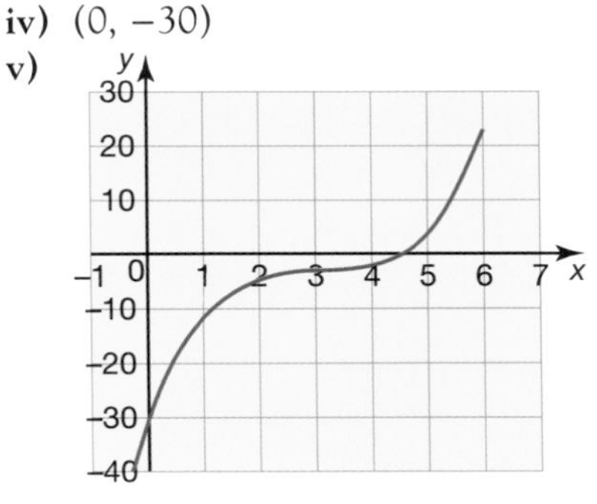

**b i)** $f'(x) = 4x^3 - 12x^2 + 16$

**ii)** $(-1, -11)$ and $(2, 16)$

**iii)** $(-1, -11)$ is a minimum point, $(2, 16)$ is a point of inflexion

**iv)** $(0, 0)$

**v)**

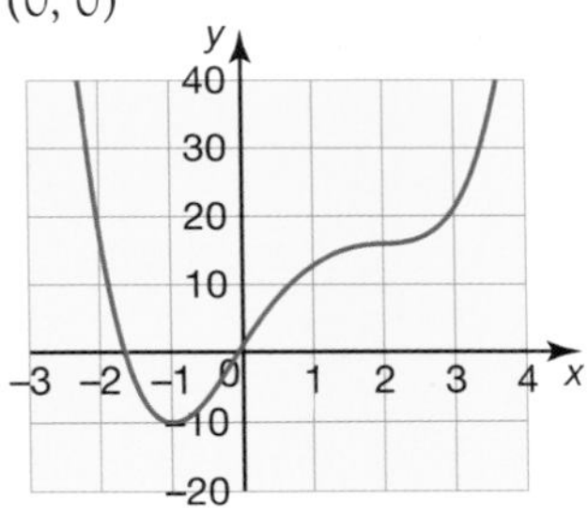

## Exercise 7.6.1

**1 a** $500 - 2x$

**b** $A = 500x - 2x^2$

**c** $500 - 4x$

**d** 125 m

**e** $31\,250\,\text{m}^2$

**f**

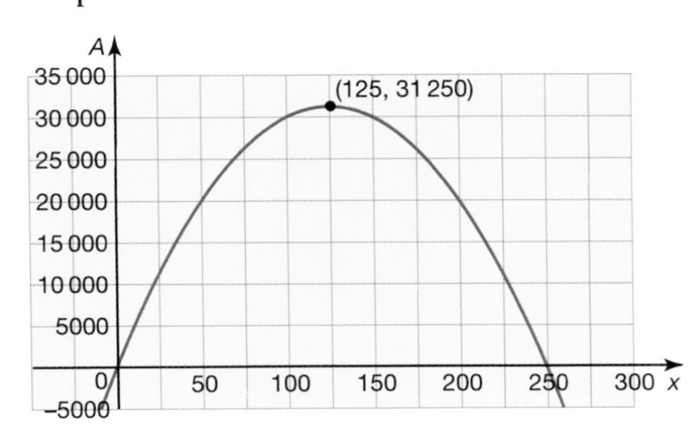


**2 a** \$40

**b** $40 + x$

**c** Each $x$ increase in price leads to a $2x$ decrease in phones sold.

**d** $P = (40 + x)(150 - 2x) = 6000 + 70x - 2x^2$

**e** $70 - 4x$

**f** 17.5

**g i** \$117 or \$118

**ii** \$6612.50

**3 a** $h = \dfrac{1600}{x^2}$

**c** $16x - \dfrac{1600}{x^2}$

**d** 10 cm

**e** When $x = 9$ cm $\dfrac{dA}{dx} = -53.5$

When $x = 11$ cm $\dfrac{dA}{dx} = 43.8$

Gradient changes from negative to positive therefore the stationary point is a minimum.

**4 b** $V = 858\,387\,\text{cm}^3 = 858\,400\,\text{cm}^3$ (4 s.f.)

### Student assessment 1

**1 a** $\dfrac{dy}{dx} = 3x^2$ **b** $\dfrac{dy}{dx} = 4x - 1$

**c** $\dfrac{dy}{dx} = -x + 2$ **d** $\dfrac{dy}{dx} = 2x^2 + 8x - 1$

**2 a** $f'(x) = 2x + 2$ **b** $f'(x) = 2x - 1$

**c** $f'(x) = 2x$ **d** $f'(x) = x + 1$

**e** $f'(x) = -3x^{-2}$ **f** $f'(x) = 1 - 2x^{-2}$

**3 a** $f'(1) = 2$ **b** $f'(0) = 1$

**c** $f'\left(-\frac{1}{2}\right) = 9$ **d** $f'\left(\frac{1}{4}\right) = 5\frac{1}{2}$

**4 a** $-\frac{1}{2}$ **b** $-1$

**c** $-\frac{1}{9}$ **d** $-\frac{2}{11}$

**5 a** $v = \dfrac{ds}{dt} = 10t$ **b** $v = 30\,\text{m}\,\text{s}^{-1}$

**c i)** 4.2 seconds **ii)** 88.2 m

**Student assessment 2**

1 a $f'(x) = 3x^2 + 2x$
b $P\left(-\frac{2}{3}, -\frac{23}{27}\right)$
c $Q(0, -1)$
d P is a maximum and Q a minimum

2 a Substituting (1, 1) into the equation gives $1 = 1^3 - 1^2 + 1 \rightarrow 1 = 1$
b 2
c $y = 2x - 1$
d $x + 2y - 3 = 0$

3 a $f'(x) = 2x - 4$
b $x < 2$

4 a $f'(x) = 4x^3 - 4x$
b (0, 0), (1, −1) and (−1, −1)
c (0, 0) is a maximum, (1, −1) and (−1, −1) are both minimum points
d i) (0, 0)
ii) (0, 0), $(\sqrt{2}, 0)$ and $(-\sqrt{2}, 0)$
e

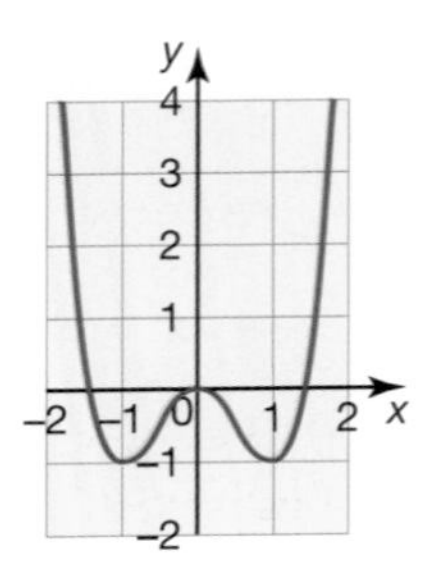


**Examination questions**

1 a $4x + 1$ b $-11$ c $-\frac{1}{4}$
2 a 32
b

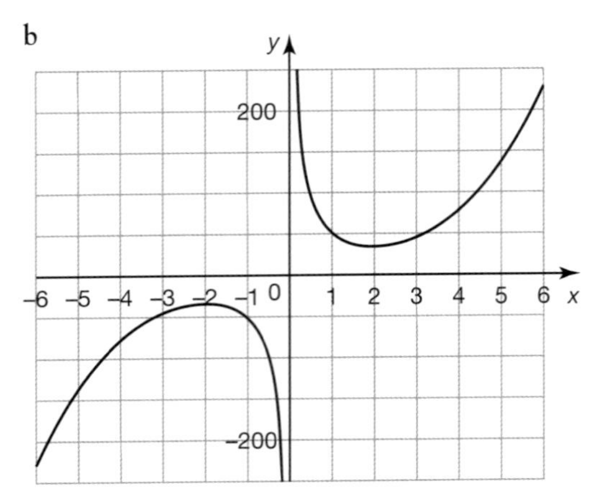


c $3x^2 - \frac{48}{x^2}$
d 0
e (−2, −32)
f $\{y \geq 32\} \cup \{y \leq -32\}$
g −45
h −1

3 **Part A**

a c

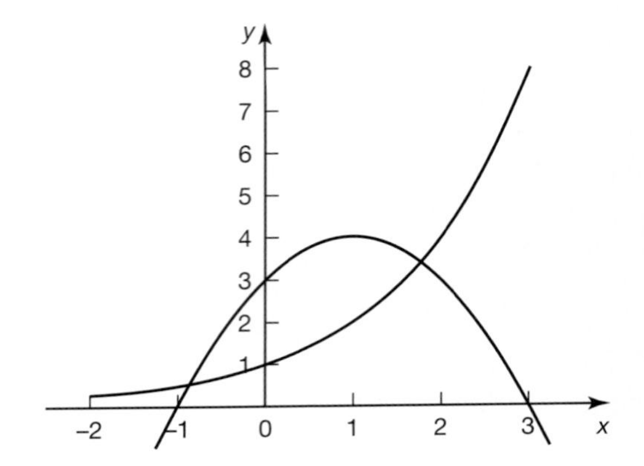


b $y = 0$
d $x = -0.857$ or $x = 1.77$
e 4
f $f'(x) = 2 - 2x$
At a maximum: $2 - 2x = 0$
$x = 1$
$f(1) = 3 + 2 \times 1 - 1^2 = 4$

**Part B**

a $4p + 2q = -6$ *or* $2p + 9q = -3$
b i $2px + q$ ii $4p + q = 1$
c $p = 2, q = -7$

# Index